**Advanced Control of Power
Converters**

Advanced Control of Power Converters

Techniques and MATLAB/Simulink Implementation

Hasan Komurcugil
Eastern Mediterranean University, Turkey

Sertac Bayhan
Hamad Bin Khalifa University, Qatar

Ramon Guzman
Technical University of Catalonia, Spain

Mariusz Malinowski
Warsaw University of Technology, Poland

Haitham Abu-Rub
Texas A&M University, Qatar

IEEE Press Series on Control Systems Theory and Applications
Maria Domenica Di Benedetto, Series Editor

IEEE PRESS

WILEY

Published by John Wiley & Sons, Inc., Hoboken, New Jersey.
Published simultaneously in Canada.

For general information on our other products and services or for technical support, please contact our Customer Care Department within the United States at (800) 762-2974, outside the United States at (317) 572-3993 or fax (317) 572-4002.

Wiley also publishes its books in a variety of electronic formats. Some content that appears in print may not be available in electronic formats. For more information about Wiley products, visit our web site at www.wiley.com.

Library of Congress Cataloging-in-Publication Data Applied for:
Hardback: 9781119854401

Cover Design: Wiley
Cover Image: © kozirsky/Shutterstock.com

Set in 9.5/12.5pt STIXTwoText by Straive, Pondicherry, India
Printed and bound by CPI Group (UK) Ltd, Croydon, CR0 4YY

C9781119854401_080723

Contents

About the Authors

Hasan Komurcugil received the BSc, MSc, and PhD degrees from the Eastern Mediterranean University (EMU), Famagusta, North Cyprus, Via Mersin 10, Turkey, in 1989, 1991, and 1998, respectively, all in electrical engineering. Then, he was promoted to assistant professor in 1998, associate professor in 2002, and full professor in 2008. From 2004 to 2010, he was the Head of the Computer Engineering Department, EMU. In 2010, he played an active role in preparing the department's first self-study report for the use of Accreditation Board for Engineering and Technology. In 2010, he was elected as the as the Board Member of Higher Education, Planning, Evaluation, Accreditation and Coordination Council (YODAK) North Cyprus. From 2010 to 2019, he played active role in evaluating the universities in North Cyprus. In 2022, he visited Texas A&M University at Qatar as an associate research scientist in the electrical and computer engineering program. He has authored more than 70 science citation index expanded journal papers and 80 conference papers. His research interests include power electronics and innovative control methods for power converters such as sliding mode control, Lyapunov-function-based control, and model predictive control. He is a coauthor of one book (Multilevel Inverters: Introduction and Emergent Topologies) chapter. He is a member of the IEEE Industrial Electronics Society (IES) and Senior Member of the IEEE. He served as the Corresponding Guest Associate Editor of the *IEEE Transactions on Energy Conversion* and Guest Editor of the *IEEE Transactions on Industrial Informatics*. Currently, he serves as the Associate Editor of the *IEEE Transactions on Industrial Electronics* and the *IEEE Transactions on Industrial Informatics*.

Sertac Bayhan received the MS and PhD degrees in electrical engineering from Gazi University, Ankara, Turkey in 2008 and 2012, respectively. His undergraduate degree was also earned at the same university, and he graduated as valedictorian. He joined Gazi University as a lecturer in 2008 and was promoted to associate professor and full professor in 2017 and 2022, respectively. He was an associate research scientist at Texas A&M University at Qatar from 2014 to 2018. Currently, Dr. Bayhan is a senior scientist at Qatar Environment and Energy Research Institute (QEERI) and an associate professor (joint appointment) in Hamad Bin Khalifa University's Sustainable Division. His research encompasses power electronics and their applications in next-generation power and energy systems, including renewable energy integration, electrified transportation, and demand-side management.

Dr. Bayhan is the recipient of many prestigious international awards, such as the Teaching Excellence Award in recognition of outstanding teaching in Texas A&M University at Qatar in 2022, Best Paper awards in the 3rd International Conference on Smart Grid and Renewable Energy, March 20–22, 2022, Doha/ Qatar and 10th International Conference on Renewable Energy Research and Applications, September 26–29, 2021, Istanbul/Turkey, the Research Fellow Excellence Award in recognition of his research achievements and exceptional contributions to the Texas A&M University at Qatar in 2018. He has acquired $13M in research funding and published more than 170 papers in mostly prestigious IEEE journals and conferences. He is also the coauthor of three books and six book chapters.

Dr. Bayhan has been active senior member of IEEE. Because of the visibility of his research, he has been recently elected as an energy cluster delegate of the Industrial Electronics Society (IES). In 2020, he founded the IES Qatar Section Chapter and is currently its chair. Furthermore, he currently serves as associate editor of the *IEEE Transactions on Industrial Electronics*, *IEEE Journal of Emerging and Selected Topics in Industrial Electronics*, *IEEE Open Journal of the Industrial Electronics Society*, and *IEEE Industrial Electronics Technology News*.

Ramon Guzman received the BSc, MSc, and PhD degrees in communications engineering from the Universitat Politècnica de Catalunya (UPC), in 1999, 2004, and 2016, respectively. He was promoted to assistant professor in 2002 and associate professor in 2016. Currently, he is an associate professor with the department of automatic control in the Universitat Politècnica de Catalunya. He has authored more than 40 documents including science citation index expanded journal papers and conference papers. His research interests include advanced control methods for power converters such as sliding mode control, model predictive control, among others. He is a coauthor of the book (*Control Circuits in Power Electronics*) chapter and the book (*Communication in Active Distribution Networks*) chapter. He is a member of the IEEE Industrial Electronics Society (IES) and Senior Member of the IEEE. Currently, he serves as an associate editor of the *IEEE Transactions on Industrial Electronics*.

Mariusz Malinowski received a PhD degree with honors in Electrical Engineering from the Warsaw University of Technology (WUT) in 2001. He then attained a habilitation in 2012 and a full professor title in 2019.

He received the IEEE Industrial Electronics Society (IES) David Irwin Early Career Award for "Outstanding research and development of modulation and control for industrial electronics converters" in 2011, IEEE IES David Bimal Bose Award for Industrial Electronics Applications in Energy Systems for *"Contributions in control of industrial electronics converters applications in energy systems"* in 2015, and Power Electronics and Motion Control (PEMC) Istvan Nagy Award in 2021.

Mariusz Malinowski has published many journals and conference papers and is a co-author of five books. He has participated in many research and industrial projects and has been a reviewer and PhD commission member for numerous PhD theses in Germany, Spain, Denmark, Australia, India, Switzerland, and Poland.

Mariusz Malinowski's public service includes activity in IEEE, where he was editor-in-chief of *IEEE Industrial Electronics Magazine*, associate editor of *IEEE Transactions on Industrial Electronics*, and associate editor of *IEEE Transactions on Power Electronics*. Mariusz Malinowski is currently the past chair of the IEEE Poland Section and president of the IEEE Industrial Electronics Society. He has an IEEE Fellow status.

Mariusz Malinowski participated in the development of technologies that received many prizes e.g. three times recognition in the competition Polish Product of the Future organized by the Polish Agency for Enterprise Development (PARP), the Grand Prix Exhibition of Innovations in Geneva (Gold Medal) and the Exhibition in Brussels "Eureco" (Bronze Medal).

Mariusz Malinowski was visiting scholar and professor in the following institutions: Aalborg University (Denmark), University of Nevada (Reno, USA), Technical University of Berlin (Germany), Universidad Tecnica Federico Santa Maria (Valparaiso, Chile), ENSEEIHT - Laplace, Toulouse (France), and ETH Zurich (Switzerland).

Haitham Abu-Rub is a full professor holding two PhDs from Gdansk University of Technology (1995) and from Gdansk University (2004). Dr. Abu Rub has long teaching and research experiences at many universities in many countries including Qatar, Poland, Palestine, USA, and Germany.

Since 2006, Dr. Abu-Rub has been associated with Texas A&M University at Qatar, where he has served for five years as the chair of Electrical and Computer Engineering Program and has been serving as the managing director of the Smart Grid Center at the same university.

His main research interests are energy conversion systems, smart grid, renewable energy systems, electric drives, and power electronic converters.

Dr. Abu-Rub is the recipient of many prestigious international awards and recognitions, such as the American Fulbright Scholarship and the German Alexander von Humboldt Fellowship. He has co-authored around 400 journal and conference papers, five books, and five book chapters. Dr. Abu-Rub is an IEEE fellow and co-editor in chief of the IEEE Transactions on Industrial Electronics.

List of Abbreviations

AC	Alternating Current
ADC	Analog Digital Converter
BDFIG	Brushless Doubly Fed Induction Generator
CCSMPC	Continuous Control Set Model Predictive Control
CDFIG	Cascaded Doubly Fed Induction Generator
CM	Control Machine
CMV	Common Mode Voltage
CPL	Constant Power Load
CRL	Constant Resistive Load
DC	Direct Current
DG	Distributed Generation
DLQR	Discrete Linear Quadratic Regulator
DPC	Direct Power Control
DSP	Digital Signal Processor
EKF	Extended Kalman Filter
EV	Electric Vehicle
FCS	Finite Control Set
FCSMPC	Finite control set model predictive control
FPGA	Field Programmable Gate Array
HM	Hysteresis Modulation
IGBT	Insulated Gate Bipolar Transistor
ISMC	Integral Sliding Mode Control
KF	Kalman Filter
KKT	Karush–Kuhn–Tucker
MCU	Microcontroller unit
MIMO	Multi-Input Multi-Output
MIPS	Millions Instructions Per Second
MOSFET	Metal Oxide Semiconductor Field Effect Transistor

MPC	Model Predictive Control
MRAS	Model Reference Adaptive System
NPC	Neutral Point Clamped
PCC	Point of Common Coupling
PI	Proportional Integral
PID	Proportional Integral Derivative
PLL	Phase-Locked Loop
PM	Power Machine
PR	Proportional Resonant
PTC	Predictive Torque Control
PV	Photovoltaic
PWM	Pulse Width Modulation
qZS	Quasi Z Source
qZSI	Quasi Z Source Inverter
RMS	Root Mean Square
SAPF	Shunt Active Power Filter
SDA	Switching Decision Algorithm
SEPIC	Single-Ended Primary-Inductor Converters
SISO	Single-Input Single-Output
SMC	Sliding Mode Control
Space	Vector Modulation
SPWM	Sinusoidal Pulse Width Modulation
STA	Super Twisting Algorithm
TDVR	Transformerless Dynamic Voltage Restorer
THD	Total Harmonic Distortion
TSMC	Terminal Sliding Mode Control
UPFR	Unity Power Factor Rectifier
UPS	Uninterruptible Power Supply
VFD	Variable Frequency Drive
VOC	Voltage Oriented Control
VSI	Voltage Source Inverter
VSS	Variable Structure System
WF	Weighting Factor

Preface

Power electronics converters play an important role in every stage of today's modernized world including computers, smart home systems, electric vehicles, airplanes, trains, marine electrical systems, microgrids, robots, renewable energy source integration systems, residential, and many industrial applications. The main function of a power converter is to achieve DC–DC, DC–AC, AC–DC, and AC–AC power conversion with high performance in terms of efficiency, stability, robustness, reduced complexity, and low cost. A power converter consists of switching devices and diodes, which are turned on and off based on a control strategy for achieving the desired power conversion performance. Hence, the control of power converters is the key point in achieving the desired target.

The traditional control techniques that are based on the linearized model (small-signal model) of converter offer satisfactory performance around the operating point. However, these control strategies fail in achieving the desired performance away from the operating point. In the last two decades, many advanced nonlinear control strategies, such as sliding mode control, Lyapunov function-based control, and model predictive control, have received significant attention of many researchers and practicing engineers. The effectiveness of these control methods has been proved in literature. Considering the advantages and superiority of nonlinear control, it is essential and timely to write a comprehensive book to present the fundamental ideas, design guidelines, mathematical modeling, and MATLAB®\Simulink®-based simulation of these advanced control strategies for various power converters employed in many applications. Thus, we decided to write this book to cover the advanced nonlinear control methods of power electronic converters in a single source. The book provides a unique combination of the advanced nonlinear control methods mentioned above for various power converters and applications. Furthermore, each control method is supported by simulation examples along with MATLAB®\Simulink® models, which will make the book of high benefit for researchers, engineering professionals, and undergraduate/graduate students in electrical engineering and mechatronics areas.

This book has nine chapters that can be divided into four parts. In the first part, a brief introduction of sliding mode control, Lyapunov function-based control, and model predictive control methods is presented (Chapters 1 and 2). In the second part, design guidelines of these control methods are presented in detail (Chapters 3–5). Third part presents a tutorial on physical modeling and experimental verification using MATLAB®\Simulink® (Chapter 6). Finally, case studies of various power converter applications are given in the fourth part (Chapters 7–9). These case studies are mainly based on our own research work available in literature. We have carefully selected each case study to cover a wide range of converter applications. All Simulink models referred to in the case studies are provided as supplementary material to be downloaded from the web site provided by the publisher. These MATLAB®\Simulink® models will be very helpful in learning the basics of these control methods. In this respect, we believe that this book fills the gap between theory and practice and provides practical guidance to the researchers, graduate and senior undergraduate students, and practicing engineers for designing and developing these advanced control methods using MATLAB®\Simulink®.

Acknowledgment

We would like to take this opportunity to express our sincere appreciation to all the people who were directly or indirectly helpful in making this book a reality. We emphasize that portions of the book appeared in earlier forms as journal papers and conference papers with some of our students and colleagues. Due to this fact, our special thanks go to all of them.

We are grateful to the Qatar National Research Fund (a member of Qatar Foundation) for funding many of the research projects, whose outcomes helped us in preparing major part of this book. Chapters 1–4 for NPRP grant (NPRP12S-0226-190158), Chapters 7 and 8 for NPRP grant (NPRP12C-33905-SP-220), and Chapters 5, 6, and 9 for NPRP grant (NPRP12S-0214-190083). The statements made herein are solely the responsibility of the authors.

This work was supported in part by the OPUS program of the National Science Centre of Poland under Grant NCN 2018/31/B/ST7/00954.

This work is also supported in part by the R+D+i project PID2021-122835OB-C21, financed by MCIN/AEI/10.13039/501000011033 and FEDER "A way of making Europe."

Also, we appreciate the help from many colleagues for providing constructive feedback on the material and for editing. Particular appreciation goes to Dr. Naki Guler from Gazi University, Turkey, for his great help not only in drawing many of the figures but also in creating some of the MATLAB®\Simulink® models.

Finally, we are indebted to our families for their continued support, endless patience, a wonderful working environment at home, especially during the difficult time of the COVID-19 pandemic, and encouragement, without which this book would not have been completed.

Hasan Komurcugil
Sertac Bayhan
Ramon Guzman
Mariusz Malinowski
Haitham Abu-Rub

About the Companion Website

This book is accompanied by the companion website:

www.wiley.com/go/komurcugil/advancedcontrolofpowerconverters

The website includes case studies of Simulink model.

1

Introduction

1.1 General Remarks

Power electronics converters are widely utilized in almost every aspect of todays' modernized world including computers, smart home systems, electric vehicles (EVs), trains, marine, aircrafts, microgrids, robots, renewable energy conversion and integration, and many industrial applications. The main function of a power converter is to convert the electrical power from one form to the other [1]. In general, there are four different categories of power converters: DC–DC, AC–DC, DC–AC, and AC–AC. In each category, various converter topologies have been developed to meet the desired conversion and application objectives. For instance, when the photovoltaic (PV) energy is to be converted and injected into the grid, a DC–DC boost converter is connected between the PV panel and DC–AC inverter to ensure a constant inverter input voltage and to possibly track the panel's maximum power point. The necessity of DC–DC boost converter in such application arises due to the buck operation of the inverter (i.e. DC input voltage is greater than the amplitude of its AC voltage). On the other hand, an AC–DC converter (usually referred to as rectifier) is used in an electric vehicle (EV) charging system to convert the grid's AC voltage to DC such that the battery can be charged. Another example is the use of series active filter (usually referred to as dynamic voltage restorer) in the protection of sensitive loads (i.e. medical equipment in the hospitals, data centers, and so on) against voltage sags, and voltage swells in the grid voltage. When such voltage variations occur in the grid, the series active filter, which is built using a DC–AC inverter, generates and injects the required compensation voltage to the point of common coupling such that the sensitive load

Advanced Control of Power Converters: Techniques and MATLAB/Simulink Implementation, First Edition. Hasan Komurcugil, Sertac Bayhan, Ramon Guzman, Mariusz Malinowski, and Haitham Abu-Rub.
© 2023 The Institute of Electrical and Electronics Engineers, Inc.
Published 2023 by John Wiley & Sons, Inc.
Companion website: www.wiley.com/go/komurcugil/advancedcontrolofpowerconverters

voltage is always kept at the desired value. Similar examples can be given for the other converter categories.

Thus, considering the importance of today's energy demand and the need for clean and reliable resources, the use of power electronics has increased tremendously. As such, the performance of the power converters used in these applications has gained utmost importance. In most of the power electronics-related applications, closed-loop control is essential to keep the voltage or current at reference values under various conditions, which include load changes, grid voltage deterioration (voltage sags, voltage swells, and distorted grid voltages), and parameter variations, which occur because of aging and operating conditions. More importantly, the stability of closed-loop system should not be jeopardized under these situations. For this reason, the design of a closed-loop system that responds to these challenges is an essential and difficult task. First of all, it should be noted that the design of closed-loop control for power electronics converters requires a deep knowledge in many areas such as circuit analysis, advanced mathematics, modeling, control systems, and power electronics. In this regard, this chapter starts with the introduction of simplest closed-loop control for power converters. Then, mathematical modeling, basic control objectives, and performance evaluation are explained briefly. Hence, the reader is urged to refresh or gain further basic information in the areas of control theory and power electronics converters from the literature.

This book is primarily concerned with advanced nonlinear control of power converters. A closed-loop control is referred to as nonlinear control if it contains at least one nonlinear component. Nonlinear control of power converters received attention of many researchers in the last two decades. The main reason of this popularity comes from the advantages over linear control methods, which lack guaranteed stability in large operation range of the converters; facing hard nonlinearities (saturation, dead-zone, backlash, and hysteresis), which cannot be approximated linearly; and having model uncertainties, which are assumed to be known when designing the linear controller. Whereas the nonlinear control is able to cope with the problems mentioned above. In this book, sliding mode control, Lyapunov function-based control, and model predictive control methodologies are explained for power converters. Although these nonlinear control methods are not new, their application in power converter control was limited in the past due to the required extensive computations. Since last decade, the advent of fast implementation platforms such as digital signal processors and field programmable gate arrays (FPGAs) relieved the computation burden issue. Therefore, compiling the design and application of the nonlinear control methods mentioned above in a single book is very beneficial for the interested readers.

1.2 Basic Closed-Loop Control for Power Converters

A basic single input single output (SISO) closed-loop power converter control system is illustrated in Figure 1.1. Here, the main aim is to control the power converter in order to accomplish specific desired control objectives (see Section 1.4). Clearly, the output signal (i.e. voltage or current) is measured and compared with the reference one to produce an error signal. This error signal is applied to the controller. Then, the controller generates modulation signal from which the pulse width modulation (PWM) signals are generated. These signals are applied to the gates of switching devices (i.e. insulated gate bipolar transistors [IGBTs], metal oxide semiconductor field effect transistors [MOSFETs], etc.) in the power converter. Upon the application of PWM signals, the switching devices are turned on and off. The value of voltage (or current) in the converter is changed by these switching actions. If the controller is well designed, the error signal is

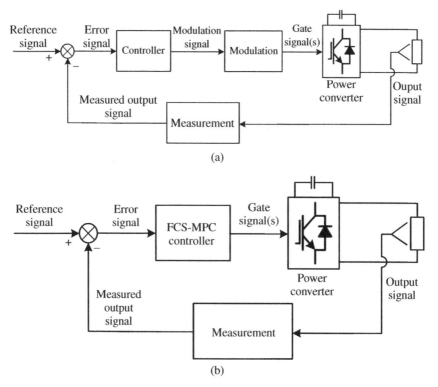

Figure 1.1 Basic closed-loop power converter control system. (a) With modulation, (b) without modulation.

continually reduced until the output signal tracks the reference signal in the steady state. On the other hand, the number of loops in a control system may be more than one depending on the converter topology and the application area.

Numerous control approaches have been developed for the power converters. Each control approach has its own advantages and disadvantages concerning with the controller complexity and cost, dynamic response, steady-state error, robustness to parameter variations, and closed-loop stability. It is worth to mention that the discussion in this book is based on the introduction, design and application of sliding mode control, Lyapunov function-based control, and model predictive control methods used in various power converters. While the sliding mode control and Lyapunov function-based control methods require a modulation block as shown in Figure 1.1a, the finite control set model predictive control (FCS-MPC) method does not require a modulation as shown in Figure 1.1b. The design of sliding mode control, Lyapunov function-based control, and model predictive control are explained in Chapters 3, 4, and 5, respectively. The following sections intend to present background information regarding the steps that should be taken into consideration when designing a controller. Even though these steps are well known in the modern control systems area, the readers, who are not fully familiar with these, will gain a knowledge before learning each of these control methods.

1.3 Mathematical Modeling of Power Converters

Usually, an accurate mathematical model of the converter is necessary when there is a need to design its controller. As it will be discussed in Chapter 3, the sliding mode control does not require mathematical modeling of the converter. Whereas the Lyapunov function-based control and model predictive control approaches rely on the mathematical model of the power converter as will be explained in Chapters 4 and 5, respectively. However, a perfect mathematical model, which represents all dynamics of the converter, is not possible in practice due to the certain noises (i.e. measurement noise) and possible failure conditions. There are two types of mathematical models in the continuous time: linear models and nonlinear models. The behavior of a linear converter system is usually described by linear differential equations written in the state-space form as follows:

$$\frac{d\mathbf{x}}{dt} = A\mathbf{x} + Bu$$
$$y = C\mathbf{x} + Du$$
(1.1)

where $\mathbf{x}$ represents the state vector, u represents the input vector, y represents the output vector, and A, B, C, and D represent the matrices with appropriate dimension. Such models are suitable to be used with the root-locus method, state-space method, and frequency domain design methods such as Bode plot and Nyquist method. As will be discussed in Chapter 4, the Lyapunov function-based control method uses the linear converter system model in (1.1). On the other hand, the model predictive control method (see Chapters 5 and 9) uses discrete-time version of the continuous-time model in (1.1) as given below:

$$\mathbf{x}(k + 1) = A_d\mathbf{x}(k) + B_d u(k)$$

$$y(k) = C_d\mathbf{x}(k) + D_d u(k), \qquad k = 0, 1, 2, 3, ... \tag{1.2}$$

where A_d, B_d, C_d, and D_d are discretized matrices of A, B, C, and D, respectively. The sampling instants are represented by k and $k + 1$. It is worth to mention that the sampling period T_s is omitted in (1.2) for brevity. In general, the discretization of (1.1) can be based on the integral approximation method or Euler's method. In the case of integral approximation method, the input is assumed to be constant between sampling instants k and $k + 1$ (i.e. $u(t) = u(kT_s)$, $kT_s \leq t \leq (k + 1)T_s$), which results in following discrete-time state equation:

$$\mathbf{x}((k + 1)T_s) = e^{AT_s}\mathbf{x}(kT_s) + \int_{kT_s}^{(k + 1)T_s} e^{A((k + 1)T_s - \tau)} Bu(kT_s) d\tau \tag{1.3}$$

Note that (1.3) is same as first equation of (1.2) if T_s is omitted. Comparing (1.2) and (1.3), it can be seen that

$$A_d = e^{AT_s}$$

$$B_d = e^{AT_s} + \int_{kT_s}^{(k + 1)T_s} e^{A((k + 1)T_s - \tau)} d\tau B \tag{1.4}$$

The discrete-time output equation can be derived in the same way. In the case of Euler's method, the first derivative of state equation at $t = kT_s$ is approximated as follows:

$$\frac{d\mathbf{x}(t)}{dt} \approx \frac{\mathbf{x}((k + 1)) - \mathbf{x}(k)}{T_s} \tag{1.5}$$

Equation (1.5) is referred to as Euler's forward approximation in literature [2], [3]. Applying (1.5) to the first equation of (1.1), one can obtain:

$$\mathbf{x}((k + 1)T_s) \approx (I + T_s A)\mathbf{x}(kT_s) + T_s Bu(kT_s) \tag{1.6}$$

where I is the identity matrix. When first equation of (1.2) and (1.6) are compared, the following relations are obtained easily:

$$A_d = I + T_s A, \quad B_d = T_s B \tag{1.7}$$

On the other hand, the behavior of a nonlinear converter can be described by using a nonlinear mathematical model of the form:

$$\frac{dx}{dt} = f(\mathbf{x}, u)$$

$$y = g(\mathbf{x}, u) \tag{1.8}$$

where $f(\mathbf{x}, u)$ and $g(\mathbf{x}, u)$ are the nonlinear functions of $\mathbf{x}$ and u. It is worth noting that the behavior of some converters can also be defined by using the following nonlinear mathematical model:

$$\frac{dx}{dt} = f(\mathbf{x}) + g(\mathbf{x})u \tag{1.9}$$

where $f(\mathbf{x})$ and $g(\mathbf{x})$ are the nonlinear functions of $\mathbf{x}$. The nonlinear functions in (1.8) and (1.9) are usually unknown. For this reason, robust control method such as the sliding mode control is emerged to obtain the desired performance.

1.4 Basic Control Objectives

Controller design is usually based on satisfying some closed-loop specifications, which can be referred to as the control objectives. These control objectives can be described as follows.

1.4.1 Closed-Loop Stability

In the control design, the closed-loop stability analysis of a converter to be controlled is the first step. The closed-loop stability of linear systems can be tested either in s-domain [4] or in z-domain [5]. When the controller is designed in continuous-time, the stability check is performed in Laplace domain. In this case, the closed-loop converter system is stable if all the roots of its characteristic equation are located in the left half of s-plane (complex plane). It should be noted that the roots of characteristic equation are the poles of closed-loop system. If any root is located in the right half of s-plane, the closed-loop converter system is unstable. The effects of the root locations on the dynamic response of the closed-loop system are depicted in Figure 1.2. Clearly, while complex conjugate roots with negative real parts cause a decreasingly oscillatory response, a root with negative real part leads to a converging smooth response. On the other hand, complex conjugate roots with positive real parts cause an increasingly oscillatory response while a root with positive real part leads to a diverging smooth response. Hence, an idea about the speed of the transient response can be obtained from the root locations. For instance, real roots located near the imaginary axis cause slow transient

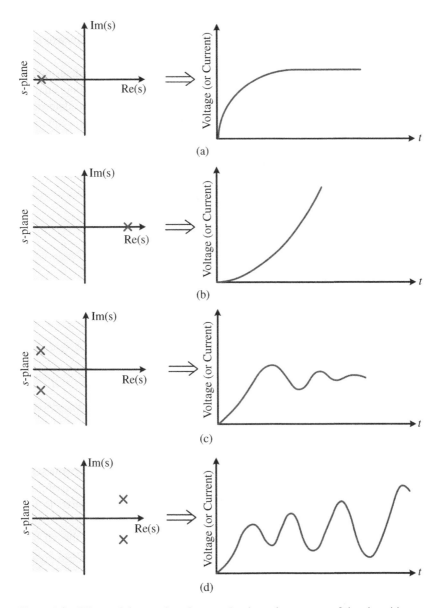

Figure 1.2 Effects of the root locations on the dynamic response of the closed-loop system. (a) Negative real root, (b) positive real root, (c) complex conjugate root with negative real parts, (d) complex conjugate root with positive real parts.

response due to the fact that $\tau_1 = 1/p_1$ where τ_1 is the closed-loop time constant and p_1 is the real root. Similarly, complex roots located close to the imaginary axis slow down the transient response.

When the order of characteristic equation is higher four, then the determining closed-loop stability is not easy due to the difficulty in solving the roots analytically. In this case, Routh–Hurwitz stability criterion can be used to determine the stability of a linear system [4]. Consider the following characteristic equation:

$$F(s) = a_n s^n + a_{n-1} s^{n-1} + a_{n-2} s^{n-2} + \dots + a_1 s + a_0 = 0 \qquad (1.10)$$

The Routh–Hurwitz stability criterion relies on ordering the coefficients of the characteristic equation in the form of an array shown below [4]:

$$
\begin{array}{c|ccc}
s^n & a_n & a_{n-2} & a_{n-4} \\
s^{n-1} & a_{n-1} & a_{n-3} & a_{n-5} \\
s^{n-2} & b_{n-1} & b_{n-3} & b_{n-5} \\
\cdot & c_{n-1} & c_{n-3} & c_{n-5} \\
\cdot & \cdot & \cdot & \cdot \\
\cdot & \cdot & \cdot & \cdot \\
\cdot & \cdot & \cdot & \cdot \\
s & \cdot & \cdot & \cdot \\
s^0 & h_{n-1} & \cdot & \cdot
\end{array}
\qquad (1.11)
$$

where $b_{n-1} = \dfrac{-1}{a_{n-1}} \det \begin{bmatrix} a_n & a_{n-2} \\ a_{n-1} & a_{n-3} \end{bmatrix}$, $b_{n-3} = \dfrac{-1}{a_{n-1}} \det \begin{bmatrix} a_n & a_{n-4} \\ a_{n-1} & a_{n-5} \end{bmatrix}$,

$c_{n-1} = \dfrac{-1}{b_{n-1}} \det \begin{bmatrix} a_{n-1} & a_{n-3} \\ b_{n-1} & b_{n-3} \end{bmatrix}$, and so on. According to the Routh–Hurwitz stability criterion, all roots of a characteristic equation are located in the left half of s-plane if and only if all first-column elements of the array in (1.11) have the same sign. However, it should be noted that the Routh–Hurwitz criterion is useful for only continuous-time linear systems and, hence, cannot be considered to test the stability of discrete-time systems.

On the other hand, when the controller is designed in discrete-time, the left-half of s-plane is mapped to the interior of a unit circle as illustrated in Figure 1.3. This implies that the closed-loop converter system is stable if all the roots of its characteristic equation are located within the unit circle (i.e. $|z| < 1$). It is worthy noting that while the entire imaginary axis in the s-plane is mapped onto the unit circle ($|z| = 1$), the right-half of s-plane is mapped to the outside of the unit circle. Thus, the closed-loop system becomes unstable when the roots are outside of the unit circle (i.e. $|z| > 1$).

In the case of difficulty in solving the roots analytically in discrete time, the Jury stability criterion can be used to test the stability of the linear system [5]. The Jury stability criterion involves determinant operations as in the Routh–Hurwitz

Figure 1.3 Stable region of the closed-loop system in discrete time.

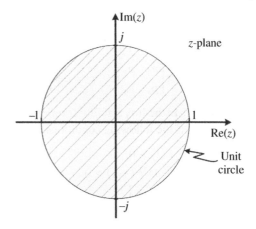

stability criterion. However, determining the closed-loop stability is more time consuming. Consider the following characteristic equation [6]:

$$F(z) = a_n z^n + a_{n-1} z^{n-1} + a_{n-2} z^{n-2} + \ldots + a_1 z + a_0 = 0 \tag{1.12}$$

Then, Table 1.1 can be constructed as follows.

The entries of table can be computed as follows:

$$b_k = \det \begin{bmatrix} a_0 & a_{n-k} \\ a_n & a_k \end{bmatrix}, \quad k = 0, 1, \ldots, n-1 \tag{1.13}$$

Table 1.1 Jury's table.

Row	z^0	z^1	z^2	...	z^{n-k}	...	z^{n-1}	z^n
1	a_0	a_1	a_2	...	a_{n-k}	...	a_{n-1}	a_n
2	a_n	a_{n-1}	a_{n-2}	...	a_k	...	a_1	a_0
3	b_0	b_1	b_2	...	b_{n-k}	...	b_{n-1}	
4	b_{n-1}	b_{n-2}	b_{n-3}	...	b_k	...	b_0	
5	c_0	c_1	c_2	...	...	c_{n-2}		
6	c_{n-2}	c_{n-3}	c_{n-4}	...	...	c_0		
.	.	.	.	...	...			
.	.	.	.	...	...			
$2n-5$	p_0	p_1	p_2	p_3				
$2n-4$	p_3	p_2	p_1	p_0				
$2n-3$	q_0	q_1	q_2					

$$c_k = \det \begin{bmatrix} b_0 & b_{n-k} \\ b_n & b_k \end{bmatrix}, \quad k = 0, 1, ..., n-2 \tag{1.14}$$

$$q_0 = \det \begin{bmatrix} p_0 & p_3 \\ p_3 & p_0 \end{bmatrix}, \quad q_1 = \det \begin{bmatrix} p_0 & p_2 \\ p_3 & p_1 \end{bmatrix}, \quad q_2 = \det \begin{bmatrix} p_0 & p_1 \\ p_3 & p_2 \end{bmatrix} \tag{1.15}$$

According to the Jury stability criterion, the roots of $F(z)$ are inside the unit circle if and only if the following conditions are held,

1. $\quad F(1) > 0$
2. $\quad (-1)^n F(-1) > 0$
3. $\quad |a_0| < a_n$
4. $\quad |b_0| > |b_{n-1}|$
5. $\quad |c_0| > |c_{n-2}|$

$\quad \cdot \qquad \cdot$

$\quad \cdot \qquad \cdot$

$n + 1 \quad |q_0| > |q_2|$

The stability analysis of Lyapunov function-based control and model predictive control methods, which will be described in Chapters 8 and 9 of this book, can be determined by using the methods described above. It is worth noting that the stability analysis of sliding mode control is based on special conditions as explained in Chapters 2 and 3. On the other hand, the stability analysis of nonlinear converter systems can be based on Lyapunov functions, which are also discussed in Chapter 2.

1.4.2 Settling Time

In the power converter control problem, the settling time is described as the time taken by a voltage (or a current) from a step variation to reach its final value and remain within a predefined tolerance band as shown in Figure 1.4. In other words, the settling time can be described as the dynamic response time of the control method. In some applications, the controller is highly desired to have a fast dynamic response (i.e. short settling time) against load variations, reference voltage (or current), and input voltage variations. In some control methods, the settling time can be adjusted to meet the desired speed by tuning the control gains. For example, the proportional gain in a proportional-integral (PI) controller has the effect on the settling time. Similarly, the sliding coefficient in the sliding mode control method plays an important role to make the settling time faster or slower. The Lyapunov function-based control and model predictive control methods have control parameters that directly affect the settling time.

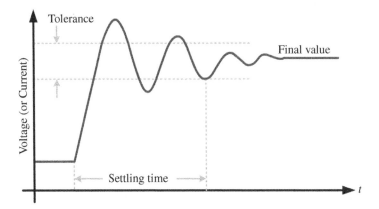

Figure 1.4 Settling time.

1.4.3 Steady-State Error

In the power converter control problem, the steady-state error is described as the difference between the actual value and reference value of a voltage (or a current) when the converter reaches steady state. Figure 1.5 shows the steady-state error graphically where the actual output is not equal to the desired output value during the steady state. When a steady-state error exists, it implies that the converter operates slightly away from its desired operating point. In such a case, the efficiency of the converter is reduced. For this reason, achievement of almost zero steady-state error is very important. For example, the integral term in the PI controller has the ability to achieve zero steady-state error. The integral gain determines how fast

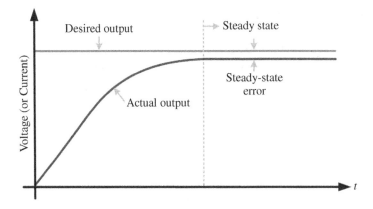

Figure 1.5 Steady-state error.

the error becomes zero in the steady state. In the case of sliding mode control, zero steady-state error can be achieved at the expense of high switching frequency. Similarly, the Lyapunov function-based control and model predictive control methods may yield zero steady-state error provided that the control parameters are selected appropriately.

1.4.4   Robustness to Parameter Variations and Disturbances

Robustness is defined as the ability of a closed-loop converter system to compensate the parameter variations and disturbances in the converter, and to maintain the stability and control objectives. The values of circuit parameters such as resistors, inductors, and capacitors vary due to aging and operating conditions. Also, the disturbances such as abrupt voltage (or current) variations may result in adverse effects in the performance and stability of the closed-loop system. For this reason, the effects of parameter variations as well as disturbances should be investigated. Among the control methods that are discussed in this book, only the sliding mode control is robust to the parameter variations and disturbances. This property is not surprising since the sliding mode control is not based on the mathematical model of the converter. As such, this property makes the sliding mode control very attractive in the applications where the robustness is important. On the other hand, the Lyapunov function-based control and model predictive control methods highly depend on the system parameters. Hence, the effect of parameter variations and disturbances on the performance of the designed controller should be investigated. In some cases, the parameters variations may endanger the closed-loop stability and result in an unstable system.

1.5   Performance Evaluation

The performance of the designed controller should be tested to see whether the desired control objectives are satisfied or not. In general, there are two methods to verify that the designed controller works properly.

1.5.1   Simulation-Based Method

In this method, a simulation model of the closed-loop converter control system is built and used to investigate the satisfaction of the control objectives. The most popular simulation platforms for power converter control are MATLAB®\Simulink, PSIM®, PLECS®, and POWERSIM®. An accurate simulation model that includes the effects of practical implementation can also be developed. Hence, the simulation-based method is very useful in enhancing the confidence that the predefined control objectives of the designed closed-loop system are achieved.

1.5.2 Experimental Method

In this method, the designed closed-loop system for the converter is implemented and tested under various operating conditions. Contrary to the simulation-based method, experimental method requires significant resources such as AC/DC power supplies, voltage/current sensors, AC/DC loads, passive components (resistors, inductors, capacitors, transformers), and measuring devices (oscilloscope, power quality analyzer, voltmeter, ammeter). More importantly, a platform is needed to realize the designed control method in real time. Digital signal processors (DSPs), field programmable gate arrays (FPGAs), and real-time simulators (e.g. OPAL-RT, Typhoon, and dSpace) are widely used in converter control problems. It is worth noting that the fixed-point DSPs are widely used in industrial applications due to their low cost. On the other hand, the computational power that is measured in terms of millions of instructions per second (MIPS) of DSPs is improved by time. Today, there exist very powerful DSPs in the market, which can handle the heavy computations needed in most of the control algorithms.

1.6 Contents of the Book

This book is organized as nine chapters. Chapter 1 presents brief information about the design of basic closed-loop controller for power converters, mathematical modeling, basic control objectives, and performance evaluation. Chapter 2 introduces the basics of advanced control methods such as sliding mode control (SMC), Lyapunov function-based control, and model predictive control (MPC). The introduction of these advanced control methods constitutes a bridge between theory and application.

Chapter 3 highlights the design of SMC for power converters in detail. For the sake of simplicity, two well-known DC–DC converters (Buck and Cuk) are selected. The simulation of these converters is presented and discussed. Thereafter, the SMC design procedure that includes sliding surface function selection, control input design, chattering mitigation techniques, and modulation techniques is presented. Finally, other types of SMC are also discussed briefly.

Chapter 4 highlights the design of Lyapunov function-based control for power converters in detail. The idea behind the selection of Lyapunov function for various converters is explained. For the sake of simplicity, two well-known DC–DC converters (Buck and Boost) are selected. The simulation of these converters is presented and discussed.

Chapter 5 highlights the design of MPC for power converters in detail. This includes the discussions about predictive control methods, finite control set MPC, continuous control set MPC, and design and implementation issues.

Chapter 6 presents MATLAB/Simulink tutorial on physical modeling and experimental setup. It includes discussions about building simulation models for power converters, modeling of sliding mode controlled single-phase grid-connected inverter, modeling of Lyapunov function-based control for three-phase rectifier, modeling of MPC-controlled quasi-Z-source three-phase four-leg inverter, modeling of distributed generations in islanded AC microgrid, real-time modeling of single-phase T-type rectifier, and building rapid control prototyping for single-phase T-type rectifier.

Chapter 7 presents the SMC of various power converters in detail with simulation as well as experimental results. The discussion starts with a single-phase grid-connected inverter with LCL filter. It includes mathematical modeling, SMC design, PWM signal generation using single- and double-band hysteresis modulations, switching frequency computation of both hysteresis modulation methods, selection of control gains, simulation and experimental results. Then, a three-phase grid-connected inverter with LCL filter is considered. The discussion includes model equations, design of SMC, stability analysis, experimental results, and computational load and performance of SMC. Then, a three-phase rectifier is considered. The discussion includes nonlinear model of the rectifier, problem formulation, axis-decoupling based on an estimator, design of SMC, and experimental results. A three-phase transformerless dynamic voltage restorer is also considered. After modeling of the system, the SMC design is presented. Then, the time-varying switching frequency equation is derived. Thereafter, constant switching frequency with boundary-layer-based SMC design is discussed. Simulation and experimental results are presented to verify the theoretical considerations. Finally, a three-phase shunt active power filter is considered. The discussion starts with the shunt active filter model and continues with problem formulation, SMC design, and experimental results.

Chapter 8 presents the Lyapunov function-based control of various power converters in detail with simulation as well as experimental results. The discussion starts with a single-phase grid-connected inverter with LCL filter. It includes mathematical modeling, controller design with capacitor voltage feedback, inverter current generation using proportional-resonant controller, grid current transfer function, harmonic impedance, simulation, and experimental results. Then, a single-phase quasi-Z-source grid-connected inverter with LCL filter is considered. The discussion includes quasi-Z-source modeling, grid-connected inverter modeling, control design of quasi-Z-source network, control design for grid-connected inverter, reference generation using cascaded proportional-resonant controller, and simulations results.

Chapter 9 presents continuous control set MPC for three-phase grid-connected voltage source inverter, MPC of single-phase three-level shunt active filter, MPC of quasi-Z-source three-phase four-leg inverter, weighting factorless MPC of DC-DC SEPIC converter, MPC droop control of distributed generation inverters in islanded AC microgrid, finite control set MPC of three-phase shunt active power filter, finite control set MPC of single-phase T-type rectifier, predictive torque control of brushless doubly fed induction generator fed by a matrix converter, and an enhanced finite control set MPC method with self-balancing capacitor voltages of three-level T-type rectifier.

References

1 M. P. Kazmierkowski, R. Krishnan, and F. Blaabjerg, *Control in Power Electronics: Selected Problems*. USA: Academic, 2002.

2 N. Guler, S. Biricik, S. Bayhan, and H. Komurcugil, "Model predictive control of DC–DC SEPIC converters with auto-tuning weighting factor," *IEEE Trans. Ind. Electron.*, vol. 68, no. 10, pp. 9433–9443, Oct. 2021.

3 N. Guler and H. Komurcugil, "Energy function based finite control set predictive control strategy for single-phase split source inverters," *IEEE Trans. Ind. Electron.*, vol. 69, no. 6, pp. 5669–5679, Jun. 2022.

4 N. S. Nise, *Control Systems Engineering*. USA: Wiley, 2004.

5 S. Buso and P. Mattavelli, *Digital Control in Power Electronics*. Switzerland: Springer-Verlag, 2015.

6 M. S. Fadali and A. Visioli, *Digital Control Engineering: Analysis and Design*. USA: Academic, 2009.

2

Introduction to Advanced Control Methods

2.1 Classical Control Methods for Power Converters

Power electronic converters are widely used in the commercial and industrial applications. The performance of these converters is crucial in achieving the required objectives in these applications. In order to achieve the required objectives which include fast transient response, reasonably low steady-state error, high stability margin, robustness to parameter variations, and small total harmonic distortion (THD), a power converter should be controlled by using a suitable control method. In literature, many control methods are developed to achieve these control objectives. In general, the control methods can be divided into two categories: linear control methods and nonlinear control methods. The most popular linear control method that is widely used in the industry is the proportional–integral–derivative (PID) control method [1]–[4]. However, the performance of the PID controller is considerably degraded when the converter is subject to parameter variations, nonlinearities, and disturbances. Thus, the researchers are focused on developing PID control methods that would improve the robustness of the system under parameter variations and disturbances [2], [4]. On the other hand, the proportional–integral (PI) controller is also popular in the control of various power converters [5]–[8]. The control technique presented in [7] is referred to as the Komurcugil–Kukrer control [9], which employs only one PI controller in controlling a three-phase active-front end rectifier and eliminates the need for measuring the grid currents contrary to the other linear controllers. Despite the developed PID and PI controllers, the performance is still not satisfactory under parameter variations, disturbances, and nonlinearities. The typical nonlinearities that exist in power converters are saturation, dead-zone, backlash, and hysteresis. In order to cope with the deficiencies of linear controllers, nonlinear control methods such

Advanced Control of Power Converters: Techniques and MATLAB/Simulink Implementation, First Edition. Hasan Komurcugil, Sertac Bayhan, Ramon Guzman, Mariusz Malinowski, and Haitham Abu-Rub.
© 2023 The Institute of Electrical and Electronics Engineers, Inc.
Published 2023 by John Wiley & Sons, Inc.
Companion website: www.wiley.com/go/komurcugil/advancedcontrolofpowerconverters

as the voltage-oriented control (VOC) [10] and direct power control (DPC) [11], [12] are emerged. Although VOC offers high dynamic performance through the use of inner current loops, it requires decoupling of active and reactive power. Also, the design of VOC is complicated due to the difficulty in controller tuning. Hence, the overall performance of VOC is degraded when the inner current loop is not designed properly. On the other hand, unlike the VOC, the DPC relies on the direct active and reactive power control by making use of a switching table, which eliminates the need for using a modulator. Also, the DPC does not require controller tuning, which leads to a simplicity in its design. Despite of employing different switching tables, no considerable improvement is achieved due to the variable switching frequency. With the aim of tackling the disadvantages of DPC, it is used by advanced nonlinear controllers, which include the sliding mode control (SMC) [13] and model predictive control (MPC) [14]. Recently, a review of DPC developed for PWM converters is summarized in [15].

In this chapter, an introduction to advanced control methods such as the SMC, Lyapunov function-based control, and MPC, which have been studied extensively for various power converter applications in the last two decades, is presented. The fundamentals of these control methods are explained in Chapters 7, 8, and 9 where the design of each control method for a specific application is discussed in detail.

2.2 Sliding Mode Control

The sliding mode control (SMC) is an effective nonlinear control approach that offers order reduction and insensitivity to parameter variations as well as disturbances and is suitable for complex systems operating under uncertainty conditions in today's modern technology [16]–[19]. Owing to these remarkable features, the interest in SMC is increased very rapidly in the last decades in many areas such as robotics, aerial vehicles, power electronics, and electrical drives. The main aim of this chapter is to explain the fundamentals of SMC theory for controlling power converters. Therefore, the other application areas of SMC are beyond the scope of this chapter and this book. It is well known that the structure of a power converter is subject to variations due to the switches and diodes connected in the power converters. Specially, the position of the switches are always changed by the binary type control signals (usually referred to as pulse width modulation signals). Hence, the power converters can be categorized in the particular class of variable structure systems (VSS). Therefore, the fundamental operating principle of SMC developed for a specific power converter can be based on by altering the dynamics of the power converter by applying a discontinuous control input that forces the error variables to slide along the predetermined surface called the sliding surface (usually referred to as sliding line, sliding manifold, or switching surface).

Thus, from this point of view, the SMC can be considered as a kind of variable structure control method that is very suitable for controlling power converters [20].

Now, let the second-order variable structure system be defined by the following differential equations:

$$\frac{dx_1}{dt} = x_2 \tag{2.1}$$

$$\frac{dx_2}{dt} = ax_2 + u \tag{2.2}$$

where x_1 and x_2 are the system states and u denotes the control input. Expressing (2.1) and (2.2) in the state-space form yields

$$\frac{d}{dt}\begin{bmatrix} x_1 \\ x_2 \end{bmatrix} = \begin{bmatrix} 0 & 1 \\ 0 & a \end{bmatrix}\begin{bmatrix} x_1 \\ x_2 \end{bmatrix} + \begin{bmatrix} 0 \\ 1 \end{bmatrix}u \tag{2.3}$$

One can easily determine that the system in (2.3) is unstable. Now, let use a linear state feedback to define the control input as follows:

$$u = -Kx_1 \tag{2.4}$$

The eigenvalues of this system can be obtained as

$$\lambda_{1,2} = \frac{a \pm \sqrt{a^2 - 4K}}{2} \tag{2.5}$$

Assuming that $a = 1$, the eigenvalues with $K = 12$ can be calculated as $\lambda_{1,2} = 0.5 \pm j3.4278$, while the eigenvalues obtained by $K = -12$ have real values $\lambda_1 = 4$ and $\lambda_2 = -3$. This implies that the system is subject to two structures. The state trajectories of these structures are shown in Figure 2.1. It is apparent from Figure 2.1a that the state trajectory diverges from the origin by making a spiral movement. This means that the equilibrium point at the origin (i.e. $x_1 = 0$ and $x_2 = 0$) is unstable for $K = 12$. On the other hand, the trajectories starting at arbitrary initial points with $K = -12$ result in a saddle point at the origin as shown in Figure 2.1b. Clearly, only one trajectory (dashed line) converges to the origin while all other trajectories diverge. Hence, the control input with $K = 12$ or $K = -12$ is not able to stabilize the system.

Now, let us consider the combination of these trajectories shown in Figure 2.2. Also, let us consider the following line from the state trajectories in Figure 2.2 as follows:

$$\sigma = \lambda x_1 + x_2 = 0, \quad 0 < \lambda < 3 \tag{2.6}$$

Hence, the control input can be structured as a combination of previous control inputs as

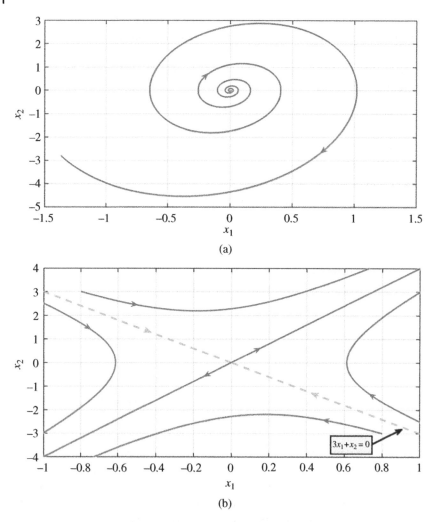

Figure 2.1 Trajectories of the system in the phase plane for two distinct K values. (a) $K = 12$, (b) $K = -12$.

$$u = \begin{cases} 12x_1 & \text{if} \quad \sigma x_1 > 0 \\ -12x_1 & \text{if} \quad \sigma x_1 < 0 \end{cases} \qquad (2.7)$$

It is apparent that the control input in (2.7) is discontinuous due to the two distinct control input values, which are needed because of two different system structures. Hence, such control is referred to as the variable structure control whose block diagram is shown in Figure 2.3.

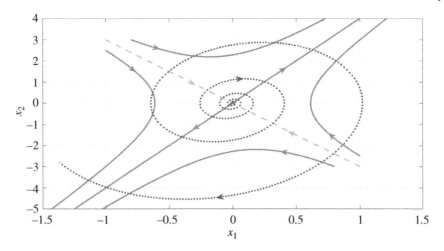

Figure 2.2 Combination of the trajectories in Figure 2.1a and Figure 2.1b.

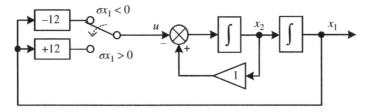

Figure 2.3 Block diagram of variable structure control.

For the purpose of explaining the effect of variable structure control on the system, the trajectories in Figure 2.2 together with lines $\lambda x_1 + x_2 = 0$ and $3x_1 + x_2 = 0$ are displayed in Figure 2.4. It is obvious that the line $\lambda x_1 + x_2 = 0$ splits the phase plane into regions in which the line has different sign. The boundaries of these regions are the x_2 axis and line $\lambda x_1 + x_2 = 0$. When the trajectory enters a region defined by these boundaries, it means that the structure of system is changed. Then, the trajectory moves inside that region until it hits the line $\lambda x_1 + x_2 = 0$. Thereafter, the applied control forces the trajectory moving on the line toward origin. Such motion of trajectories on the line is called sliding mode and the line is callsed sliding line (or sliding surface). Hence, the origin becomes an asymptotically stable equilibrium point of the controlled system. Therefore, the sliding mode control is as a kind of variable structure control. On the other hand, since the state trajectory moves on $\lambda x_1 + x_2 = 0$, system's order is reduced from two to one during

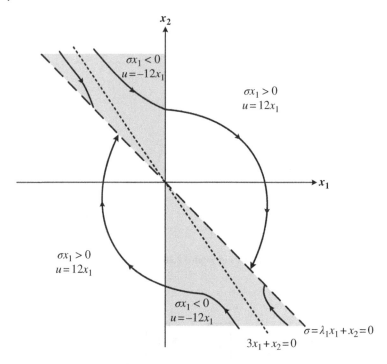

Figure 2.4 Regions of variable structure control that establish sliding mode.

the sliding mode. More importantly, the motion during the sliding mode is insensitive to system parameters and disturbance.

Generally, the state trajectory passes through two phases. While the first movement is known as the reaching phase, the latter is called the sliding phase. Figure 2.5 shows the typical reaching and sliding phases of a second-order system. The discontinuous control inputs are denoted by u^+ and u^-. When the state trajectory reaches the sliding surface, the sliding surface function satisfies $\sigma = 0$, which means that the sliding mode exists. On the other hand, the zigzag movements on the sliding line denote chattering that is the main problem of sliding mode control.

2.3 Lyapunov Function-Based Control

Stability theory is crucial in determining the behavior of linear as well as nonlinear control systems. The most widely used method to evaluate the stability of a nonlinear control system is the Lyapunov stability theory invented by the Russian

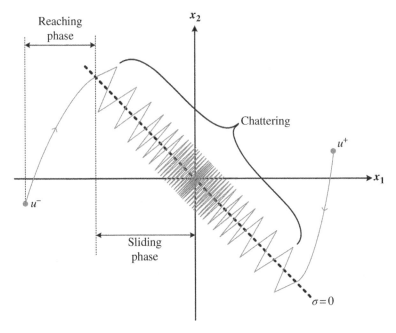

Figure 2.5 Reaching and sliding phases of a second-order system in the phase plane.

mathematician Alexander Mikhailovich Lyapunov in the nineteenth century. However, this theory became popular in the early 1960s. Basically, Lyapunov stability theory contains two methods, known as the linearization and direct methods, for analyzing the stability of a system [21]. While the linearization method relies on the linear approximation of the nonlinear system, the direct method is based on the energy-like function (i.e. Lyapunov function) of the system. At this stage, it is worthy to remark that while the linearization method can be used to investigate the local stability, the direct method provides information about local and global stability without doing linear approximation.

2.3.1 Lyapunov's Linearization Method

Now, consider the following time-invariant nonlinear system:

$$\frac{d\mathbf{x}}{dt} = \mathbf{f}(\mathbf{x}) \tag{2.8}$$

where $\mathbf{x}$ denotes the $n \times 1$ state variable vector and $\mathbf{f}(\mathbf{x})$ is the $n \times 1$ vector function. Assuming that $\mathbf{f}(\mathbf{x})$ is continuous, then (2.8) is expressed as

$$\frac{dx}{dt} = \left(\frac{\partial \mathbf{f}}{\partial \mathbf{x}}\right)\bigg|_{\mathbf{x}=0} \mathbf{x} = \mathbf{A}\mathbf{x} \tag{2.9}$$

where $\mathbf{A}$ is the Jacobian matrix. Eq. (2.9) is linearization of nonlinear system in (2.8) at $\mathbf{x} = 0$. For a second-order system, Eq. (2.9) can be written as

$$\frac{d}{dt}\begin{bmatrix} x_1 \\ x_2 \end{bmatrix} = \begin{bmatrix} \dfrac{\partial f_1}{\partial x_1} & \dfrac{\partial f_1}{\partial x_2} \\ \dfrac{\partial f_2}{\partial x_1} & \dfrac{\partial f_2}{\partial x_2} \end{bmatrix}\Bigg|_{\substack{x_1=0 \\ x_2=0}} \begin{bmatrix} x_1 \\ x_2 \end{bmatrix} \tag{2.10}$$

The linearization method is useful to investigate the behavior of a nonlinear system under its linearized approximation for small variations around the equilibrium point (i.e. $\mathbf{x} = \mathbf{x}^*$, where $\mathbf{x}^*$ is the reference of $\mathbf{x}$). The stability test of the linearized system provides knowledge about the stability of nonlinear system locally. For example:

i) If all eigenvalues of $\mathbf{A}$ are located in the left-half plane, then the linearized system is strictly stable.
ii) If at least one eigenvalue of $\mathbf{A}$ is located in the right-half plane, then the linearized system is unstable.
iii) If all eigenvalues of $\mathbf{A}$ are located in the left-half plane with at least one of them is located on the imaginary axis, then the linearized system is marginally stable.

Therefore, the information about the stability of linearized system can be obtained from above conditions. For instance, if the linearized system is strictly stable, then the nonlinear system is locally stable and vice versa. Even though Lyapunov's linearization method provides information about the stability of nonlinear system, it does not give any information about the extent of stability.

2.3.2 Lyapunov's Direct Method

Lyapunov's direct method relies on observing the total energy of a nonlinear system as well as a linear system. If the total energy of a system is dissipated continuously, then the system should settle down to its equilibrium point (i.e. the operating point of the system in the steady state). Hence, the variation of energy can be examined to obtain information about the stability of the system. In the case of power converters, the following relations can be obtained between the total energy and converter's stability:

i) When the total energy is settled down to zero, it means that the converter operates at its equilibrium point.

ii) When the total energy is settled down to zero, it means that asymptotic stability of converter is assured.

iii) When the total energy grows, it means that the converter is unstable.

Therefore, the stability of converter can be characterized by the variation of its total energy. According to Lyapunov's direct method, a scalar energy-like function (usually referred to as the Lyapunov function) possessing the following features should be selected:

i) $V(\mathbf{x}) = 0$ when $\mathbf{x} = \mathbf{x}^*$

ii) $V(\mathbf{x}) > 0$ when $\mathbf{x} \neq \mathbf{x}^*$

iii) $V(\mathbf{x}) \to \infty$ when $\sqrt{x_1^2 + x_2^2 + \ldots + x_n^2} \to \infty$

iv) $\dfrac{dV(\mathbf{x})}{dt} \leq 0$ for all $\mathbf{x}$ in the state space

As can be seen from the first two conditions, the Lyapunov function should be strictly positive for nonzero state vector $\mathbf{x}$ and be zero when the state vector equals to its reference (i.e. the system is in its equilibrium point). Also, it should converge to infinity as the magnitude of state vector tends to infinity. The last condition dictates global asymptotic stability by achieving negative derivative of Lyapunov function. Hence, the main idea behind Lyapunov's direct method is to guarantee the global asymptotic stability of a system by maintaining the derivative of Lyapunov function always negative. The first property of Lyapunov function (i.e. $V(\mathbf{x}) = 0$) means that when the Lyapunov function is assessed at the equilibrium point ($\mathbf{x} = \mathbf{x}^*$), the Lyapunov function should possess zero energy. According to the second property (i.e. $V(\mathbf{x}) > 0$), the Lyapunov function should be always positive when the state vector is away from the equilibrium point ($\mathbf{x} \neq \mathbf{x}^*$). The geometrical view of three different Lyapunov functions with $n = 3$ is depicted in Figure 2.6. The Lyapunov functions $V_1(\mathbf{x})$ and $V_2(\mathbf{x})$ are the constant energy curves with $V_1(\mathbf{x}) > V_2(\mathbf{x}) > V_3(\mathbf{x})$. When the state vector $\mathbf{x}$ moves toward the equilibrium point ($\mathbf{x} = \mathbf{x}^*$), the energy changes from high to low level. Eventually, the Lyapunov function $V_3(\mathbf{x})$ possess zero energy at the equilibrium point. The control inputs should be selected carefully to force the system to move to the lower energy level. It is worth to remark that the derivative of $V(\mathbf{x})$ along the system trajectory provides useful information about the stability. In this respect, $\dfrac{dV(\mathbf{x})}{dt} \leq 0$ for a direction of $\mathbf{x}^*$ means that state variables tend to the lower energy level. Hence, the condition $\dfrac{dV(\mathbf{x})}{dt} \leq 0$ is sufficient to guarantee the asymptotic stability.

Now, let us consider the following first-order system:

$$\frac{dx}{dt} = ax + u, \quad a < 0 \tag{2.11}$$

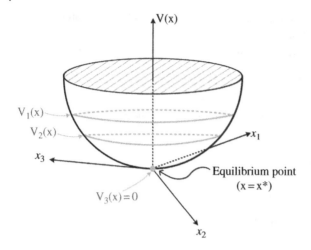

Figure 2.6 Geometrical view of three different Lyapunov functions for *n* = 3.

where u is the control input. Assuming that the control input is designed as

$$u = -Kx, \quad \frac{dK}{dt} = \alpha x^2, \quad \alpha > 0 \tag{2.12}$$

and letting $x_1 = x$ and $x_2 = K$, the equations of closed-loop system is expressed as

$$\frac{dx_1}{dt} = -x_1(x_2 - a) \tag{2.13}$$

$$\frac{dx_2}{dt} = \alpha x_1^2 \tag{2.14}$$

The equilibrium point is $x_1 = 0$. In other words, the control objective is to obtain $x_1 = 0$ as time tends to infinity. In this case, the controller regulates x to zero. Now, let us take into consideration the following Lyapunov function candidate:

$$V(\mathbf{x}) = \frac{1}{2}x_1^2 + \frac{1}{2\alpha}x_2^2 \tag{2.15}$$

Close inspection to (2.15), one can see that the first three conditions related with Lyapunov function mentioned before hold. Now, taking derivative of $V(\mathbf{x})$, we obtain

$$\frac{dV(\mathbf{x})}{dt} = x_1\frac{dx_1}{dt} + \frac{1}{\alpha}x_2\frac{dx_2}{dt} = ax_1^2 \le 0 \tag{2.16}$$

Since $a < 0$, the condition $\dfrac{dV(\mathbf{x})}{dt} \le 0$ is also achieved, which means that the closed-loop system is stable around $x = x_1 = 0$.

2.4 Model Predictive Control

The idea of model predictive control (MPC) was first introduced in the 1970s by industrial circles (rather than control theorists). In the 1980s, it steadily gained popularity. At present, it has become the most widely studied multivariable control technique in chemical process industries, as well as in other fields. MPC principles are particularly promising in power electronics applications for controlling power converters and electrical drives. Several advantages of predictive control algorithms make them beneficial for the control of power converters, including:

1) There is an intuitive and easy-to-understand structure to the concepts.
2) Multiple control objectives can be handled by MPC.
3) It is easy to incorporate constraints and nonlinearities.
4) In general, the controller can be implemented easily.

In this sub-chapter, we will review the basic theoretical underpinnings of MPC and demonstrate their application to power converters.

2.4.1 Functional Principle

MPC controller can be explained using the structure shown in Figure 2.7. An integral part of the MPC is the model that is used to predict the future behavior of the system. Future error is calculated as the difference between the future reference value and the precalculated actual value that is obtained through the model of the system. By taking system constraints and cost functions into account, an optimization algorithm (optimizer) determines an optimum set of future actuating values. Every sampling cycle, the whole procedure of prediction, optimization, and control, is repeated.

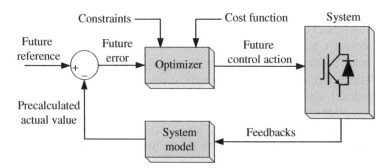

Figure 2.7 Typical structure of an MPC controller.

Due to its closeness to human instinct, MPC based on receding horizon control can be called a sort of "natural" predictive control. As an example, when a driver drives, he does not look directly in front of his car, but far ahead and changes the actuating variables, such as steering wheel position, gas pedal, and brake positions, before he approaches a junction or a traffic light. Further, he determines the behavior of the car over a specified distance in front of him until a finite horizon based on future values of the actuating variables. Based on the optimization criteria, he has set for this distance, he determines the amount of acceleration or braking to be used, and then acts accordingly. Like MPC, there are different optimization criteria that can be used, which can result in different results. For example, it is more likely that a driver will accelerate and brake more rapidly if he is optimizing for the shortest possible duration of his trip instead of a reduction in fuel consumption [22].

2.4.2 Basic Concept

Typically, MPC algorithms are formulated in discrete time with a constant sampling interval. Furthermore, the system inputs can change their values only at discrete sampling intervals, i.e. at time $t = k$. The method of discrete-time state-space representation is convenient for power electronics applications since nonlinear dynamic relations are frequently involved:

$$x(k + 1) = f(x(k), u(k)) \tag{2.17}$$

where $x(k)$ and $u(k)$ are the state value at time k and the system input, respectively. The system state may consist of capacitor voltages, inductor and load currents, and fluxes, depending on the application.

Among the interesting features of the MPC framework is its ability to incorporate state and input constraints. The state constraints can, for instance, be imposed on the capacitor voltage in neutral point clamped converters or T-type converters. It is also possible to model load current constraints as state constraints. On the other hand, the input constraints are based on the switch positions during the interval. A converter that uses a modulator should be constrained to operate within a bounded continuous space. It may be possible to use duty cycles ($d(k)$) or PWM reference signals for the components of the system input ($u(k)$) (see Figure 2.8). Indirect control applications, however, where no modulator is used, require $u(k)$ to be part of a finite set of switch combinations. There has been considerable interest in these approaches in the power electronics community, often referred to as Finite Control Set Model Predictive Control (FCS-MPC). This predictive control strategy has the main advantage of directly accounting for switching actions ($S(k)$) as input constraints in the optimization (see Figure 2.9).

3

Design of Sliding Mode Control for Power Converters

3.1 Introduction

With the development of technology and various power converter topologies, researchers have become interested in designing nonlinear control methodologies that are not merely valid around small operating point but also provide excellent performance (i.e. fast dynamic response, zero steady-state error, ensured closed-loop stability) away from the operating point without being sensitive to the parameter variations and disturbances. Sliding mode control (SMC) is one of the powerful nonlinear control methods suitable for controlling power converters.

Figure 3.1 reveals the design of SMC for controlling power converters [1]. It should be noted that the discrete-time SMC design is omitted here since it is not widely reported yet. Clearly, the design of continuous-time SMC is based on the sliding surface function design, sliding coefficient selection, control input selection, chattering reduction, and modulation technique. For the sake of simplicity, the design of SMC will be explained for the second-order DC–DC buck converter and fourth-order DC–DC Cuk converter. The SMC of other converter topologies will be discussed in the case studies in Chapter 7.

3.2 Sliding Mode Control of DC–DC Buck and Cuk Converters

With the aim of explaining SMC concept in power converters, let us consider a traditional DC–DC buck converter illustrated in Figure 3.2.

Advanced Control of Power Converters: Techniques and MATLAB/Simulink Implementation, First Edition. Hasan Komurcugil, Sertac Bayhan, Ramon Guzman, Mariusz Malinowski, and Haitham Abu-Rub.
Published 2023 by John Wiley & Sons, Inc.
Companion website: www.wiley.com/go/komurcugil/advancedcontrolofpowerconverters

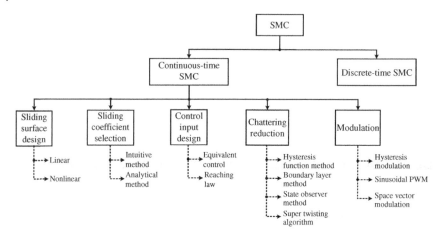

Figure 3.1 The design of SMC for controlling power converters. *Source:* Komurcugil et al. [1]/with permission of IEEE.

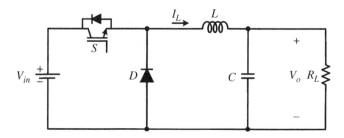

Figure 3.2 DC–DC buck converter.

The differential equations of the converter are written as [2]

$$\frac{dI_L}{dt} = \frac{V_{in}}{L}u - \frac{V_o}{L} \tag{3.1}$$

$$\frac{dV_o}{dt} = \frac{I_L}{C} - \frac{V_o}{R_L C} \tag{3.2}$$

where I_L is the inductor current, V_o is the capacitor voltage (i.e. load voltage), V_{in} is the DC input voltage, and u is the discontinuous control input that takes 1 when the switch (S) is on and 0 when the switch is off. The fundamental control objective in a DC–DC converter is to achieve the load voltage control under load and reference variations. Now, let the error variables be given as follows:

$$x_1 = V_o - V_o^* \tag{3.3}$$

$$x_2 = \frac{dx_1}{dt} = \frac{dV_o}{dt} - \frac{dV_o^*}{dt} \tag{3.4}$$

where V_o^* is the reference for V_o. Since the derivative of V_o^* is zero, Eq. (3.4) can be rewritten as

$$x_2 = \frac{dV_o}{dt} = \frac{I_C}{C} \tag{3.5}$$

Taking derivative of (3.5) yields

$$\frac{dx_2}{dt} = \frac{1}{C}\frac{dI_L}{dt} - \frac{1}{R_L C}x_2 \tag{3.6}$$

Now, substituting (3.1) in (3.6) gives

$$\frac{dx_2}{dt} = \frac{V_{in}}{LC}u - \frac{x_2}{R_L C} - \frac{V_o}{LC} \tag{3.7}$$

Substitution of $V_o = x_1 + V_o^*$ in (3.7) results in

$$\frac{dx_2}{dt} = \frac{V_{in}}{LC}u - \frac{x_1}{LC} - \frac{x_2}{R_L C} - \frac{V_o^*}{LC} \tag{3.8}$$

Equations (3.4) and (3.8) can be put into state-space form as follows:

$$\frac{d}{dt}\begin{bmatrix} x_1 \\ x_2 \end{bmatrix} = \begin{bmatrix} 0 & 1 \\ -\dfrac{1}{LC} & -\dfrac{1}{R_L C} \end{bmatrix}\begin{bmatrix} x_1 \\ x_2 \end{bmatrix} + \begin{bmatrix} 0 \\ \dfrac{V_{in}}{LC} \end{bmatrix}u + \begin{bmatrix} 0 \\ -\dfrac{V_o^*}{LC} \end{bmatrix} \tag{3.9}$$

In order to demonstrate the behavior of the system in the phase-plane (i.e. x_1 versus x_2), the buck converter is simulated for $u = 0$ and $u = 1$ using various initial values of x_1 and x_2 for each switching case. Figure 3.3 depicts these trajectories obtained by using $V_{in} = 30\,\text{V}$, $V_o^* = 12\,\text{V}$, $L = 1\,\text{mH}$, $C = 500\,\mu\text{F}$ and $R = 10\,\Omega$. An insight into the behavior of the system can be gained by observing these trajectories. It is clear from Figure 3.3a that the trajectories converge to $x_1 = -V_o^* = -12\,\text{V}$ and $x_2 = 0$ for $u = 0$. It is remarkable to note that when I_L becomes zero, the capacitor is discharged through the load resistance and cannot take negative values due to diode D. Similarly, the trajectories converge to $x_1 = V_{in} - V_o^* = 18\,\text{V}$ and $x_2 = 0$ for $u = 1$ as shown in Figure 3.3b. However, these trajectories should be combined in order to be able to exploit the behavior of these trajectories in the design of the SMC. Figure 3.3c reveals the combination of the trajectories for both $u = 0$ and $u = 1$. In order to be able to display the behavior of the trajectories in the vicinity of origin ($x_1 = 0$ and $x_2 = 0$), the trajectories in Figure 3.3c are magnified as shown in Figure 3.3d. It is evident that these trajectories belong to a line (called sliding line) defined by

$$x_2 = \frac{dx_1}{dt} = -\lambda x_1 \qquad (3.10)$$

where λ represents the positive sliding coefficient. It can be noticed that λ constitutes the slope of the sliding line. The solution of first-order differential equation in (3.10) is given by

$$x_1(t) = x_1(0)e^{-\lambda t} \qquad (3.11)$$

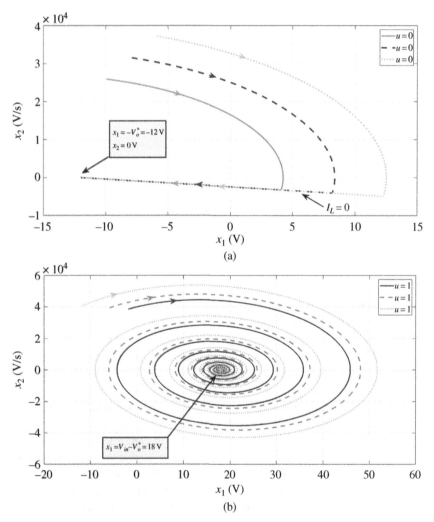

Figure 3.3 Phase-plane trajectories obtained by various initial conditions for: (a) $u = 0$, (b) $u = 1$, (c) combination of both $u = 0$ and $u = 1$, and (d) magnified trajectories in (c).

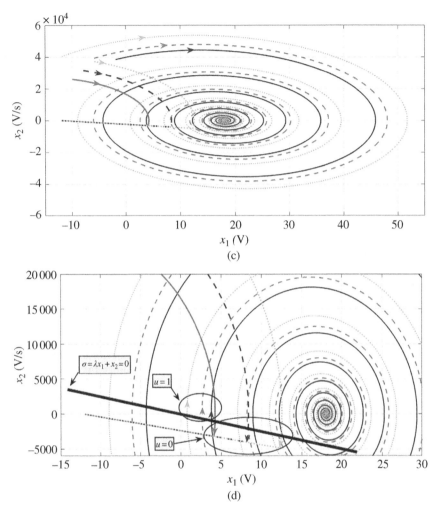

Figure 3.3 (Continued)

Clearly, $x_1(t)$ decays to zero exponentially as $t \to \infty$. Equation (3.10) can also be written as

$$\sigma = \lambda x_1 + x_2 = 0 \qquad (3.12)$$

where σ denotes the sliding surface function. Equation (3.12) describes the converter dynamics in the sliding mode. Hence, Eq. (3.12) is valid when the converter operates in the sliding mode. For this reason, the derivative of sliding surface function ($d\sigma/dt$) should also be zero during the sliding mode. Based on these facts, we can write the following statements while the system operates in the sliding mode:

i) The convergence to the origin does not depend on the circuit parameters and disturbances as shown in Eq. (3.11). The rate of convergence can be adjusted by selecting suitable λ.

ii) The order of system is reduced by one since the error variables are constrained by $\sigma = \lambda x_1 + x_2 = 0$.

Thus, it is reasonable to formulate the sliding surface function by linear combination of x_1 and x_2. Also, it can be seen from Figure 3.3d that the sliding line divides the phase-plane into two areas. These areas are below and above the sliding line. It can be observed from Figure 3.3d that when the trajectory is above the sliding line ($\sigma > 0$), the value of control input should be 0 ($u = 0$) so that the trajectory is forced to move toward the sliding line. Similarly, when the trajectory is below the sliding line ($\sigma < 0$), the value of control input should be 1 ($u = 1$) so that the trajectory is forced to move toward the sliding line. Hence, the control input can be expressed as follows:

$$u = \frac{1}{2}(1 - sign(\sigma)) = \begin{cases} 0 & \text{if} \quad \sigma > 0 \\ 1 & \text{if} \quad \sigma < 0 \end{cases} \tag{3.13}$$

When the control in (3.13) is applied to the converter, the trajectories are forced to move toward the sliding line. However, the control input in (3.13) is not able to maintain the trajectories on this line. The condition to maintain the trajectories on the line will be discussed in Section 3.3.

Figure 3.4 shows the phase-plane trajectory of the buck converter simulated with the parameters mentioned above with different λ values and V_o initial points

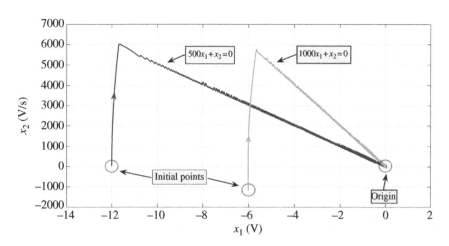

Figure 3.4 Phase-plane trajectories of buck converter with different λ values.

(0 and 6 V). It can be seen that the trajectory with $\lambda = 500$ starts at an initial point ($x_1 = -12$ V and $x_2 = 0$ V/s) and reaches the sliding line after passing the reaching phase. When the trajectory reaches the sliding line ($500x_1 + x_2 = 0$), the sliding mode operation starts and trajectory moves along the sliding line until origin is reached. Similarly, the trajectory with $\lambda = 1000$ starts at an initial point ($x_1 = -6$ V and $x_2 \simeq -1000$ V/s) and reaches the sliding line after passing the reaching phase. Comparing these trajectories, one can see that both trajectories have similar behavior and the trajectory obtained with $\lambda = 1000$ reaches the origin faster. This means that the sliding time of trajectory with $\lambda = 1000$ on the sliding line is shorter than the other trajectory. Despite the trajectories start at different initial positions, both of them successfully reach the origin. On the other hand, the chattering is discernible on the trajectories. In order to be able to alleviate chattering, the sign function in (3.13) can be replaced by a hysteresis function. The detailed discussion on the chattering mitigation techniques will be presented in Section 3.4.

The responses of output voltage attained with two distinct λ values corresponding to Figure 3.4 are shown in Figure 3.5.

As a second example, let us consider a more complicated converter such as a fourth-order DC–DC Cuk converter feeding a resistive load (R_L) as shown in Figure 3.6. It is worth noting that Cuk converter has the ability to operate as buck converter (output voltage is smaller than input voltage) as well as boost converter (output voltage is greater than input voltage) by varying the duty ratio. However, unlike the traditional DC–DC converters, the polarities of the output capacitor voltage (V_{C2}) are reversed.

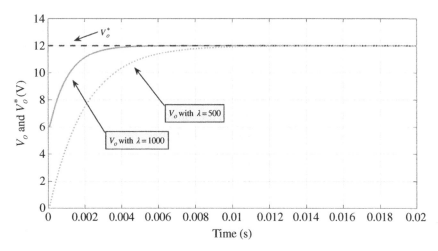

Figure 3.5 Responses of output voltage with different λ values.

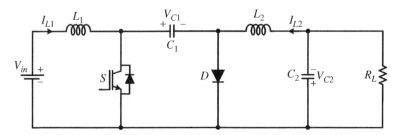

Figure 3.6 DC–DC Cuk converter.

The differential equations of the converter are written as [3]

$$\frac{dI_{L1}}{dt} = \frac{1}{L_1}(V_{in} - V_{C1} + V_{C1}u) \tag{3.14}$$

$$\frac{dV_{C1}}{dt} = \frac{1}{C_1}(I_{L1} - I_{L1}u - I_{L2}u) \tag{3.15}$$

$$\frac{dI_{L1}}{dt} = \frac{1}{L_2}(V_{C1}u - V_{C2}) \tag{3.16}$$

$$\frac{dV_{C2}}{dt} = \frac{1}{C_2}\left(I_{L2} - \frac{V_{C2}}{R_L}\right) \tag{3.17}$$

where u is the discontinuous control input defined by

$$u = \begin{cases} 1 & \text{if} \quad S \text{ is closed} \\ 0 & \text{if} \quad S \text{ is open} \end{cases} \tag{3.18}$$

Unlike the differential equations of buck converter in which the control input exists in the derivative of inductor current equation (see Eq. (3.1)) only, the differential equations of Cuk converter involves control input as shown in (3.14)–(3.16). This makes the control of Cuk converter more challenging than the control of buck converter. Hence, it is reasonable to design the control input such that the inductor current follows its reference. In this case, the control input design can be based on the sign of inductor current tracking error as follows:

$$u = \frac{1}{2}(1 - sign(\sigma)) = \begin{cases} 0 & \text{if} \quad \sigma > 0 \\ 1 & \text{if} \quad \sigma < 0 \end{cases} \tag{3.19}$$

where $\sigma = I_{L1} - I_{L1}^*$ is the inductor current tracking error, which is also considered as the sliding surface. By making use of power balance equation of the converter, the reference of inductor current is derived as

$$I_{L1}^* = \frac{(V_{C2}^*)^2}{R_L V_{in}} \tag{3.20}$$

When the converter operates in the sliding mode, then the continuous control input can easily be deduced by solving $d\sigma/dt = 0$. Now, taking the derivative of sliding surface function and equating it to zero yields

$$\frac{1}{L_1}(V_{in} - V_{C1} + V_{C1} u_{ss}) = 0 \tag{3.21}$$

where u_{ss} is the control input (usually referred to as the equivalent control that will be explained in the next section) needed in the steady state. Solving for u_{ss} gives

$$u_{ss} = \frac{V_{C1} - V_{in}}{V_{C1}} \tag{3.22}$$

The fundamental control objective is to accomplish output voltage control in the steady state. Now, let the output voltage tracking error be given as follows:

$$x_1 = V_{C2} - V_{C2}^* \tag{3.23}$$

where V_{C2}^* is the reference of V_{C2}. Taking derivative of x_1 one can obtain the following first-order linear differential equation

$$\frac{dx_1}{dt} + \frac{1}{R_L C_2} x_1 = 0 \tag{3.24}$$

Apparently, the solution of (3.24) is straight forward and is expressed as

$$x_1(t) = x_1(0)e^{-\frac{1}{R_L C_2} t} \tag{3.25}$$

Hence, $x_1(t) \rightarrow 0$ as $t \rightarrow \infty$. Figure 3.7 reveals the responses of output voltage and inductor current for an abrupt change in V_{in} from 80 to 40 V obtained by using $V_{C2}^* = 50$ V, $L_1 = 10$ mH, $L_2 = 10$ mH, $C_1 = 0.1$ μF, $C_2 = 0.1$ μF, and $R_L = 50\,\Omega$. Initial values of V_{C2} and I_{L1} are taken as zero. It can be noticed that when the converter operates in the buck mode ($V_{in} = 80$ V), both V_{C2} and I_{L1} track their references successfully and converge to 50 V and 0.625 A, respectively. When the input voltage is suddenly changed to 40 V, the converter also changes its operation mode from buck mode to the boost mode. In this case, again, both V_{C2} and I_{L1} track their references and converge to 50 V and 1.25 A, respectively. These results clearly demonstrate that the fourth-order DC–DC Cuk converter can be easily controlled by using SMC technique that does not depend on converter's parameters.

Figure 3.8 shows the trajectory of error variables (i.e. σ and x_1) in phase plane corresponding to Figure 3.7. The trajectory for the buck mode operation is depicted in Figure 3.8a. It is obvious that the trajectory starts at an initial point

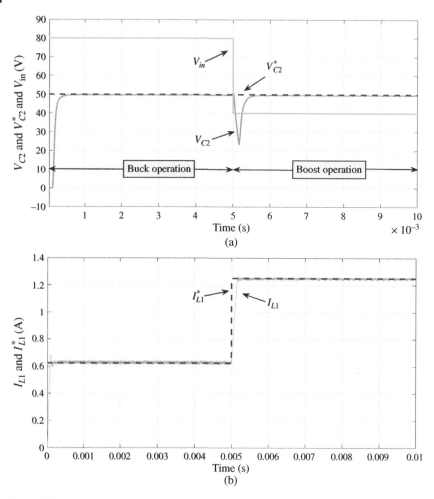

Figure 3.7 Responses of V_{C2} and I_{L1} for an abrupt change in V_{in} from 80 to 40 V. (a) V_{in} and V_{C2}, (b) I_{L1}.

where $\sigma = -0.625$ A and $x_1 = -50$ V. Then, the trajectory reaches to $\sigma = 0$ very fast and moves along it with zigzag motions toward the origin ($\sigma = 0$ and $x_1 = 0$). It can be noticed that the frequency of zigzag motions increases as the trajectory moves toward the origin. On the other hand, the trajectory for the boost mode operation is depicted in Figure 3.8b. Unlike the trajectory in Figure 3.8a, the trajectory starts at origin and reaches to $\sigma = 0$. Then, it moves along $\sigma = 0$ with zigzag motions until it reaches to origin. The controller maintains the trajectory at origin unless there is no change in the operating point.

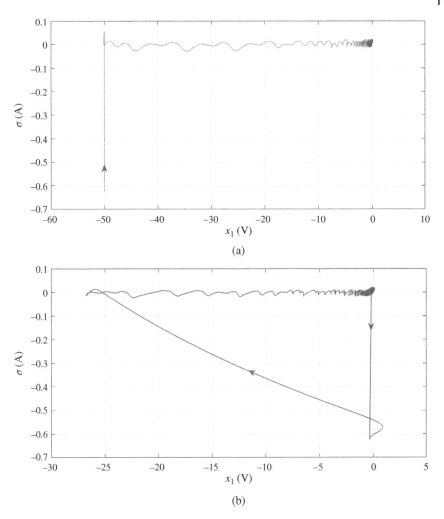

Figure 3.8 Trajectory of σ and x_1 in phase-plane corresponding to Figure 3.7. (a) Trajectory in the buck mode, (b) trajectory in the boost mode.

Because of using SMC in controlling both DC–DC buck and Cuk converters, it is observed that the control objective (regulation of output voltage) is achieved successfully without depending on the variations in system parameters. However, two main problems (chattering and variable switching frequency) remain unresolved. The techniques to mitigate chattering effect and make the switching frequency constant will be discussed in Section 3.4.

3.3 Sliding Mode Control Design Procedure

Generally, the SMC design relies on two significant steps.

Step 1. Selection of sliding surface function σ.
Step 2. Design of control input **u**.

While selection of sliding surface function should be done such that the stable operation is guaranteed, the control input design should ensure that the state trajectory moves toward the selected sliding surface and move along the surface until the steady-state point (equilibrium point) is reached. Once the designs of sliding surface function and control input for a specific power converter are performed properly, then the closed-loop control system is tested by simulation as well as by implementation in real time. The generalized block diagram of a sliding mode controlled power converter is illustrated in Figure 3.9.

3.3.1 Selection of Sliding Surface Function

Depending on the converter topology, the scalar sliding surface function is generally selected in terms of the error variables in the following form:

$$\sigma = \begin{cases} \lambda_1 x_1 + \lambda_2 x_2 + \ldots + \lambda_m x_m & \Rightarrow \quad \text{SMC} \\ \lambda x_1 + x_2 & \Rightarrow \quad \text{SMC} \\ \lambda_1 x_1 + x_2 + \lambda_2 \int x_1 dt & \Rightarrow \quad \text{ISMC} \\ x_1 & \Rightarrow \quad \text{SMC} \\ \lambda x_1^{\gamma} + x_2 & \Rightarrow \quad \text{TSMC} \end{cases} \tag{3.26}$$

where x_1, x_2, and x_m are the error variables, $\lambda_1 > 0$, $\lambda_2 > 0$, and $\lambda_m > 0$ are the sliding coefficients, and γ is the constant which should satisfy $0 < \gamma < 1$. It should be noted that γ is chosen as $\gamma = q/p$ in which p and q are positive odd integers, which are selected by using trial-and-error method so as to obtain the desired performance. The first type of sliding surface function formation in (3.26) is generally used for high-order converter systems such as fourth-order DC–DC Cuk converter [3] and DC–DC SEPIC converter [4]. Clearly, this type of sliding surface function

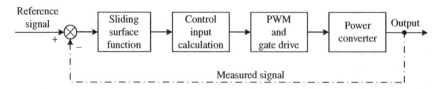

Figure 3.9 Generalized block diagram of a sliding mode controlled power converter.

formation relies on linear combination of the error variables. The second type of sliding surface function formation in (3.26) is utilized for the second-order converters such as DC–DC buck converter [2], single-phase UPS inverter with LC filter [5], and single-phase dynamic voltage restorer [6]. The third type of sliding surface function is used in the integral SMC (ISMC) method where the steady-state error is reduced [7], [8]. The fourth type of sliding surface function is the simplest one which is based on an error variable only [9]. Finally, the last type of sliding surface function formation in (3.26) is used for the terminal SMC (TSMC) methods developed for various converters such as DC–DC buck converter [10] and DC–DC boost converter [11], [12].

Generally, the selection of sliding surface function relies on the complexity of converter under consideration. That is to say, the order of converter, number of controlled variables, and number of measured variables should be taken into consideration before the sliding surface function is formulated. On the other hand, 2nd, 3rd, and 5th sliding surface functions in (3.26) are used for the converters whose mathematical modeling is defined in the following state-space form:

$$\frac{dx_1}{dt} = x_2$$

$$\frac{dx_2}{dt} = w + D(t) \qquad\qquad (3.27)$$

$$\frac{dx_3}{dt} = x_1$$

where w denotes system dynamics and $D(t)$ denotes the disturbance. In this case, 2nd, 3rd, and 5th sliding surface functions in (3.26) can yield a defined convergence time. For instance, let us consider the 2nd sliding surface function in the control of a power converter. Assume that the trajectory passes the reaching phase and reaches to the sliding mode with appropriate control input. During the sliding mode, the trajectory is forced to slide along the $\sigma = 0$. This means that the sliding mode exists and the sliding surface function can be equated to zero as follows:

$$\sigma = \lambda_1 x_1 + \frac{dx_1}{dt} = 0 \quad \Rightarrow \quad \frac{dx_1}{dt} = -\lambda_1 x_1 \qquad\qquad (3.28)$$

Obviously, the term on the right-hand side of (3.28) is first-order differential equation whose solution is expressed as

$$x_1(t) = x_1(0)e^{-\lambda t} \qquad\qquad (3.29)$$

where $x_1(0)$ is the initial value of $x_1(t)$. It is evident from (3.29) that $x_1(t) \rightarrow 0$ if and only if $t \rightarrow \infty$. This means that $x_1(t)$ cannot converge to zero in finite time. Also, it is worth noting that $x_1(t)$ converges to zero independent of system parameters and disturbances, which clearly shows the robustness feature of the SMC. In addition,

Eq. (3.28) shows that the system dynamics can be represented with lower order than the converter under consideration. On the other hand, when the 3rd sliding surface function is employed in the control of a power converter, the following equation can be written in the sliding mode:

$$\frac{d^2 x_1}{dt^2} + \lambda_1 \frac{dx_1}{dt} + \lambda_2 x_1 = 0 \tag{3.30}$$

It is evident from (3.30) that $x_1(t)$ converges to zero.

On the other hand, when the 5th sliding surface function is employed in the control of power converter, the sliding surface function can be equated to zero if the sliding mode exists [11]:

$$\sigma = \lambda x_1^{\gamma} + \frac{dx_1}{dt} = 0 \quad \Rightarrow \quad \frac{dx_1}{dt} = -\lambda x_1^{\gamma} \tag{3.31}$$

It is worthy to remark that (3.31) reduces to (3.28) for $\gamma = 1$. Now, let us rewrite (3.31) as follows:

$$dt = -\frac{dx_1}{\lambda x_1^{\gamma}} \tag{3.32}$$

Integrating both sides of (3.32) and evaluating the result of integration on the closed interval (i.e. $x_1(0) \neq 0$, $x_1(t_s) = 0$) results in [10]:

$$t_s = -\frac{1}{\lambda} \int_{x_1(0)}^{0} \frac{1}{x_1^{\gamma}} dx_1 = \frac{|x_1(0)|^{1-\gamma}}{\lambda(1-\gamma)} \tag{3.33}$$

Equation (3.33) implies that $x_1(t) \rightarrow 0$ in finite time. Since the equilibrium $x_1(t) = 0$ is finite-time reachable, it is referred to as terminal attractor [13]. Clearly, the desired reaching time t_s can be obtained by using λ and γ.

3.3.2 Control Input Design

The control input that can achieve the required motion during the sliding phase is referred to as equivalent control u_{eq}. The equivalent control should satisfy the following condition:

$$0 < u_{eq} < 1 \tag{3.34}$$

In order to achieve this condition, u_{eq} is derived from $\frac{d\sigma}{dt} = 0$. Now, let us consider the sliding surface functions of buck and Cuk converters described in Section 3.2. Now, taking derivative of each sliding surface function and equating the resulting equation to zero renders

$$\frac{d\sigma}{dt} = 0 = \begin{cases} \lambda\dfrac{dx_1}{dt} + \dfrac{V_{in}}{LC}u_{eq} - \dfrac{x_2}{R_LC} - \dfrac{(x_1 + V_o^*)}{LC} & \Rightarrow \quad \text{DC-DC Buck Converter} \\[4mm] \dfrac{1}{L_1}\left(V_{in} - V_{C1} + V_{C1}u_{eq}\right) & \Rightarrow \quad \text{DC-DC Cuk Converter} \end{cases}$$

$$(3.35)$$

Solving u_{eq} from (3.35), one can derive the following expression for each converter

$$u_{eq} = \begin{cases} \dfrac{1}{V_{in}}\left(V_o^* + x_1 - LC\lambda\dfrac{dx_1}{dt} + \dfrac{Lx_2}{R_L}\right) & \Rightarrow \quad \text{DC-DC Buck Converter} \\[4mm] \dfrac{V_{C1} - V_{in}}{V_{C1}} & \Rightarrow \quad \text{DC-DC Cuk Converter} \end{cases}$$

$$(3.36)$$

Hence, one can easily see that the condition in (3.34) is satisfied in the steady state for both converters

$$\begin{aligned} 0 < \frac{V_o^*}{V_{in}} < 1 &\quad \Rightarrow \quad \text{DC-DC Buck Converter} \\[3mm] 0 < \frac{V_{C2}^*}{V_{in} + V_{C2}^*} < 1 &\quad \Rightarrow \quad \text{DC-DC Cuk Converter} \end{aligned}$$

$$(3.37)$$

It is worth to note that $V_{C1} = V_{in} + V_{C2}^*$ in the steady state for the Cuk converter. On the other hand, when the converter is controlled with the control input deduced from the equivalent control, the switching signals needed to drive the switch in the converter can be attained by using traditional pulse width modulation (PWM) method. The simplest and easiest way of generating PWM signal is possible by comparing the equivalent control signal (i.e. low-frequency modulating signal) with the high-frequency triangular carrier signal. Even though this enables the converter to work with constant switching frequency, the performance of the controller becomes sensitive to the variations in the system parameters. Thus, the equivalent control is not robust against variations in the system parameters.

Alternatively, the control input design can be based on the relay control. The idea of relay control relies on the discontinuous control defined in (3.38) as follows:

$$u = \frac{1}{2}(1 - sign(\sigma)) \tag{3.38}$$

where *sign* function is expressed as

$$sign = \begin{cases} 1 & \text{when} \quad \sigma > 0 \\ 0 & \text{when} \quad \sigma = 0 \\ -1 & \text{when} \quad \sigma < 0 \end{cases} \tag{3.39}$$

However, when the converter is controlled with the control input in (3.38), it is subjected to high and uncontrolled switching frequency, which is referred to as chattering. Since high and uncontrolled switching frequency is not desired in practice, the effect of chattering should be reduced to a tolerable level. The chattering mitigation techniques will be discussed in the next section.

3.4 Chattering Mitigation Techniques

Chattering is the major shortcoming of applying SMC to real applications in industry. When the chattering occurs, the state trajectory moves along the sliding line with low-amplitude and high-frequency oscillations (see Figure 2.5). Usually, the main source of chattering is due to:

i) the practical limitations of switching devices and components employed in the converter,
ii) the unmodeled dynamics (i.e. unstructured uncertainties) with small time constants.

Chattering causes steady-state errors in the state variables, high heat losses, time-varying switching frequency, and instability when the unmodeled high-order dynamics of the system are excited. Hence, the chattering alleviation is essential in sliding mode controlled power converters. The chattering mitigation techniques are discussed in the next sections.

3.4.1 Hysteresis Function Technique

The hysteresis function technique is one of the widely used chattering mitigation techniques. It is based on replacing the *sign* function by a hysteresis function that has a band width h as shown in Figure 3.10.

When the state trajectory reaches the sliding line, it is enforced to move along the sliding line within the hysteresis band. It is possible to adjust this band such that a reduction is obtained in the chattering and switching frequency. However, it is worth to note that the levels of chattering and switching frequency are inversely

Figure 3.10 Replacing *sign* function with the hysteresis function.

proportional with the value of hysteresis band. A detailed analysis of chattering and switching frequency will be discussed in Chapter 7. Since the switching signals needed for driving the switches in a power converter are obtained by using hysteresis band, this method is referred to as hysteresis modulation in literature. Various hysteresis modulation methods have been developed, which employ single [14], double [7], [8], and multi-hysteresis bands [15].

3.4.2 Boundary Layer Technique

The idea behind the boundary layer technique is to alleviate the discontinuities in the vicinity of sliding surface function [16]. In order to achieve this, a boundary layer with thickness Φ_t and width Φ_w is used such that the undesired discontinuities in the sliding surface function are smoothed out. The visualization of boundary layer method for a second-order system in the phase plane is shown in Figure 3.11. When the state trajectory is not inside the boundary layer, the boundary layer is made attractive by satisfying (3.38). However, when the state trajectory is inside the boundary layer, the control input u should be interpolated inside the boundary layer such that the trajectory remains inside the boundary layer. In order

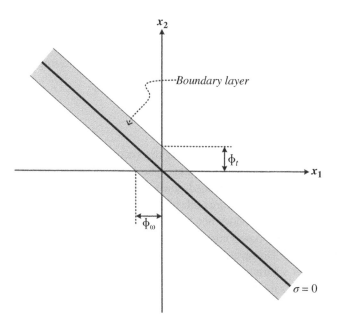

Figure 3.11 Visualization of boundary layer method for a second-order system in phase plane.

Figure 3.12 Replacing *sign* function with the saturation function.

to achieve this, the discontinuous control input $u = sign\,(\sigma)$ is replaced by σ/Φ_t as shown in Figure 3.12. Hence, the boundary layer method is considered as the continuous approximation of the *sign* function by using the saturation function defined below:

$$u \cong sat\left(\frac{\sigma}{\Phi_t}\right) = \begin{cases} sign\left(\dfrac{\sigma}{\Phi_t}\right) & \text{for} \quad |\sigma| \geq 1 \\[2mm] \dfrac{\sigma}{\Phi_t} & \text{for} \quad |\sigma| < 1 \end{cases} \qquad (3.40)$$

As a result of boundary layer method, the control input is a continuous signal inside the boundary layer that alleviates chattering. The boundary layer method is successfully applied to mitigate the chattering in sliding mode controlled uninterruptible power supply (UPS) inverters [17], dynamic voltage restorers [18], and quasi-Z-source inverters [19], [20]. However, the boundary thickness value (Φ_t) leads to steady-state error when it is not selected appropriately. Hence, the boundary thickness value is to be selected to make a compromise between chattering mitigation and steady-state error.

3.4.3 State Observer Technique

The state observer technique is based on estimating the system states required in the sliding mode controller in an observer loop. The block diagram of state observer method in a sliding mode controlled power converter is illustrated in Figure 3.13. Clearly, the estimated states are utilized in the sliding mode controller. This implies that the observer loop does not include unmodeled dynamics that exist due to the aggregated use of power converter and sensors. Thus, the existence of chattering can be avoided. However, the system may become sensitive to the parameter variations under a parameter mismatch between the observer and system states. The state observer method is successfully applied to mitigate the chattering in sliding-mode-controlled DC–DC buck converters [21], DC–DC boost converters [22], three-phase grid-connected inverters [23], and fault detection in modular multilevel converters [24].

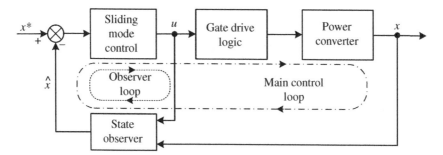

Figure 3.13 Block diagram of state observer method in a sliding mode controlled power converter.

3.5 Modulation Techniques

A modulation technique produces the gate signals needed for the semiconductor switches in a power converter leading to a switched output waveform so as to adjust the amplitude, frequency, and phase as desired. The widely used modulation techniques in the sliding mode controlled power converters are discussed in the next sections.

3.5.1 Hysteresis Modulation Technique

The hysteresis modulation (HM) technique relies on using single and multi-band hysteresis switching logic. Let us define the control input by using a single hysteresis band h as follows:

$$u = \begin{cases} u^+ & \text{if} \quad \sigma > +h \\ u^- & \text{if} \quad \sigma < -h \\ \text{Previous value} & \text{if} \quad \text{Otherwise} \end{cases} \tag{3.41}$$

The implementation of (3.41) and its switching logic for a single-phase H-bridge grid-connected inverter with switches S_1, S_2, S_3, and S_4 is shown in Figure 3.14. Clearly, the switches S_1 and S_4 are switched on while the switches are switched off and vice versa. Although the use of single-band hysteresis modulation is useful in alleviating the effect of chattering, it yields time-varying switching frequency that is not suggested in power converters. However, the use of double-band-based hysteresis modulation method provides better performance in mitigating the switching frequency. The implementation of single-band and double-band hysteresis modulations together with their switching logics and switching frequency computations will be discussed in more detail in Chapter 7.

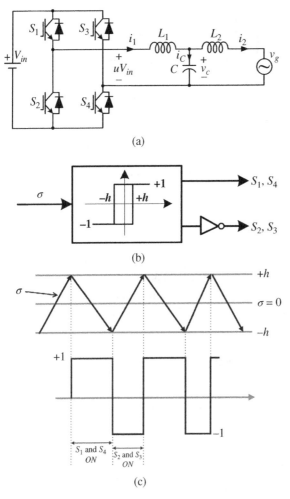

Figure 3.14 (a) Single-phase H-bridge grid-connected inverter, (b) implementation of (3.41), and (c) switching logic.

3.5.2 Sinusoidal Pulse Width Modulation Technique

Unlike the hysteresis modulation technique discussed above, the sinusoidal pulse width modulation (SPWM) technique uses a continuous modulation signal. The gate drive signals needed for the switches in the inverter are produced by comparing the continuous modulation signal with the carrier signal that has fixed frequency. In this case, the switching frequency of converter is always fixed. The basic SPWM scheme for a single-phase H-bridge inverter is illustrated in

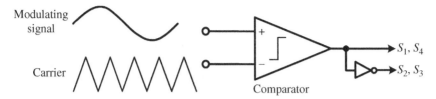

Figure 3.15 Basic SPWM scheme for a single-phase H-bridge grid-connected inverter.

Figure 3.15. The low-frequency modulating signal comes from either equivalent control in (3.36) or boundary layer method in (3.40).

3.5.3 Space Vector Modulation Technique

The space vector modulation (SVM) technique is preferable to be utilized in three-phase power converters. It is well known that the mathematical modeling of these converters is usually performed in the synchronously rotating dq frame. In the sliding mode controlled three-phase power converters, the sliding surface functions of corresponding phases should also be computed in the dq frame. The required control inputs, which are the outputs of the SMC method, are computed in the dq frame. Then, these control inputs are transformed into the abc frame. Finally, the required switching pulses are extracted from the space vector table, which includes all switching states of the converter under consideration. The block diagram of sliding mode controlled three-phase two-level power converter (see Chapter 7) with SVM is shown in Figure 3.16. It is worthy to remark that the SVM can be used in three-phase multilevel power converters as well.

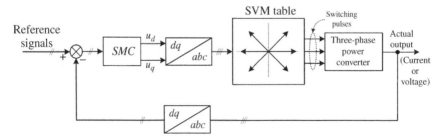

Figure 3.16 Block diagram of sliding mode controlled two-level three-phase converter with SVM.

3.6 Other Types of Sliding Mode Control

The SMC method described so far suffers from chattering. In addition, the error variables cannot converge to the equilibrium point in finite time, which leads to steady-state errors in the controlled variables. As a remedy to these disadvantages, various SMC methods such as terminal SMC and second-order SMC approaches are emerged. These approaches will be introduced in the following sections.

3.6.1 Terminal Sliding Mode Control

The TSMC approach relies on formulating the sliding surface function by using the fractional power [25]. The block diagram of TSMC-based control of a power converter is illustrated in Figure 3.17. In [11], a fast TSMC is developed to control a DC–DC boost converter. In [26], a continuous output feedback-based TSMC is applied to control a single-phase DC–AC inverter.

3.6.2 Second-Order Sliding Mode Control

A second-order SMC should satisfy the following condition:

$$\sigma = \frac{d\sigma}{dt} = 0 \tag{3.42}$$

Various second-order SMC algorithms such as twisting algorithm, sub-optimal algorithm, drift algorithm, and super twisting algorithm (STA) are developed. The STA is more popular and received much attention than the other algorithms since it yields continuous control, which leads to reduction in the chattering. However, it can be applied to the systems with relative degree one. Relative degree is described as the number of times one has to differentiate the output in order to have the control input appearing explicitly. The control input designed by using the STA contains two terms. While one of them is the continuous-time function

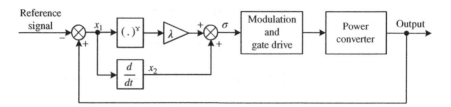

Figure 3.17 Block diagram of TSMC-based control of a power converter.

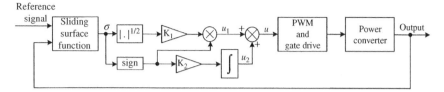

Figure 3.18 Block diagram of STA-based SMC method for controlling a power converter.

of the sliding surface function, the other one is the integral of the sliding surface function as shown below:

$$u = u_1 + u_2$$
$$u_1 = -K_1|\sigma|^{1/2}sign(\sigma)$$
$$\frac{du_2}{dt} = -K_2sign(\sigma)$$

(3.43)

where $K_1 > 0$ and $K_2 > 0$ represent gains, which play significant role in improving the performance of the converter under consideration regarding chattering, steady-state error, and dynamic response. For instance, while K_1 has the ability to adjust the dynamic response, K_2 can alleviate the steady-state error. The block diagram of STA-based SMC method applied to control a power converter is illustrated in Figure 3.18.

The STA-based SMC method is applied to control power converters such as three-phase AC–DC rectifiers [27], single-phase grid-connected inverters [28], three-phase shunt active power filters [29], and single-phase dynamic voltage restorers [30].

References

1 H. Komurcugil, S. Biricik, S. Bayhan, and Z. Zhang, "Sliding mode control: Overview of its applications in power converters," *IEEE Ind. Electron. Mag.*, vol. 15, no. 1, pp. 40–49, Mar. 2021.

2 S. C. Tan, Y. M. Lai, M. K. H. Cheung, and C. K. Tse, "On the practical design of a sliding mode voltage controlled buck converter," *IEEE Trans. Power Electron.*, vol. 20, no. 2, pp. 425–437, Mar. 2005.

3 Z. Chen, "PI and sliding mode control of a cuk converter," *IEEE Trans. Power Electron.*, vol. 27, no. 8, pp. 3695–3703, Aug. 2012.

4 E. Mamarelis, G. Petrone, and G. Spagnuolo, "Design of a sliding-mode-controlled SEPIC for PV MPPT applications," *IEEE Trans. Ind. Electron.*, vol. 61, no. 7, pp. 3387–3398, Jul. 2014.

5 H. Komurcugil, "Rotating-sliding-line-based sliding-mode control for single-phase UPS inverters," *IEEE Trans. Ind. Electron.*, vol. 59, no. 10, pp. 3719–3726, Oct. 2012.

6 S. Biricik and H. Komurcugil, "Optimized sliding mode control to maximize existence region for single-phase dynamic voltage restorers," *IEEE Trans. Ind. Informat.*, vol. 12, no. 4, pp. 1486–1497, Aug. 2016.

7 O. Kukrer, H. Komurcugil, and A. Doganalp, "A three-level hysteresis function approach to the sliding mode control of single-phase UPS inverters," *IEEE Trans. Ind. Electron.*, vol. 56, no. 9, pp. 3477–3486, Sep. 2009.

8 H. Komurcugil, S. Ozdemir, I. Sefa, N. Altin, and O. Kukrer, "Sliding-mode control for single-phase grid-connected LCL-filtered VSI with double-band hysteresis scheme," *IEEE Trans. Ind. Electron.*, vol. 63, no. 2, pp. 864–873, Feb. 2016.

9 H. Komurcugil, S. Biricik, and N. Guler, "Indirect sliding mode control for DC-DC SEPIC converters," *IEEE Trans. Ind. Informat.*, vol. 16, no. 6, pp. 4099–4108, Jun. 2020.

10 H. Komurcugil, "Adaptive terminal sliding-mode control strategy for DC–DC buck converters," *ISA Trans.*, vol. 51, pp. 673–681, Nov. 2012.

11 I. Yazici and E. K. Yaylaci, "Fast and robust voltage control of DC–DC boost converter by using fast terminal sliding mode controller," *IET Power Electron.*, vol. 9, no. 1, pp. 120–125, Feb. 2016.

12 Z. Wang, S. Li, and Q. Li, "Continuous nonsingular terminal sliding mode control of DC–DC boost converters subject to time-varying disturbances," *IEEE Trans. Circuits Syst. II Express Briefs*, vol. 67, no. 11, pp. 2552–2556, Nov. 2020.

13 M. Zak, "Terminal attractors for addressable memory in neural network," *Physical Letters A*, vol. 133, no. 1/2, pp. 18–22, Oct. 1988.

14 W. Yan, J. Hu, V. Utkin, and L. Xu, "Sliding mode pulsewidth modulation," *IEEE Trans. Power. Electron.*, vol. 23, no. 2, pp. 619–626, Mar. 2008.

15 S. Gautam and R. Gupta, "Switching frequency derivation for the cascaded multilevel inverter operating in current control mode using multiband hysteresis modulation," *IEEE Trans. Power Electron.*, vol. 29, no. 3, pp. 1480–1489, Mar. 2014.

16 J. J. E. Slotine and W. Li, Applied Nonlinear Control. Englewood Cliffs, NJ: Prentice-Hall, 1991.

17 A. Abrishamifar, A. A. Ahmad, and M. Mohamadian, "Fixed switching frequency sliding mode control for single-phase unipolar inverters," *IEEE Trans. Power Electron.*, vol. 27, no. 5, pp. 2507–2514, May 2012.

18 H. Komurcugil and S. Biricik, "Time-varying and constant switching frequency-based sliding-mode control methods for transformerless DVR employing half-bridge VSI," *IEEE Trans. Ind. Electronics*, vol. 64, no. 4, pp. 2570–2579, Apr. 2017.

19 S. Bayhan and H. Komurcugil, "A sliding-mode controlled single-phase grid-connected quasi-Z-source NPC inverter with double-line frequency ripple suppression," *IEEE Access*, vol. 7, pp. 160004–160016, Oct. 2019.

20 F. Bagheri, H. Komurcugil, O. Kukrer, N. Guler, and S. Bayhan, "Multi-input multi-output based sliding-mode controller for single-phase quasi-Z-source inverters," *IEEE Trans. Ind. Electronics*, vol. 67, no. 8, pp. 6439–6449, Sept. 2020.

21 J. Wang, S. Li, J. Yang, B. Wu, and Q. Li, "Extended state observer-based sliding mode control for PWM-based DC–DC buck converter systems with mismatched disturbances," *IET Control Theory Applications*, vol. 9, no. 4, pp. 579–586, Feb. 2015.

22 S. Oucheriah and L. Guo, "PWM-based adaptive sliding-mode control for boost DC–DC converters," *IEEE Trans. Ind. Electron.*, vol. 60, no. 8, pp. 3291–3294, Aug. 2013.

23 R. Guzman, L. G. de Vicuna, M. Castilla, J. Miret, and J. de la Hoz, "Variable structure control for three-phase LCL-filtered inverters using a reduced converter model," *IEEE Trans. Ind. Electron.*, vol. 65, no. 1, pp. 5–15, Jan. 2018.

24 S. Shao, P. W. Wheeler, J. C. Clare, and A. J. Watson, "Fault detection for modular multilevel converters based on sliding mode observer," *IEEE Trans. Power Electron.*, vol. 28, no. 11, pp. 4867–4872, Nov. 2013.

25 X. Yu, Y. Feng, and Z. Man, "Terminal sliding mode control – An overview," *IEEE Open J. Ind. Electron. Soc.*, vol. 2, pp. 36–52, Nov. 2021.

26 Z. Zhao, J. Yang, S. Li, X. Yu, and Z. Wang, "Continuous output feedback TSM control for uncertain systems with a DC–AC inverter example," *IEEE Trans. Circuits Syst. II Express Briefs*, vol. 65, no. 1, pp. 71–75, Jan. 2018.

27 J. Liu, S. Vazquez, L. Wu, A. Marquez, H. Gao, and L. G. Franquelo, "Extended state observer-based sliding-mode control for three-phase power converters," *IEEE Trans. Ind. Electron.*, vol. 64, no. 1, pp. 22–31, Jan. 2017.

28 J. Xia, Y. Guo, X. Zhang, J. Jatskevich, and N. Amiri, "Robust control strategy design for single-phase grid-connected converters under system perturbations," *IEEE Trans. Ind. Electron.*, vol. 66, no. 11, pp. 8892–8901, Nov. 2019.

29 S. Ouchen, M. Benbouzid, F. Blaabjerg, A. Betka, and H. Steinhart, "Direct power control of shunt active power filter using space vector modulation based on super twisting sliding mode control," *IEEE J. Emerging Sel. Top. Power Electron.*, vol. 9, no. 3, pp. 3243–3253, Jun. 2021.

30 S. Biricik, H. Komurcugil, H. Ahmed, and E. Babaei, "Super twisting sliding mode control of DVR with frequency-adaptive Brockett oscillator," *IEEE Trans. Ind. Electron.*, vol. 68, no. 11, pp. 10730–10739, Nov. 2021.

4

Design of Lyapunov Function-Based Control for Power Converters

4.1 Introduction

With the development of technology and various power converter topologies, researchers have become interested in designing nonlinear control methodologies that are not merely valid around small operating point, but also provide excellent performance away from the operating point. Lyapunov function-based control method is one of the nonlinear control methods which guarantees the stability of the converter and provides excellent performance under large transients [1], [2]. Figure 4.1 shows the steps involved in the Lyapunov function-based control approach. The discussion of linearization method is omitted since it is briefly discussed in Chapter 2. The design of Lyapunov function-based control using direct method will be explained in the next sub-sections.

4.2 Lyapunov-Function-Based Control Design Using Direct Method

Figure 4.2 depicts the typical energy distribution in basic power converters [3]–[6]. When the input energy (E_{in}) is delivered to the converter, part of this energy is consumed by the resistor of inductance and switching devices in the converter. While E_s denotes the energy dissipated by the switches, E_L and E_C denote the energy stored in the inductor and capacitor, respectively. Since capacitors and inductors do not consume energy, the rest of input energy is continuously exchanged between capacitors and inductors in a bidirectional manner until this energy settles down to the equilibrium point of the converter. This implies that the rest of input energy denoted by E_o is transferred to the load or utility grid for a

Advanced Control of Power Converters: Techniques and MATLAB/Simulink Implementation,
First Edition. Hasan Komurcugil, Sertac Bayhan, Ramon Guzman, Mariusz Malinowski, and Haitham Abu-Rub.
© 2023 The Institute of Electrical and Electronics Engineers, Inc.
Published 2023 by John Wiley & Sons, Inc.
Companion website: www.wiley.com/go/komurcugil/advancedcontrolofpowerconverters

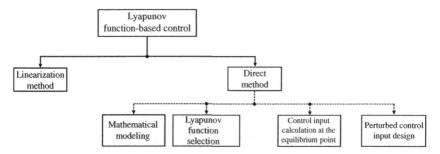

Figure 4.1 Steps of Lyapunov function-based control design.

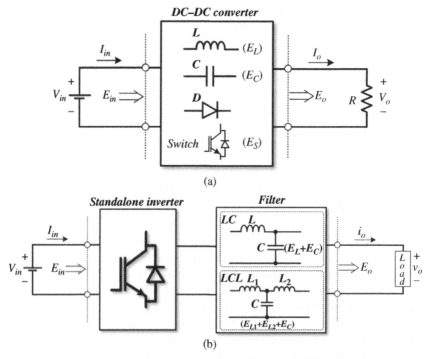

Figure 4.2 Energy distribution in basic power converters. (a) DC–DC converter, (b) DC–AC standalone inverter, (c) DC–AC grid-connected inverter, (d) AC–DC rectifier.

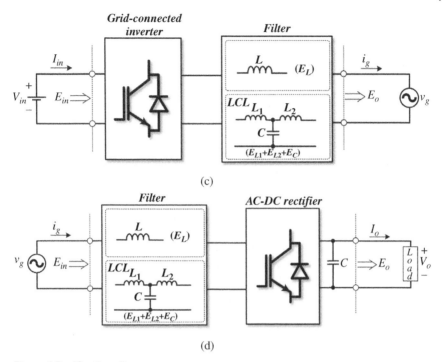

Figure 4.2 (Continued)

grid-connected inverter case. Even though Lyapunov's direct method is useful to investigate the global stability of the converter under consideration, it does not provide any guideline to select Lyapunov function. It is worth noting that Lyapunov function is formulated by considering the energies stored in capacitors and inductors connected in the converter.

As mentioned before, the Lyapunov function should be an energy-like function. This implies that the Lyapunov function can be formulated by considering the energy stored in the inductors and capacitors connected in a power converter. The typical Lyapunov function candidate for various power converters can be selected as follows

$$
V(\mathbf{x}) = \begin{cases}
K_L L x_1^2 + K_C C x_2^2 & \Rightarrow \begin{array}{l} \text{DC–DC converter/single-phase} \\ \text{inverter with LC filter} \end{array} \\[2mm]
K_{L1} L_1 x_1^2 + K_C C x_2^2 + K_{L2} L_2 x_3^2 & \Rightarrow \begin{array}{l} \text{Single-phase inverter/rectifier} \\ \text{with LCL filter} \end{array} \\[2mm]
\begin{aligned} & K_{L1d} L_1 x_1^2 + K_{L1q} L_1 x_2^2 + K_{Cd} C x_3^2 \\ & + K_{Cq} C x_4^2 + K_{L2d} L_2 x_5^2 + K_{L2q} L_2 x_6^2 \end{aligned} & \Rightarrow \begin{array}{l} \text{Three-phase inverter/rectifier} \\ \text{in dq frame with LCL filter} \end{array}
\end{cases}
$$

$$(4.1)$$

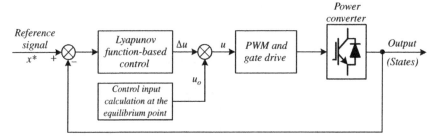

Figure 4.3 Generalized block diagram of the Lyapunov function-based control of a power converter.

where $K_L > 0$, $K_C > 0$, $K_{L1} > 0$, $K_{L2} > 0$, $K_{L1d} > 0$, $K_{L1q} > 0$, $K_{L2d} > 0$, and $K_{L2q} > 0$ are positive constants and x_1, x_2, x_3, x_4, x_5, and x_6 are the error variables. It is obvious from (4.1) that the Lyapunov function candidate should involve some terms which are in the form of energy stored in the inductor and capacitor, respectively. The number of these terms depends on the mathematical modeling of the converter under consideration. While traditional DC–DC buck and boost converters with single switch and LC filter require two energy terms, the three-phase inverters/rectifiers modeled in the rotating dq frame with an LCL filter requires six energy terms. In the case of a single-phase inverter (or rectifier) with LCL filter, three energy terms are needed. On the other hand, the values of constants in the Lyapunov function for DC–DC converters and single-phase inverters/rectifiers with LCL filter are usually selected as $K_L = K_C = K_{L1} = K_C = K_{L2} = 1/2$. The other constants in the Lyapunov function formulated for three-phase inverters/rectifiers are to be determined in the controller design step. The generalized block diagram of the Lyapunov function-based control of a power converter is illustrated in Figure 4.3. Control input calculation at the equilibrium point and perturbed control input design will be explained together with the mathematical modeling of DC–DC buck and DC–DC boost converters in Sections 4.3 and 4.4, respectively.

4.3 Lyapunov Function-Based Control of DC–DC Buck Converter

Let us rewrite the differential equations of the DC–DC buck converter shown in Figure 3.2 again

$$L\frac{dI_L}{dt} = V_{in}u - V_o \tag{4.2}$$

$$C\frac{dV_o}{dt} = I_L - \frac{V_o}{R_L} \tag{4.3}$$

where u is the control input. The control objective is to achieve the regulation of load voltage (V_o) against variations in load (R_L), reference load voltage (V_o^*) and input voltage (V_{in}). Now, define the state variables and control input as follows

$$x_1 = V_o - V_o^* \tag{4.4}$$

$$x_2 = I_L - I_L^* \tag{4.5}$$

$$u = U_o + \Delta u \tag{4.6}$$

where V_o^* represents the reference of V_o, I_L^* represents the reference of I_L and Δu represents the perturbation away from the nominal control input (U_o). Assuming that the actual variables follow their references in the steady-state (i.e. $V_o = V_o^*$, $I_L = I_L^*$, and $u = U_o$), one can derive the following equations

$$L\frac{dI_L^*}{dt} = V_{in}U_o - V_o^* \tag{4.7}$$

$$C\frac{dV_o^*}{dt} = I_L^* - \frac{V_o^*}{R_L} \tag{4.8}$$

Considering $V_o = x_1 + V_o^*$, $I_L = x_2 + I_L^*$ and $u = U_o + \Delta u$ in Eqs. (4.2) and (4.3) yields

$$L\frac{dx_2}{dt} + L\frac{dI_L^*}{dt} = V_{in}U_o + V_{in}\Delta u - x_1 - V_o^* \tag{4.9}$$

$$C\frac{dx_1}{dt} + C\frac{dV_o^*}{dt} = x_2 + I_L^* - \frac{x_1}{R_L} - \frac{V_o^*}{R_L} \tag{4.10}$$

Taking into consideration the steady-state Eqs. (4.7) and (4.8), the error dynamics of the converter can be obtained as

$$L\frac{dx_2}{dt} = V_{in}\Delta u - x_1 \tag{4.11}$$

$$C\frac{dx_1}{dt} = x_2 - \frac{x_1}{R_L} \tag{4.12}$$

Now, let us define a Lyapunov function as follows

$$V(x) = \frac{1}{2}Cx_1^2 + \frac{1}{2}Lx_2^2 \tag{4.13}$$

It should be noted that Lyapunov function is in the form of total energy stored in the converter. While the first term represents the energy stored in the capacitor C, the latter is the energy stored in the inductor L. Clearly, the Lyapunov function satisfies the conditions ($V(0) = 0$, $V(x) > 0$ when $x \neq 0$, and $V(x) \to \infty$ when

$\sqrt{x_1^2 + x_2^2} \to \infty$). As can be seen from Chapter 2, only $\dfrac{dV(x)}{dt} < 0$ condition is not satisfied. Now, let us deign the control input that satisfies $\dfrac{dV(x)}{dt} < 0$. The derivative of Lyapunov function can be written as

$$\frac{dV(x)}{dt} = x_1 C \frac{dx_1}{dt} + x_2 L \frac{dx_2}{dt} \tag{4.14}$$

Substituting (4.11) and (4.12) into (4.14) gives

$$\frac{dV(x)}{dt} = -\frac{x_1^2}{R_L} + V_{in} x_2 \Delta u \tag{4.15}$$

The stability of the converter can be achieved if $\dfrac{dV(x)}{dt} < 0$ holds. It is obvious that the first term in (4.15) is always negative. Hence, in order to achieve $\dfrac{dV(x)}{dt} < 0$, the perturbed control input should be selected as

$$\Delta u = \alpha x_2, \quad \alpha < 0 \tag{4.16}$$

The block diagram of Lyapunov function-based control of buck converter is shown in Figure 4.4.

The performance of closed-loop system can be investigated by MATLAB/Simulink simulations where the system and control parameters are selected as

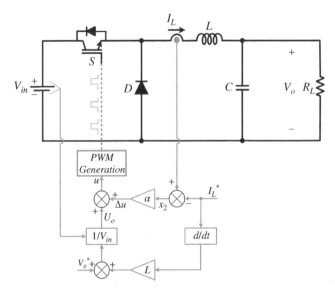

Figure 4.4 Block diagram of Lyapunov function-based control for DC–DC buck converter.

$V_{in} = 100$ V, $V_o^* = 50$ V, $L = 5$ mH, $C = 470$ µF and $R_L = 10\,\Omega$. The switching frequency was selected as 10 kHz. The reference inductor current is generated by using a PI regulator as follows

$$I_L^* = K_p\left(V_o^* - V_o\right) + K_i \int \left(V_o^* - V_o\right) dt \tag{4.17}$$

In the simulation study, while K_p is selected as 0.1, the value of K_i was 70. Finally, the current control gain was selected to be $\alpha = -2$. Figure 4.5 shows the responses of output voltage and inductor current for an abrupt change in V_{in} from 100 to 70 V.

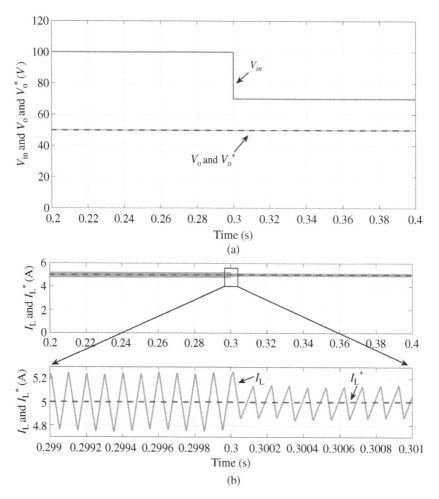

Figure 4.5 Responses of output voltage and inductor current for an abrupt change in V_{in} from 100 to 70 V. (a) Input voltage, output voltage and its reference, (b) Inductor current and its reference.

It is obvious from Figure 4.5a that the output voltage is not affected from the variation in V_{in}. On the other hand, the inductor current and its reference are depicted in Figure 4.5b. It can be seen that this variation in V_{in} reduces the amount of ripple in the inductor current. However, the average value of the inductor current is not changed since there is no change in the output voltage. Since output voltage is constant, the average of capacitor current becomes zero which implies that $I_L^* = V_o^*/R_L = 5\,A$.

Figure 4.6 shows the responses of output voltage and inductor current for an abrupt change in R_L from 10 to $5\,\Omega$ under $V_{in} = 100\,V$. Despite the output voltage

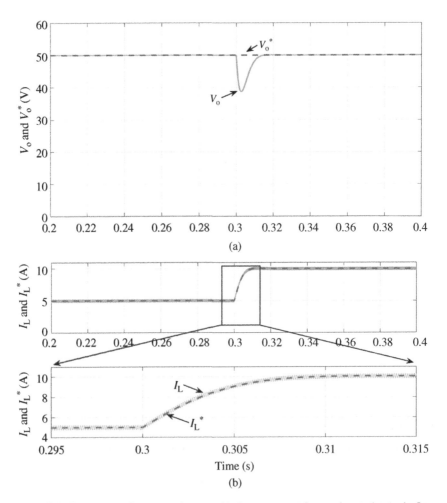

Figure 4.6 Responses of output voltage and inductor current for an abrupt change in R_L from 10 to $5\,\Omega$. (a) Output voltage and its reference, (b) Inductor current and its reference.

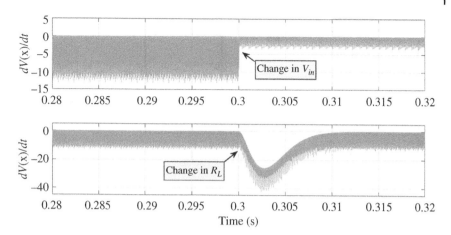

Figure 4.7 Responses of Lyapunov function derivative corresponding to the changes in V_{in} and R_L.

exhibits an undershoot due to this sudden change in R_L, the controller forces the output voltage to follow its reference and, thereafter, the output voltage is regulated at 50 V as shown in Figure 4.6a. The responses of inductor current and its reference are shown in Figure 4.6b. It is evident that both inductor current and its reference react to this sudden drop in R_L by gradually changing their values from 5 to 10 A. It is worth to point out that the inductor current follows its reference during this transition and thereafter.

Figure 4.7 shows the responses of Lyapunov function derivative for the abrupt changes in V_{in} and R_L corresponding to Figures 4.5 and 4.6, respectively. It is clear that the derivative of Lyapunov function remains always negative against the variations in V_{in} and R_L. Hence, the stability of the buck converter against such disturbances is guaranteed.

4.4 Lyapunov Function-Based Control of DC–DC Boost Converter

The circuit diagram of a DC–DC boost converter is shown in Figure 4.8. The main goal of this converter is to achieve a regulated output voltage which is always greater than the input voltage.

Now, let us write the differential equations of this converter

$$L\frac{dI_L}{dt} = V_{in} + V_o(u-1) \tag{4.18}$$

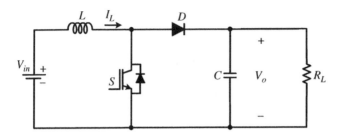

Figure 4.8 Circuit diagram of a DC–DC boost converter.

$$C\frac{dV_o}{dt} = I_L(1-u) - \frac{V_o}{R_L} \tag{4.19}$$

It is worth to note that since u is the control variable, then Eqs. (4.18) and (4.19) are nonlinear. The state variables and control input are same as in (4.4)–(4.6). Assuming that the actual variables follow their references in the steady-state (i.e. $V_o = V_o^*$, $I_L = I_L^*$, and $u = U_o$), one can derive the following equations

$$L\frac{dI_L^*}{dt} = V_{in} + V_o^*(U_o - 1) \tag{4.20}$$

$$C\frac{dV_o^*}{dt} = I_L^*(1 - U_o) - \frac{V_o^*}{R_L} \tag{4.21}$$

Substituting $V_o = x_1 + V_o^*$, $I_L = x_2 + I_L^*$ and $u = U_o + \Delta u$ in Eqs. (4.18) and (4.19) yields

$$L\frac{dx_2}{dt} + L\frac{dI_L^*}{dt} = V_{in} + x_1 U_o + x_1 \Delta u + V_o^* U_o + V_o^* \Delta u - x_1 - V_o^* \tag{4.22}$$

$$C\frac{dx_1}{dt} + C\frac{dV_o^*}{dt} = x_2 + I_L^* - x_2 U_o - x_2 \Delta u - I_L^* U_o - I_L^* \Delta u - \frac{x_1}{R_L} - \frac{V_o^*}{R_L} \tag{4.23}$$

Considering the steady-state Eqs. (4.20) and (4.21), the error dynamics of the converter becomes

$$L\frac{dx_2}{dt} = x_1 U_o + x_1 \Delta u - x_1 + V_o^* \Delta u \tag{4.24}$$

$$C\frac{dx_1}{dt} = x_2 - x_2 U_o - x_2 \Delta u - \frac{x_1}{R_L} - I_L^* \Delta u \tag{4.25}$$

The same Lyapunov function in Eq. (4.13) can be considered for the boost converter as well. In this case, the derivative of Lyapunov function in Eq. (4.14) can also be used for the boost converter. Thus, our aim is to obtain the perturbed control input that satisfies $\dfrac{dV(x)}{dt} < 0$. Substituting (4.24) and (4.25) in (4.13), we obtain

$$\frac{dV(x)}{dt} = -\frac{x_1^2}{R_L} - \left(I_L^* x_1 - V_o^* x_2\right)\Delta u \tag{4.26}$$

It is evident that the first term in (4.26) is always negative. Therefore, in order to guarantee $\dfrac{dV(x)}{dt} < 0$, the perturbed control input should be selected as

$$\Delta u = \beta\left(I_L^* x_1 - V_o^* x_2\right), \quad \beta > 0 \tag{4.27}$$

The block diagram of Lyapunov function-based control of boost converter is shown in Figure 4.9.

The performance of closed-loop system can be investigated by MATLAB/ Simulink simulations where the system and control parameters are selected as $V_{in} = 50\,\text{V}$, $V_o^* = 100\,\text{V}$, $L = 5\,\text{mH}$, $C = 470\,\mu\text{F}$ and $R_L = 20\,\Omega$. The switching frequency was selected as 10 kHz. The reference inductor current (I_L^*) is generated by using the PI regulator in (4.17) with $K_p = 0.5$ and $K_i = 100$. The current control gain was selected to be $\beta = 0.08$. Figure 4.10 shows the responses of output voltage and inductor current for an abrupt change in V_{in} from 50 to 70 V. It is obvious that the output voltage exhibits a small overshoot and settles down at 100 V after a short transition. On the other hand, the inductor current and its reference are well

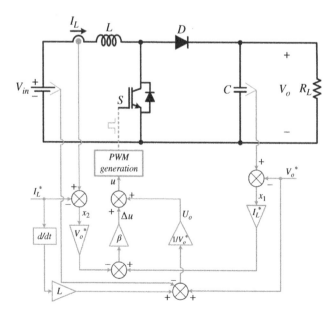

Figure 4.9 Block diagram of Lyapunov function-based control for DC–DC boost converter.

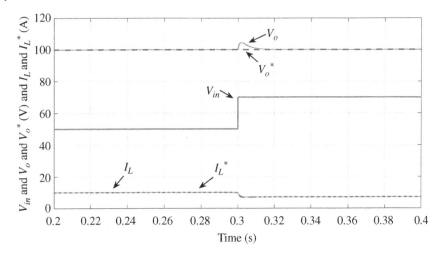

Figure 4.10 Responses of output voltage and inductor current for an abrupt change in V_{in} from 50 to 70 V.

regulated at 10 A before V_{in} is changed. However, it can be seen that the inductor current and its reference settle down around 7.14 A after V_{in} is changed. These results clearly demonstrate that the controller works correctly to achieve the regulation of output voltage and inductor current at their reference values.

Figure 4.11 shows the responses of output voltage and inductor current for an abrupt change in R_L from 20 to 10 Ω under $V_{in} = 50$ V. Despite the output voltage exhibits an undershoot due to the sudden change in R_L, the controller forces the

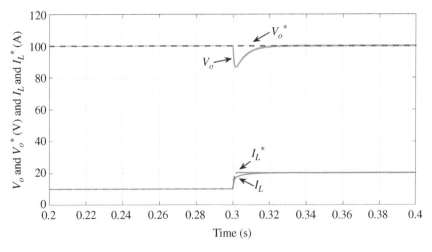

Figure 4.11 Responses of output voltage and inductor current for an abrupt change in R_L from 20 to 10 Ω.

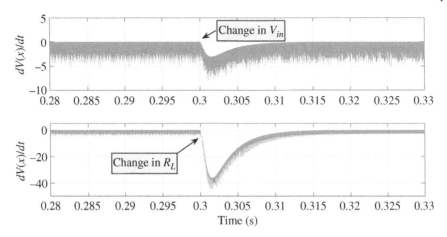

Figure 4.12 Responses of Lyapunov function derivative corresponding to the changes in V_{in} and R_L.

output voltage to follow its reference and as a consequence of this, the output voltage is regulated at 50 V. Also, it can be seen that both inductor current and its reference react to this sudden drop in R_L by gradually changing their values from 10 to 20 A. It is worth to point out that the dynamic response of output voltage and inductor current can be improved by tuning the controller gains.

Figure 4.12 shows the responses of Lyapunov function derivative for the abrupt changes in V_{in} and R_L corresponding to Figures 4.10 and 4.11, respectively. It is clear that the derivative of Lyapunov function remains always negative against the variations in V_{in} and R_L. Hence, the stability of the boost converter against such disturbances is guaranteed.

References

1 J. J. E. Slotine and W. Li, *Applied Nonlinear Control*. Englewood Cliffs, NJ: Prentice-Hall, 1991.

2 S. R. Sanders and G. C. Verghese, "Lyapunov-based control for switched power converters," *IEEE Trans. Power Electron.*, vol. 7, no. 1, pp. 17–24, Jan. 1992.

3 H. Komurcugil, N. Altin, S. Ozdemir, and I. Sefa, "An extended Lyapunov-function-based control strategy for single-phase UPS inverters," *IEEE Trans. Power Electron.*, vol. 30, no. 7, pp. 3976–3983, Jul. 2015.

4 H. Komurcugil, N. Altin, S. Ozdemir, and I. Sefa, "Lyapunov-function and proportional-resonant-based control strategy for single-phase grid-connected VSI with LCL filter," *IEEE Trans. Ind. Electron.*, vol. 63, no. 5, pp. 2838–2849, May 2016.

5 I. Sefa, S. Ozdemir, H. Komurcugil, and N. Altin, "Comparative study on Lyapunov-function-based control schemes for single-phase grid-connected voltage-source inverter with LCL filter," *IET Renewable Power Gener.*, vol. 11, no. 11, pp. 1473–1482, Aug. 2017.

6 I. Sefa, S. Ozdemir, H. Komurcugil, and N. Altin, "An enhanced Lyapunov-function based control scheme for three-phase grid-tied VSI with LCL filter," *IEEE Trans. Sustainable Energy*, vol. 10, no. 2, pp. 504–513, Apr. 2019.

5

Design of Model Predictive Control

5.1 Introduction

This chapter is focused on the design and implementation of different model predictive control strategies. The model predictive control is an advanced control technique that has been mainly used in industrial processes. This technique is characterized by using a predictive model to predict the output signal and a cost function, which contains the error between those predicted output value and their references. The cost function is minimized inside a predictive window to obtain the optimum control signal. The chapter will present two different strategies based on the model predictive control method. First, the finite control set model predictive control is presented. This method has been used in several applications where the number of the system states is finite. Then, it has a straightforward application on the control of three-phase power converters. An important feature of this control method is that no modulator is needed since the control signals for the switches are directly obtained. The other technique is called continuous control set model predictive control. Unlike the previous technique, a duty cycle is obtained from the minimization procedure, leading to the use of a modulator to obtain the control signals. The chapter also will introduce how constraints can be added to the cost function, and how the minimization procedure can be solved via quadratic programming. Several examples will be presented to make this complex control method more comprehensive.

5.2 Predictive Control Methods

Model predictive control (MPC) is an advanced and powerful control method, which has been used in industrial and chemical processes during the last two

Advanced Control of Power Converters: Techniques and MATLAB/Simulink Implementation, First Edition. Hasan Komurcugil, Sertac Bayhan, Ramon Guzman, Mariusz Malinowski, and Haitham Abu-Rub.
© 2023 The Institute of Electrical and Electronics Engineers, Inc.
Published 2023 by John Wiley & Sons, Inc.
Companion website: www.wiley.com/go/komurcugil/advancedcontrolofpowerconverters

decades [1], [2]. MPC has emerged and has had a significant impact on the industrial processes because it can be applied either in single-input-single-output (SISO) systems or multiple-input-multiple-output (MIMO) systems. Another important feature of the MPC technique is that various constraints can be included in the formulation of the control law in a very easy way.

All the MPC algorithms are based on a predictive model of the system, which is used to obtain a prediction of the output variable. The other important feature is that a cost function is always used to obtain the optimum control signal. Then, the objective is to find the control signal that minimizes the error between that predicted value and its reference. There are known in the literature two classes of MPC algorithms regarding the nature of the control variable. The first one is known as finite control set model predictive control (FCSMPC). The FCSMP is usually applied to systems that have a finite number of states. In this technique, a cost function is calculated for every state, and the objective of the control algorithm is to find the state that minimizes the error between the output variable and its reference. For instance, in three-phase two-level power converters, only eight possible voltage vectors are possible depending on the state of the switches. If the objective of the controller is to track a desired current, then the cost function will be selected as the difference between the measured current and the reference. The combination of the switches, which minimizes the error between the measured current and its reference will be used as the control signal. This combination will generate the optimum voltage vector in the converter. Note that it will not be necessary for the use of modulator since the obtained control signal can be directly send to the drivers of the switches.

The continuous control set model predictive control (CCSMPC) is an alternative to the FCSMPC. The control algorithm in this method is based on the prediction of state variables from a discrete model of the system. To obtain the vector of future control actions, a cost function is evaluated over a prediction horizon using the predicted state variables. The first value of this vector is usually used when considering a receding horizon. Unlike the FCS-MPC, a continuous duty cycle is obtained. In the case of three-phase power converters, a modulator is required, usually a space vector modulator (SVM). If no constraints are included in the cost function, this method is similar to a classic state feedback controller, and the optimization problem can be solved offline. Contrary, in case of constraints, a quadratic programming algorithm is usually required to solve the optimization problem.

Different MPC algorithms has been applied to power converters: for instance, in three-phase grid-connected voltage source inverters [3], [4], three-phase rectifiers [5], [6], and three-phase shunt active power filters [7], [8]. Also, this technique has been introduced to the control of microgrids with excellent results [9]. A distributed MPC has also been presented in [10] in a secondary control for frequency regulation.

5.3 FCS Model Predictive Control

In this section, first, the main idea of the FCS-MPC for power electronic converters is summarized, then, the detailed design steps will be explained. Although FCS-MPC is naturally well suited for the control of all power electronic converters, due to simplicity and ease of analysis, traditional three-phase voltage source inverter (VSI) will be used in this chapter as a base system to explain its implementation for power converters.

Figure 5.1 depicts the block diagram of the FCS-MPC technique for controlling the load current. It is clearly seen that the requirement for the modulation stage and linear PI controller is eliminated by this control approach. Another important point is that FCS-MPC implements preventive control actions before a large current error occurs, unlike the other control techniques that take necessary control actions after the error has occurred. This control technique is generally a combination of three main subsystems, namely extrapolation, predictive model, and cost function. The extrapolation stage is often needed to estimate a value based on extending a known sequence of values; in this case the reference current is extrapolated from kth to $(k + 1)$ sampling instant as seen in Figure 5.1. The predictive model is the core of FCS-MPC, which calculates the future value of the control variable (load currents). The finite number of switching states allows these values to be enumerated for all feasible switching states. In the final stage, the cost function is employed to evaluate the predictions and generate the gate signals for the power electronic converter. The cost function is usually defined as an absolute error between the reference and predicted variable; in this case, the control variable is load current and the cost function can be represented as

$$g(k) = |i^*(k + 1) - i(k + 1)| \qquad (5.1)$$

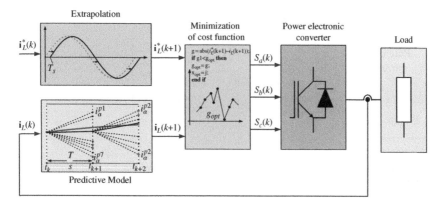

Figure 5.1 The simplified block diagram of the FCS-MPC for controlling load currents.

Switching states that minimize the current error in the next sampling instant are selected as optimal actuation and directly applied to the converter. In comparison to classical control techniques, FCS-MPC offers a faster dynamic response as well as comparable steady-state reference tracking. It is more flexible to select optimal switching signals in FCS-MPC than in deadbeat predictive control, and it is possible to incorporate multiobjective control requirements (e.g. nonlinearity and restrictions in the system). Using measurements and system parameters, FCS-MPC calculates the reference vector directly (without any online optimizations). Due to the nature of the online optimizations, FCS-MPC increases the computational burden significantly, but the current industry-standard digital control platforms are able to handle the computational capacity required by FCS-MPC.

5.3.1 Design Procedure

In this section, a step-by-step procedure for the design of the FCS-MPC implementation is presented. The detailed block diagram of the load current control technique based on FCS-MPC of the 3ph VSI is illustrated in Figure 5.2. To simplify the analysis, three-phase currents are represented in the *abc* reference frame.

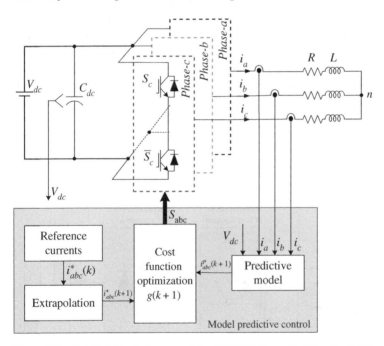

Figure 5.2 Detailed block diagram of the FCS-MPC for controlling the load currents of 3ph VSI.

The implementation of FCS-MPC is achieved in five steps, namely (i) reference currents generation, (ii) extrapolation of reference currents, (iii) measurement of feedback signals, (iv) predictive model, and (v) cost function minimization. The main control objective of this control approach is to force load currents to track the reference currents during all operating conditions.

Step 1: Determination of reference currents: The first step of the FCS-MPC implementation is to determine the reference current signals based on the application. For example, the reference currents are obtained through a maximum power point tracking algorithm for photovoltaic systems, whereas the machine model and PI controller can be used to generate the reference current in motor drive applications. On the other hand, without loss of generality, here, the reference current signals are generated simply by the user-defined three-phase sinusoidal sources as seen in Figure 5.3.

Step 2: Extrapolation of reference currents: The need for the extrapolation of reference signals plays a critical role when the sampling time (T_s) is not sufficiently small as the output of the cost function is the absolute error at the next sampling instant ($k + 1$). For that reason, the reference currents, which are generated in Step 1, are required to be extrapolated from the kth instant to the ($k + 1$) sampling instant. Although there are many extrapolation methods available in the literature, in this specific example, one of the basic polynomial extrapolation methods, namely the third-order Lagrange extrapolation technique, which is represented in (5.2), has been used due to its simplicity and accuracy. The third-order Lagrange extrapolation uses present (k) and three past samples (($k - 1$), ($k - 2$), and ($k - 3$)) to calculate the future reference currents [11].

$$i^*(k + 1) = 4i^*(k) - 6i^*(k - 1) + 4i^*(k - 2) - i^*(k - 3) \qquad (5.2)$$

Step 3: Measurement of feedback signals: Like other control techniques, the performance of the FCS-MPC technique not only depends on the predictive model and cost function but also on the accuracy of the feedback signals. The required feedback signals can be measured through the current and voltage sensors. As seen in Figure 5.2, the DC-link voltage and the load currents are measured, and these signals are used in the predictive model and the cost function. It is worthy to mention that the accuracy of the feedback signal plays a critical role in the experiments, and this accuracy directly has an impact on the performance of the MPC. For that reason, the control board (MCU, FPGA, DSP, etc.) should have at least 12-bit ADC resolution. The reference design of the current and voltage sensing circuits can be found in Chapter 6.

Step 4: Predictive model: As mentioned previously, the predictive model of the system is the core of the FCS-MPC, and the digital implementation of this control

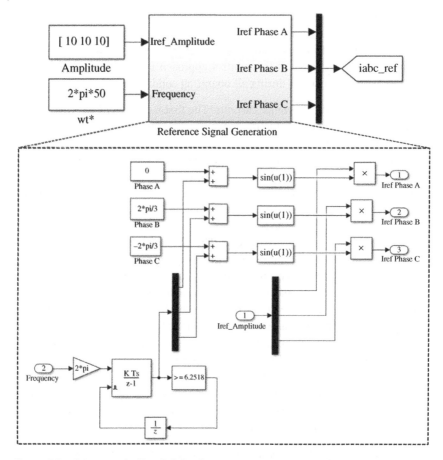

Figure 5.3 Subsystem in Simulink for three-phase current generation.

technique requires a discrete-time model of the system. In this specific example, since we are just controlling the output current, then predictive model is based on a single discrete-time equation. The load current model can be written as:

$$\frac{d\mathbf{i}(t)}{dt} = \frac{1}{L}[\mathbf{v}(t) - R\mathbf{i}(t)] \tag{5.3}$$

where $\mathbf{v}$ is the inverter output voltage vector and $\mathbf{i}$ is the load current vector.

$$\left.\frac{dx(t)}{dt}\right|_{t=k} \approx \frac{x(k+1) - x(k)}{T_s} \tag{5.4}$$

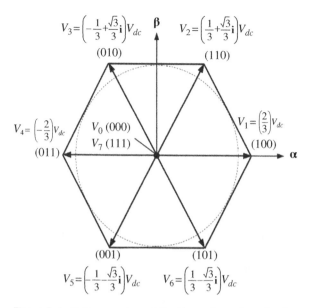

Figure 5.4 Voltage vectors of the three-phase three-leg inverter.

By using the forward Euler approximation in (5.4), the load current model in (5.3) is mapped to the discrete time as follows:

$$\mathbf{i}(k+1) = \left(1 - \frac{RT_s}{L}\right)\mathbf{i}(k) + \left(\frac{T_s}{L}\right)\mathbf{v}(k) \tag{5.5}$$

where $\mathbf{i}(k+1)$ is the predicted load current vector at the next sampling time. The model in (5.5) indicates that the future load currents are calculated based on the prediction of converter voltages $\mathbf{v}(k)$. The converter voltages $\mathbf{v}(k)$ are related to the switching states $S(k) = [S_a\ S_b\ S_c]^T$. Therefore, identifying the possible switching state combinations for the converter under study is important. It is therefore important to identify the possible switching state combinations (S_a, S_b, S_c) for the converter under study. As shown in Figure 5.4, there are eight possible combinations of switches, resulting in eight voltage vectors; there are six active vectors (v_1–v_6), while there are two zero vectors (v_0, v_7). Figure 5.4 illustrates seven different voltage vectors obtained as a result.

As FCS-MPC only has a finite number of switching state combinations, which simplifies the number of online iterations. Based on the eight possible switching state combinations, eight different predictions can be made for the converter output voltage, which then leads to eight different predictions for the load current.

Step 5: Cost function minimization:

A cost function $g(k)$ is used as a final step in the design process to identify the absolute error between the eight predicted currents $\mathbf{i}(k+1)$ and the extrapolated reference currents $\mathbf{i}^*(k+1)$.

$$g(k) = |\mathbf{i}^*(k+1) - \mathbf{i}(k+1)|$$

$$g(k) = \left|i_a^*(k+1) - i_a^p(k+1)\right| + \left|i_b^*(k+1) - i_b^p(k+1)\right| + \left|i_c^*(k+1) - i_c^p(k+1)\right|$$
$$= |9.42 - 9.98| + |4.43 - 6.51| + |-13.85 + 14.24| = 3.031$$

$$(5.6)$$

It is ideal to have a value of zero for the cost function, which represents perfect regulation of load currents. Zero cost function values are impossible, however, in practice. As a result of the cost evaluation in (5.6), eight different cost functions are obtained ($g_0 \sim g_7$). The goal of the cost function minimization subsystem is to identify the minimum value of the cost function and corresponding switching state combinations as shown in Figure 5.5 [11].

As an example, if the cost function g_5 has the minimum value among the pool, the corresponding switching state S_5 with $S_a =$ '0', $S_b =$ '1', and $S_c =$ '1' is chosen as an optimal actuation and applied directly to the converter. A switch state is selected at the next sampling instant based on the results of the whole design procedure during the kth sampling interval.

5.3.2 Tutorial 1: Implementation of FCS-MPC for Three-Phase VSI

The FCS-MPC is basically an optimization problem, and it is easy to implement the FCS-MPC algorithm in the digital control platforms such as MCU, DSP, and FPGA. In the first part of this section, the implementation stage of the FCS-MPC will be described by using a flowchart. Then, the implementation stage is going to be explained in detail through MATLAB/Simulink software.

The high-level flowchart of the FCS-MPC algorithm is depicted in Figure 5.6. As shown in the flowchart, the core of the FCS-MPC algorithm is the for-loop that predicts each voltage vector, evaluates the cost function, and stores the minimum value and the index value of the corresponding switching state. The control algorithm can be summarized in the next steps:

1) Measurement of the feedback signals (such as DC-link voltage and output currents),
2) Calculate back-emf as seen in Figure 5.6,
3) Initializing the FCS-MPC algorithm; set the counter (j) value to zero and the optimal cost function (g_{op}) value to infinity (practically very high value),

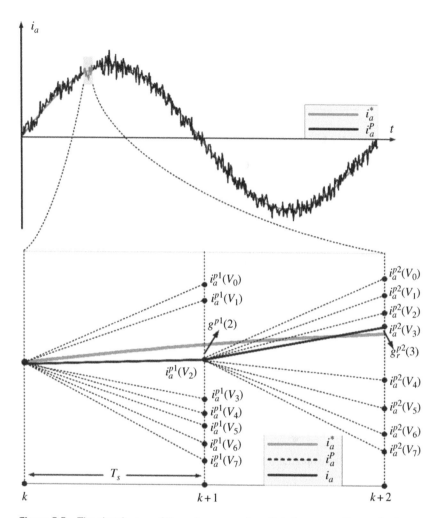

Figure 5.5 The sketch map of the reference and predicted currents. *Source:* Bayhan and Abu-Rub [11]/with permission of Elsevier.

4) Then the FCS-MPC algorithm enters the for-loop to predict the inverter voltage vector ($\mathbf{v}(k)$) and based on the $\mathbf{v}(k)$, the load currents are predicted by using the load current model,

5) A cost function is then used to evaluate the predicted load currents,

6) During any iteration, if $g < g_{op}$, the minimum g value is stored as an optimal value g_{op} and the corresponding switching state combination is stored as j_{op}. The switching state combination number j is counted by the algorithm. When

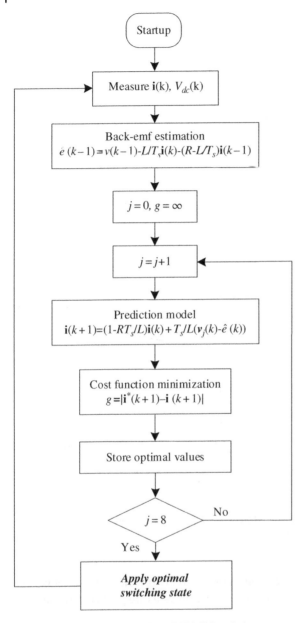

Figure 5.6 The flowchart of the FCS-MPC technique.

the loop runs for eight times $(0 \sim 7)$, the optimal j_{op} number is released by the algorithm.

7) Based on the j_{op} value, inverters are switched based on a predefined list of switching states.

It is worthy to mention that this flowchart can be applied to all power converters. However, some minor changes need to be considered based on the converter topology and application. First, the predefined switching-state table and the counter need to be updated according to the power converter topology. Second, the prediction model needs to be updated accordingly as well. Third, the cost function can be modified to meet different control objectives in addition to the basic function of the load current control. The flowchart in Figure 5.6 can easily be implemented in simulation environment such as MATLAB/Simulink or embedded systems such as MCUs and FPGAs. The m-file of FCS-MPC for three-phase VSI is given in Algorithm 5.1 [12]. The lines from 1 to 25 of the code correspond to the algorithm initialization. After that, the feedback signals are measured and the back-emf is estimated. Then, the line 40 defines the counter operation and the maximum count limit for the loop. The algorithm enters for loop at line 42 and exists at line 50. The load current is predicted based on measured currents, predicted inverter voltages, and estimated back-emf. The absolute error in the alpha beta components of the currents is evaluated in line 46 by cost function. To select the optimal cost function value and switching state combination, lines 48–54 are implemented. Finally, in lines 56–59, the optimal switching signals that minimize the cost function are generated based on the x_opt value.

Algorithm 5.1   FCS-MPC Algorithm for Three-Phase VSI

```
1 % Define variable for the control algorithm [12]
2 global Ts R L v states
3 % Setting up the parameters of the system
4 Ts = 40e-6; %Sampling time [s]
5 R = 5; % Resistance [Ohm]
6 L = 10e-3; % Inductance [H]
7 e = 150; % Back-EMF peak amplitude [V]
8 f_e = 50*(2*pi); % Nominal frequency [rad/s]
9 Vdc = 500; % DC-link voltage [V]
10 % The definition of the reference current amplitude
and frequency
11 I_ref_peak = 10; % Peak amplitude [A]
12 f_ref = 50*(2*pi); % Frequency [rad/s]
13 % Define the voltage vectors based on Figure 4
14 v0 = 0;
15 v1 = 2/3*Vdc;
```

```
16 v2 = 1/3*Vdc + 1j*sqrt(3)/3*Vdc;
17 v3 = -1/3*Vdc + 1j*sqrt(3)/3*Vdc;
18 v4 = -2/3*Vdc;
19 v5 = -1/3*Vdc - 1j*sqrt(3)/3*Vdc;
20 v6 = 1/3*Vdc - 1j*sqrt(3)/3*Vdc;
21 v7 = 0;
22 v = [v0 v1 v2 v3 v4 v5 v6 v7];
23 % Define the switching states based on Figure 4
24 states = [0 0 0;1 0 0;1 1 0;0 1 0;0 1 1;0 0 1;1 0 1;1 1 1];
25 function [Sa,Sb,Sc] = control(I_ref,I_meas)
26 % Optimum vector and measured current at instant k-1
27 persistent x_old i_old
28 % Initialize values
29 if isempty(x_old), x_old = 1; end
30 if isempty(i_old), i_old = 0+1j*0; end
31 g_opt = 1e10;
32 % Read current reference inputs at sampling instant k
33 ik_ref = I_ref(1) + 1j*I_ref(2);
34 % Read current measurements at sampling instant k
35 ik = I_meas(1) + 1j*I_meas(2);
36 % Back-EMF estimate
37 e = v(x_old) - L/Ts*ik - (R - L/Ts)*i_old;
38 % Store the measured current for the next iteration
39 i_old = ik;
40 for i = 1:8
41 % i-th voltage vector for current prediction
42 v_o1 = v(i);
43 % Current prediction at instant k+1
44 ik1 = (1 - R*Ts/L)*ik + Ts/L*(v_o1 - e);
45 % Cost function
46 g = abs(real(ik_ref - ik1)) + abs(imag(ik_ref - ik1));
47 % Selection of the optimal value
48 if (g<g_opt)
49 g_opt = g;
50 x_opt = i;
51 end
52 end
53 % Store the present value of x_opt
54 x_old = x_opt;
55 % Output switching states
56 Sa = states(x_opt,1);
57 Sb = states(x_opt,2);
58 Sc = states(x_opt,3);
```

Example 5.1 Consider a predictive current controlled 2L-VSI as shown in Figure 4.3. The system parameters are as follows: $R = 10\,\Omega$ and $L = 10$ mH. For a given sampling instant, the instantaneous feedback measurements are as follows: $v_{dc}(k) = 400$ V, $i_a(k) = 9.61$ A, $i_b(k) = 5.75$ A, and $i_c(k) = -12.86$ A. The extrapolated reference currents are $i_a^*(k + 1) = 9.42$ A, $i_b^*(k + 1) = 4.43$ A, and $i_c^*(k + 1) = -13.85$ A. The following is calculated with a sampling time $T_s = 100\,\mu$s: (a) predicted converter voltages, (b) predicted load currents, (c) cost function values, and (e) optimal switching signals for the next sampling interval [13].

a) The converter voltages with respect to the load neutral are predicted by using the measured DC-link voltage and eight possible switching states. The prediction of converter output voltages for a switching state combination $s_2 \in [`1', `1', `0']$ is obtained as follows:

$$
\mathbf{v}_2^p(k) = v_{dc}(k)
\begin{bmatrix}
\frac{2}{3} & -\frac{1}{3} & -\frac{1}{3} \\
-\frac{1}{3} & \frac{2}{3} & -\frac{1}{3} \\
-\frac{1}{3} & -\frac{1}{3} & \frac{2}{3}
\end{bmatrix}
\begin{bmatrix}
S_a(k) \\
S_b(k) \\
S_b(k)
\end{bmatrix}
= 400
\begin{bmatrix}
\frac{2}{3} & -\frac{1}{3} & -\frac{1}{3} \\
-\frac{1}{3} & \frac{2}{3} & -\frac{1}{3} \\
-\frac{1}{3} & -\frac{1}{3} & \frac{2}{3}
\end{bmatrix}
\begin{bmatrix}
1 \\
1 \\
0
\end{bmatrix}
$$

$$
=
\begin{bmatrix}
133.33 \\
133.33 \\
-266.67
\end{bmatrix}
\text{V}
$$

b) The prediction for load currents is obtained for $\mathbf{v}_2^p(k)$ from the model in (5.5):

$$
\mathbf{i}_2^p(k + 1) = \left(1 - \frac{RT_s}{L}\right)\mathbf{i}(k) + \left(\frac{T_s}{L}\right)\mathbf{v}_2^p(k) = \left(1 - \frac{10 * 0.0001}{0.01}\right)
\begin{bmatrix}
9.61 \\
5.75 \\
-12.86
\end{bmatrix}
$$

$$
+ \left(\frac{0.0001}{0.01}\right)
\begin{bmatrix}
133.33 \\
133.33 \\
-266.67
\end{bmatrix}
=
\begin{bmatrix}
9.98 \\
6.51 \\
-14.24
\end{bmatrix}
\text{A}
$$

c) The cost function value for switching state combination S_2 is

$$
g(k) = \left|i_a^*(k + 1) - i_a^p(k + 1)\right| + \left|i_b^*(k + 1) - i_b^p(k + 1)\right| + \left|i_c^*(k + 1) - i_c^p(k + 1)\right|
$$
$$
= |9.42 - 9.98| + |4.43 - 6.51| + |-13.85 + 14.24| = 3.031
$$

By repeating the calculations for all other combinations of switching states, the predicted converter output voltages and load currents, as well as cost function values, are evaluated. A minimum cost function value of 3.031 is found in this case for the switching state combination S_2. Therefore, the converter should be switched on with the following optimal switching signals at the next sampling interval: $S_a(k) = $ '1', $S_b(k) = $ '1', and $S_c(k) = $ '0'.

5.4 CCS Model Predictive Control

This section is dedicated to the CCSMPC technique. First, a model with an embedded integrator will be presented. This model will be used to predict the output of the system. The second stage is to use this prediction in a cost function which is minimized with the objective of obtaining the optimum control signal that makes the error between the output and its reference minimum. An important feature of this control method is that different constraints can be added to the cost function. This part will be also presented and explained, and a quadratic programing method to deal with the optimum. Finally, a brief explanation about stability analysis will be presented.

5.4.1 Incremental Models

The use of incremental models is a useful technique to eliminate the steady-state errors and to increase the system robustness against disturbances. The main idea is to express the model using incremental variables that also allow to embed an integrator to the system. For the sake of simplicity, an SISO model is considered in the analysis.

Let's consider a system defined by the following sate-space model where an external disturbance has been considered:

$$x_m(k + 1) = A_m x_m(k) + B_m u(k) + D_m q(k) \tag{5.7}$$

$$y(k) = C_m x_m(k). \tag{5.8}$$

In the aforementioned model A_m, B_m are the system matrices, D_m is the disturbance matrix, $x_m(k)$ is the state vector, $y(k)$ is the output of the system, $u(k)$ is the control variable, and $q(k)$ is an external disturbance. Note that the vector C_m is used to determine which variable is selected as the output. The model can be expressed in an incremental form by taking the difference operation on both sides of (5.7):

$$x_m(k + 1) - x_m(k) = A_m(x_m(k) - x_m(k - 1)) + B_m(u(k) - u(k - 1)) + D_m(q(k) - q(k - 1))$$

$$\tag{5.9}$$

After rearranging some terms, this expression can be rewritten as follows:

$$\Delta x_m(k+1) = A_m \Delta x_m(k) + B_m \Delta u(k) + D_m \Delta q(k). \tag{5.10}$$

Note that the input of the system is now $\Delta u(k)$, where Δ denotes the incremental operator.

Equation (5.10) defines the incremental model of the system, and as it has been stated before, the effect of the perturbation is removed. However, an integrator can be embedded in the system if the incremental operation is applied to the output equation. Then based on (5.8), one has:

$$y(k+1) - y(k) = C_m(x_m(k+1) - x_m(k)) \tag{5.11}$$

or equivalently:

$$y(k+1) = C_m \Delta x(k+1) + y(k) = C_m A_m \Delta x_m(k) + C_m B_m \Delta u(k) \\ + C_m D_m \Delta q(k) + y(k) \tag{5.12}$$

Equation (5.12) shows the expression of an integrator. This integrator is embedded in the system, removing the steady state error in the output variable $y(k)$.

The next step is to link the incremental variables with the output. At this point, a new space-state vector can be defined, which contains the incremental variables and the output:

$$x(k) = [\Delta x_m(k)\, y(k)] \tag{5.13}$$

Then, according to the definition of the new state-space vector (5.13), and using (5.10) and (5.12), the new state-space model is defined as follows:

$$\underbrace{\begin{bmatrix} \Delta x_m(k+1) \\ y(k+1) \end{bmatrix}}_{x(k+1)} = \underbrace{\begin{pmatrix} A_m & 0_n^T \\ C_m A_m & 1 \end{pmatrix}}_{A} \underbrace{\begin{bmatrix} \Delta x_m(k) \\ y(k) \end{bmatrix}}_{x(k)}$$
$$+ \underbrace{\begin{pmatrix} B_m \\ C_m B_m \end{pmatrix}}_{B} \Delta u(k) + \underbrace{\begin{pmatrix} D_m \\ C_m D_m \end{pmatrix}}_{D} \Delta q(k) \tag{5.14}$$

$$y(k) = \underbrace{[0_n\ 1]}_{C} \begin{bmatrix} \Delta x_m(k) \\ y(k) \end{bmatrix} \tag{5.15}$$

where 0_n^T represents a vector of zeros, the size of which is the same as the number of state variables. Note that the new space-state model can be expressed in a more compact form as follows:

$$x(k + 1) = Ax(k) + B\Delta u(k) + D\Delta q(k) \tag{5.16}$$

$$y(k) = Cx(k) \tag{5.17}$$

where the new system matrices are expressed as a function of the original system matrices:

$$A = \begin{pmatrix} A_m & 0_n^T \\ C_m A_m & 1 \end{pmatrix} \tag{5.18}$$

$$B = \begin{pmatrix} B_m \\ C_m B_m \end{pmatrix} \tag{5.19}$$

$$C = [0_n \ 1] \tag{5.20}$$

$$D = \begin{pmatrix} D_m \\ C_m D_m \end{pmatrix} \tag{5.21}$$

Example 5.2 A system is defined by the following space-state model:

$$x_m(k + 1) = A_m x(k) + B_m u(k) \tag{5.22}$$

$$y(k) = C_m x_m(k) \tag{5.23}$$

Find the incremental model of the system considering that $A_m = \begin{pmatrix} 1 & 0.5 \\ 0 & 1 \end{pmatrix}$, $B_m = \begin{bmatrix} 0.5 \\ 0 \end{bmatrix}$, and $C_m = [1 \ 0]$.

The original system has two state variables, $n = 2$. The system should be expressed in the form of ((5.16) and (5.17)). The new state vector is determined by the two incremental state variables and the output, $x = [\Delta x_{m1} \ \Delta x_{m2} \ y]$. By using the relations ((5.18), (5.17), (5.20)), and considering that the original system has two state variables, $n = 2$, the incremental model is obtained as follows:

$$A = \begin{pmatrix} 1 & 0.5 & 0 \\ 0 & 1 & 0 \\ 1 & 0.5 & 1 \end{pmatrix}; B = \begin{bmatrix} 0.5 \\ 0 \\ 0.5 \end{bmatrix}; C = [0 \ 0 \ 1]$$

5.4.2 Predictive Model

The predictive model is essential in any MPC algorithm, which is used to predict the output variable of the system. The past values and an actual measure are used to predict the system behavior within a prediction window, usually called

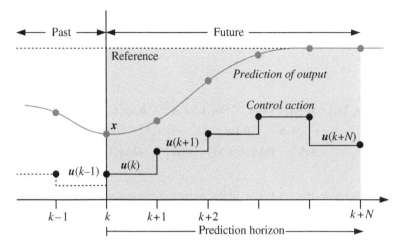

Figure 5.7 Prediction of the system output based on the past values and a measure.

prediction horizon. The incremental model defined in the last section can be used as a prediction model as it will be shown later. The MPC strategy is shown in Figure 5.7. As shown in the figure, the algorithm uses the past values of the system output and an actual measure to predict the future behavior of the output variable. Let's suppose that at the sampling instant k_i, with $k_i > 0$, the system state is sampled. The information of the sate vector $x(k)$ obtained from the measures is then known. The state $x(k_i)$ will provide the actual information of the system. It will be possible to predict the system behavior according to the actual information and the information given by the past values of the system states.

Then, denoting the future incremental control trajectory as:

$$\Delta u(k_i), \Delta u(k_i + 1), \dots, \Delta u(k_i + N_c - 1) \tag{5.24}$$

where N_c is the control horizon, which determines the number of values of the control signal used to obtain the control trajectory. On the other hand, the predictions of the system states are expressed by the following:

$$\Delta x(k_i + 1 \mid k_i), \Delta x(k_i + 2 \mid k_i), \dots, \Delta x(k_i + N_p \mid k_i) \tag{5.25}$$

where $N_p \geq N_c$ is the so-called prediction horizon and $\Delta x(k_i + m \mid k_i)$ is the prediction of the sate vector at the instant $k_i + m$ according to the system information $x(k_i)$. Note that it always be accomplished that $N_p \geq N_c$.

From the last definitions, and according to the model defined in (5.16), the following predictions for the system states can be written:

$$x(k_i + 1 \mid k_i) = Ax(k_i) + B\Delta u(k_i) + D\Delta q(k) \tag{5.26}$$

$$x(k_i + 2 \mid k_i) = Ax(k_i + 1) + B\Delta u(k_i + 1) + D\Delta q(k + 1)$$
$$= A^2 x(k_i) + AB\Delta u(k_i) + AD\Delta q(k_i) + B\Delta u(k_i + 1) \tag{5.27}$$

$$\vdots$$

$$x(k_i + N_p \mid k_i) = A^{N_p} x(k_i) + A^{N_p-1} Bu(k_i) + A^{N_p-2} B\Delta u(k_i + 1)$$
$$+ \dots + A^{N_p-N_c} B\Delta u(k_i + N_c - 1) + A^{N_p-1} D\Delta q(k_i)$$
$$+ A^{N_p-2} D\Delta q(k_i + 1) + \dots + A^{N_p-N_c} D\Delta q(k_i + N_c - 1) \tag{5.28}$$

In a similar manner, the prediction of the output of the system can be obtained as follows:

$$y(k_i + 1 \mid k_i) = CAx(k_i) + CB\Delta u(k_i) + CD\Delta q(k) \tag{5.29}$$
$$y(k_i + 2 \mid k_i) = CA^2 x(k_i) + CAB\Delta u(k_i) + CB\Delta u(k_i + 1) + CAD\Delta q(k_i) \tag{5.30}$$

$$\vdots$$

$$y(k_i + N_p \mid k_i) = CA^{N_p} x(k_i) + CA^{N_p-1} B\Delta u(k_i) + CA^{N_p-2} B\Delta u(k_i + 1)$$
$$+ \dots + CA^{N_p-N_c} B\Delta u(k_i + N_c - 1) + CA^{N_p-1} D\Delta q(k_i)$$
$$+ CA^{N_p-2} D\Delta q(k_i + 1) + \dots + CA^{N_p-N_c} D\Delta q(k_i + N_c - 1) \tag{5.31}$$

From the last expressions, one can observe that all the predictions of the output of the system are expressed in terms of the actual state $x(k_i)$ and from the future control trajectory $\Delta u(k_i + j)$, where $j = 0, 1, 2, \dots, N_c - 1$. The vectors containing all the control trajectory and the predicted system output quantities at instant k_i are respectively defined as follows:

$$\Delta U = [\Delta u(k_i) \quad \Delta u(k_i + 1) \quad \cdots \quad \Delta u(k_i + N_c - 1)]^T \tag{5.32}$$

$$Y = [y(k_i + 1) \quad y(k_i + 2) \quad \cdots \quad y(k_i + N_p)]^T \tag{5.33}$$

$$\Delta Q = [\Delta q(k_i) \quad \Delta q(k_i + 1) \quad \cdots \quad \Delta q(k_i + N_c - 1)]^T \tag{5.34}$$

Then, according to (5.28), and from (5.31)–(5.33), the prediction model, which includes the disturbance term, can be expressed in a compacted matrix form, yielding:

$$Y = Fx(k_i) + G_b \Delta U + G_d \Delta Q \tag{5.35}$$

where

$$F = \begin{pmatrix} CA \\ CA^2 \\ \vdots \\ CA^{N_p} \end{pmatrix} \tag{5.36}$$

$$G_b = \begin{pmatrix} CB & 0 & 0 & 0 & \cdots & 0 \\ CAB & CB & 0 & 0 & \cdots & 0 \\ CA^2B & CAB & CB & 0 & \cdots & 0 \\ CA^3B & CA^2B & CAB & CB & \cdots & 0 \\ \vdots & \vdots & \vdots & \vdots & \vdots & \vdots \\ CA^{N_p-1}B & CA^{N_p-2}B & CA^{N_p-3}B & CA^{N_p-4}B & \cdots & CA^{N_p-N_c}B \end{pmatrix} \tag{5.37}$$

$$G_d = \begin{pmatrix} CD & 0 & 0 & 0 & \cdots & 0 \\ CAD & CD & 0 & 0 & \cdots & 0 \\ CA^2D & CAD & CD & 0 & \cdots & 0 \\ CA^3D & CA^2D & CAD & CD & \cdots & 0 \\ \vdots & \vdots & \vdots & \vdots & \vdots & \vdots \\ CA^{N_p-1}D & CA^{N_p-2}D & CA^{N_p-3}D & CA^{N_p-4}D & \cdots & CA^{N_p-N_c}D \end{pmatrix} \tag{5.38}$$

Note that the number of columns is equal to the prediction horizon, N_p, while the number of rows is the difference between the prediction and horizon control, respectively.

Example 5.3 Find the predictions of the system defined in the Example 5.2 with $N_p = 3$ and $N_c = 2$.

The matrices of the incremental model are $A = \begin{pmatrix} 1 & 0.5 & 0 \\ 0 & 1 & 0 \\ 1 & 0.5 & 1 \end{pmatrix}$;

$B = \begin{bmatrix} 0.5 \\ 0 \\ 0.5 \end{bmatrix}$; $C = [0 \quad 0 \quad 1]$.

Using the expression in (5.36) and (5.34), with $N_p = 3$ and $N_c = 2$, the expressions for F and G can be obtained as follows:

$$F = \begin{pmatrix} CA \\ CA^2 \\ CA^3 \end{pmatrix} = \begin{pmatrix} 1 & 0.5 & 1 \\ 2 & 1.5 & 1 \\ 3 & 3 & 1 \end{pmatrix}; \quad G_b = \begin{pmatrix} CB & 0 \\ CAB & CB \\ CA^2B & CAB \end{pmatrix} = \begin{pmatrix} 0.5 & 0 \\ 1 & 0.5 \\ 1.5 & 1 \end{pmatrix}$$

Now, according to (5.35), one has:

$$Y = Fx(k_i) + G_b \Delta U = \begin{pmatrix} y(k_i + 1 \mid k_i) \\ y(k_i + 2 \mid k_i) \\ y(k_i + 3 \mid k_i) \end{pmatrix} = \begin{pmatrix} 1 & 0.5 & 1 \\ 2 & 1.5 & 1 \\ 3 & 3 & 1 \end{pmatrix} \begin{pmatrix} \Delta x_{m1}(k_i) \\ \Delta x_{m2}(k_i) \\ y(k_i) \end{pmatrix}$$

$$+ \begin{pmatrix} 0.5 & 0 \\ 1 & 0.5 \\ 1.5 & 1 \end{pmatrix} \begin{pmatrix} \Delta u(k_i) \\ \Delta u(k_i + 1) \end{pmatrix}$$

5.4.3 Cost Function in CCSMPC

The cost function is used in a predictive control system to obtain the future control vector, which minimizes the error between the output of the system and its reference. The function is always minimized inside a prediction horizon. Inside this window, the reference can be assumed constant, which can be expressed as follows:

$$Y_{ref}^T = \underbrace{\begin{bmatrix} 1 & 1 & \cdots & 1 \end{bmatrix}}_{N_p} y^*(k_i) = R_r y^*(k_i) \tag{5.39}$$

where $R_r = [1 \ 1 \ \cdots \ 1]$ is a vector of length N_p and $y^*(k_i)$ is the reference at instant k_i.

The cost function is usually defined as

$$J = (Y_{ref} - Y)^T (Y_{ref} - Y) + \Delta U^T R \Delta U \tag{5.40}$$

where R is a diagonal weight matrix, which dimension is N_c

$$R = \begin{pmatrix} r_\omega & 0 & \cdots & 0 \\ 0 & r_\omega & 0 & 0 \\ 0 & 0 & \ddots & 0 \\ 0 & 0 & \cdots & r_\omega \end{pmatrix} \tag{5.41}$$

This weight matrix can be used as a tuning parameter. In particular, R is the control effort and is used to modify the closed-loop performances of the system. Small values of R leads to a faster closed-loop dynamic, while with a large value of R, the dynamics becomes slower.

As explained before, the goal of a predictive control algorithm is to obtain the future control trajectory ΔU, which minimizes the error between the output and its reference.

5.4.4 Cost Function Minimization

The next step in a CCSMPC algorithm is to minimize the cost function defined in (5.40). For this purpose, the prediction of the output signal, (5.35), is used in (5.40), yielding

$$J = \left(Y_{ref} - Fx(k_i) - G_b\Delta U - G_d\Delta Q\right)^T \left(Y_{ref} - Fx(k_i) - G_b\Delta U - G_d\Delta Q\right) + \Delta U^T R\Delta U \tag{5.42}$$

that after some manipulations the following expression is obtained:

$$\begin{aligned} J = &\left(Y_{ref} - Fx(k_i) - G_d\Delta Q\right)^T \left(Y_{ref} - Fx(k_i) - G_d\Delta Q\right) \\ &- 2\Delta U^T G_b^T \left(Y_{ref} - Fx(k_i) - G_d\Delta Q\right) + \Delta U^T \left(G_b^T G_b + R\right)\Delta U \end{aligned} \tag{5.43}$$

Now we need to obtain the vector of the future control trajectory, ΔU, that minimizes the cost function (5.43). The condition for obtaining the minimum of J is expressed as:

$$\frac{\partial J}{\partial \Delta U} = 0$$

Applying this condition to the cost function J leads to the following condition:

$$- 2G_b^T \left(Y_{ref} - Fx(k_i) - G_d\Delta Q\right) + 2\left(G_b^T G_b + R\right)\Delta U = 0$$

where the optimal control signal can be obtained by solving the expression for ΔU, yielding:

$$\Delta U = \left(G_b^T G_b + R\right)^{-1} G_b^T \left(Y_{ref}^T - Fx(k_i) - G_d\Delta Q\right) \tag{5.44}$$

According to the definitions in (5.15) and (5.39), the last expression can be rewritten as follows:

$$\Delta U = \left(G_b^T G_b + R\right)^{-1} G_b^T \left(R_r y^*(k_i) - F\begin{bmatrix} \Delta x_m(k_i) \\ y(k_i) \end{bmatrix} - G_d\Delta Q\right) \tag{5.45}$$

Which after some manipulations can be separated into three different terms depending on the reference signal, the incremental state variables, and the incremental perturbations, respectively:

$$\Delta U = -K_r(y(k_i) - y^*(k_i)) - K_c \Delta x_m(k_i) - K_d \Delta Q \qquad (5.46)$$

where K_r is obtained by taking the last column of the resulting matrix $(G_b^T G_b + R)^{-1} G_b^T F$, while K_c are the remaining columns of $(G_b^T G_b + R)^{-1}$ $G_b^T F$. Note that $K_d = (G_b^T G_b + R)^{-1} G_b^T G_d$ is the gain of the disturbance term. As shown in [4], the disturbance term can be used as a feedforward signal, giving a robust transient performance regardless of loading dynamics and variations.

Example 5.4   Find the incremental control signal vector at time $k_i = 1$, for the system described in the Example 5.3, considering that the space vector is $x(1) = [0.993\ 0.297\ 0.993]$, the reference signal is $y^*(1) = 1$, and the matrix R is set to zero. The prediction and control horizons are $N_p = 3$ and $N_c = 2$, respectively.

The incremental control signal vector is defined by Eq. (5.44). Considering the matrices obtained in the last example with matrix R the identity matrix, the following expression can be written:

$$\Delta U = (G_b^T G_b + R)^{-1} G_b^T \left(Y_{ref}^T - Fx(k_i) \right)$$

$$= \left(\begin{pmatrix} 0.5 & 0 \\ 1 & 0.5 \\ 1.5 & 1 \end{pmatrix}^T \begin{pmatrix} 0.5 & 0 \\ 1 & 0.5 \\ 1.5 & 1 \end{pmatrix} \right)^{-1} \begin{pmatrix} 0.5 & 0 \\ 1 & 0.5 \\ 1.5 & 1 \end{pmatrix}^T$$

$$\left(\begin{pmatrix} 1 \\ 1 \\ 1 \end{pmatrix} \cdot 1 - \begin{pmatrix} 1 & 0.5 & 1 \\ 2 & 1.5 & 1 \\ 3 & 3 & 1 \end{pmatrix} \begin{pmatrix} 0.993 \\ 0.297 \\ 0.993 \end{pmatrix} \right)$$

and after solving the expression, the control signal vector results in:

$$\Delta U = (-2.2195\ 0.5090)^T$$

Note that as stated before, it is accomplished that:

$$(G_b^T G_b + R)^{-1} G_b^T R_r = \left(\begin{pmatrix} 0.5 & 0 \\ 1 & 0.5 \\ 1.5 & 1 \end{pmatrix}^T \begin{pmatrix} 0.5 & 0 \\ 1 & 0.5 \\ 1.5 & 1 \end{pmatrix} \right)^{-1} \begin{pmatrix} 0.5 & 0 \\ 1 & 0.5 \\ 1.5 & 1 \end{pmatrix}^T$$

$$\begin{pmatrix} 1 \\ 1 \\ 1 \end{pmatrix} \cdot 1 = \begin{pmatrix} 2 \\ -2 \end{pmatrix}$$

which is the last column of

$$\left(G_b^T G_b + R\right)^{-1} G_b^T F = \left(\begin{pmatrix} 0.5 & 0 \\ 1 & 0.5 \\ 1.5 & 1 \end{pmatrix}^T \begin{pmatrix} 0.5 & 0 \\ 1 & 0.5 \\ 1.5 & 1 \end{pmatrix}\right)^{-1} \begin{pmatrix} 0.5 & 0 \\ 1 & 0.5 \\ 1.5 & 1 \end{pmatrix}^T \begin{pmatrix} 1 & 0.5 & 1 \\ 2 & 1.5 & 1 \\ 3 & 3 & 1 \end{pmatrix}$$

$$= \begin{pmatrix} 2 & 0.833 & 2 \\ 0 & 1.667 & -2 \end{pmatrix}$$

and then the controller gains can be obtained according to (5.46):

$$K_r = \left(G_b^T G_b + R\right)^{-1} G_b^T R_r = \begin{pmatrix} 2 \\ -2 \end{pmatrix}$$

$$K_c = \begin{pmatrix} 2 & 0.833 \\ 0 & 1.667 \end{pmatrix}$$

Note that the control signal vector ΔU could be also obtained using the aforementioned control gains.

As shown in the example, the incremental control signal vector has two components, which is in accordance to the control horizon defined in the Example 5.3, $N_c = 2$.

To show the effect of the matrix R, the incremental control signal vector is computed for $r_\omega = 10$. In this case, the state-space vector at time $k_i = 1$ is defined by $x(1) = [0.993 \ 0.297 \ 0.993]$. The incremental control signal vector will be

$$\Delta U = \left(G_b^T G_b + R\right)^{-1} G_b^T \left(Y_{ref}^T - Fx(k_i)\right)$$

$$= \left(\begin{pmatrix} 0.5 & 0 \\ 1 & 0.5 \\ 1.5 & 1 \end{pmatrix}^T \begin{pmatrix} 0.5 & 0 \\ 1 & 0.5 \\ 1.5 & 1 \end{pmatrix} + \begin{pmatrix} 10 & 0 \\ 0 & 10 \end{pmatrix}\right)^{-1} \begin{pmatrix} 0.5 & 0 \\ 1 & 0.5 \\ 1.5 & 1 \end{pmatrix}^T$$

$$\left(\begin{pmatrix} 1 \\ 1 \\ 1 \end{pmatrix} \cdot 1 - \begin{pmatrix} 1 & 0.5 & 1 \\ 2 & 1.5 & 1 \\ 3 & 3 & 1 \end{pmatrix} \begin{pmatrix} 0.993 \\ 0.297 \\ 0.993 \end{pmatrix}\right) = (0.86 \ 0.5)^T$$

As it can be seen, with a larger value of r_ω leads to a smaller value of the control signal vector. In fact, it is equivalent to a slower transient response in the output signal. In Figure 5.8, the system transient response is shown for both values of r_ω.

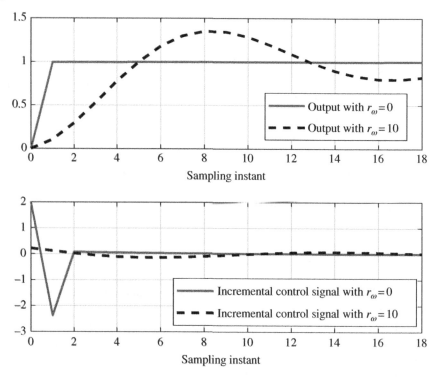

Figure 5.8 Comparison between the output signal of the system and the incremental control signal for different values of the control effort r_ω.

5.4.5 Receding Control Horizon Principle

As shown in the last section, the incremental control signal vector has a length equal to the desired prediction horizon N_c. In fact, in the Example 5.4, a vector of two components has been obtained. The receding horizon principle is based on using the first value of the vector while avoiding the remaining values. Although the control signal vector is fully computed at each sampling instant, only the first value of this vector is sent to the plant. However, this value could be different depending on the length of the control horizon. For instance, in the Example 5.4, with a value of $r_\omega = 10$, the first value for a control horizon $N_c = 2$ is 0.1182. It should be noted that the first value of the incremental control vector could be different for a different prediction and control horizons. With a prediction and control horizons $N_p = N_c = 5$, the incremental control vector is now

$$\Delta U = \left(G^T G + R\right)^{-1} G^T \left(Y_{ref}^T - Fx(k_i)\right) = [0.075 \quad 0.006 \quad -0.022 \quad -0.023 \quad -0.011]$$

Then, the value of the control signal that is sent to the plant will be 0.075.

The incremental control signal vector defined in (5.44) can be separated in two terms. One term depends on the state vector $x(k_i)$ while the other term on the reference signal $y^*(k_i)$. The first element of the control vector ΔU can be written as:

$$\Delta u(k_i) = -\left(G_b^T G_b + R\right)^{-1} G_b^T Fx(k_i) + W\left(G_b^T G_b + R\right)^{-1} G_b^T Y_{ref}^T \quad (5.47)$$

where $W = [1 \quad 0 \quad \cdots \quad 0]$ which length is N_c.

In a similar way as explained in the last section, the expression of $\Delta u(k_i)$ can be expressed as two terms with different gains K_x and K_y:

$$\Delta u(k_i) = -K_c x(k_i) + K_y y^*(k_i) \quad (5.48)$$

where K_c is the first row of $\left(G_b^T G_b + R\right)^{-1} G_b^T F$ and k_y is the first element of $\left(G_b^T G_b + R\right)^{-1} G_b^T R_r$.

As shown in Expression (5.48), the solution of an unconstrained MPC problem is similar to a classical problem of a state-feedback control.

5.4.6 Closed-Loop of an MPC System

In case of an unconstrained problem, the closed-loop of an MPC system can be regarded as a state-feedback controller. Using (5.48) with index k in (5.16), the following closed-loop system equation is obtained:

$$x(k+1) = Ax(k) - BK_x x(k) + BK_y y^*(k) \quad (5.49)$$

or equivalently

$$x(k+1) = (A - BK_x)x(k) + BK_y y^*(k) \quad (5.50)$$

The poles of the closed-loop system are given by the eigenvalues of $(A - BK_x)$, by solving de determinant $|A - BK_x - \lambda I| = 0$. It should be noticed that the eigenvalues have a dependence of the control gain K_x, and as a consequence a dependence on the control effort r_ω. The closed-loop system will be stable if eigenvalues are inside the unit circle.

5.4.7 Discrete Linear Quadratic Regulators

The cost function presented in (5.40) is based on the error between the output of the system and its reference. This is a simple way to express the objective function for a simpler minimization procedure. However, the cost function can be reformulated in terms of a discrete linear quadratic regulator (DLQR) as follows:

$$J = \sum_{m=1}^{N_p} x(k_i + m \mid k_i)^T Q x(k_i + m \mid k_i) + U^T R_q U \tag{5.51}$$

where Q is a positive definite matrix which dimension matches with the number of state variables and R_q is a positive define matrix that has the same dimension as the control vector L.

As shown previously, the control signal vector defined in (5.48) can be rewritten in a matrix form, and after rearranging some terms yields:

$$\Delta u(k_i) = -\begin{bmatrix} K_x & K_r \end{bmatrix} \begin{bmatrix} \Delta x(k_i) \\ y(k_i) \end{bmatrix} + K_r y^*(k_y) = -K_x \Delta x(k_i) - K_r(y(k_i) - y^*(k_i))$$

$$\tag{5.52}$$

which allows to redefine the incremental control vector as follows:

$$\Delta u(k_i) = -K\tilde{x}(k_i) \tag{5.53}$$

being the new state-space vector $\tilde{x}(k_i) = [\Delta x(k_i) \ y(k_i) - y^*(k_i)]^T$.

Then, the prediction of the state vector $\tilde{x}(k_i + m \mid k_i)$ can be defined:

$$\tilde{x}(k_i + m \mid k_i) = [\Delta x(k_i + m \mid k_i) \ y(k_i + m \mid k_i) - y^*(k_i)] \tag{5.54}$$

and using the expression of the DLQR defined in (5.51), the following cost function can be derived:

$$J = \sum_{m=1}^{N_p} \tilde{x}^T(k_i + m \mid k_i) Q\tilde{x}(k_i + m \mid k_i) + U^T R U \tag{5.55}$$

where selecting $Q = C^T C$, the expression is equivalent to the cost function defined in (5.40). It implies that to minimize (5.40) is similar to minimize the discrete linear quadratic regulator defined in (5.55). This fact allows to redefine the problem of the closed-loop system stability according to an adequate Lyapunov function. As shown in [14], the stability analysis of an MPC system can be performed by using a Lyapunov function that is exactly the cost function of the DLQR. Then, the cost function (5.51) can be rewritten as follows:

$$J = \sum_{m=1}^{N_p} \tilde{x}^T(k_i + m \mid k_i) Q\tilde{x}(k_i + m \mid k_i) + \sum_{m=0}^{N_c} \Delta u^T(k_i + m \mid k_i) R\Delta u(k_i + m \mid k_i)$$

$$\tag{5.56}$$

The aforementioned expression is a quadratic function, and it is a Lyapunov function candidate to analyze the stability.

Let's define the optimum of the cost function, J_{opt}, as a Lyapunov function V:

$$V(\tilde{x}(k_i), k_i) = J_{opt}$$

$$= \sum_{m=1}^{N_p} \tilde{x}_{opt}^T(k_i + m \mid k_i) Q \tilde{x}_{opt}(k_i + m \mid k_i) \tag{5.57}$$

$$+ \sum_{m=0}^{N_c} \Delta u_{opt}^T(k_i + m \mid k_i) R \Delta u_{opt}(k_i + m \mid k_i)$$

In a classical predictive control problem, the stability is ensured if the Lyapunov function decreases along its trajectory. Then, the following condition must be hold:

$$V(\tilde{x}(k_i + 1), k_i + 1) - V(\tilde{x}(k_i), k_i) < 0 \tag{5.58}$$

By using (5.57) in (5.58), the following inequality can be derived:

$$\tilde{x}_{opt}^T(k_i + m \mid k_i) Q \tilde{x}_{opt}(k_i + m \mid k_i) - \tilde{x}_{opt}^T(k_i \mid k_i) Q \tilde{x}_{opt}(k_i \mid k_i)$$
$$- \Delta u_{opt}^T(k_i \mid k_i) R \Delta u_{opt}(k_i \mid k_i) < 0 \tag{5.59}$$

In the last expression, and according to (5.53), the optimum incremental control signal $\Delta u_{opt}(k_i \mid k_i)$ can be substituted by $\Delta u_{opt}(k_i \mid k_i) = -K_c \tilde{x}_{opt}^T(k_i \mid k_i)$, which leads to the following condition to ensure asymptotic stability:

$$(A - BK)^T (A - BK) - K^T RK - Q < 0 \tag{5.60}$$

The last expression shows clearly that the stability of the closed-loop system depends on the system matrices, the control gain and also of the control effort. This way is interesting since it allows to analyze the stability in case of system parameters deviation. For instance, if the system has small deviations, it can be formulated as $(A + \Delta A)$ and $(B + \Delta B)$, where ΔA and ΔB contains the system parameters deviations. Then, to include these deviations, the inequality (5.60) can be rewritten as:

$$((A + \Delta A) - (B + \Delta B)K)^T ((A + \Delta A) - (B + \Delta B)K) - K^T RK - Q < 0 \tag{5.61}$$

Regarding the last expression, the system stability is clearly affected by the system uncertainties that could bring the system to the instability.

5.4.8 Formulation of the Constraints in MPC

The use of constraints in a model predictive control system is one of the reasons why this technique has become more attractive in recent years. With the development of digital signal processors, MPC algorithms have become a promising control method. The inclusion of constraints is very useful in cases where either the control variable or the output of the system should be kept in a specific range. This section will address how to include constraints in different variables of the system.

The easiest case is the inclusion of constraints in the control or in the incremental control signal. The mathematical formulation is simple, and it can be expressed as follows:

$$u_{\min} \leq u(k_i) \leq u_{\max} \tag{5.62}$$

$$\Delta u_{\min} \leq \Delta u(k_i) \leq \Delta u_{\max} \tag{5.63}$$

In these cases, the limitation is applied to the input of the system or in the increment of the input of the system after each sampling period. As an example, we can consider a digital control system where the digital signal processor only accepts ± 5 V at the input. Then, to avoid damages in the signal processor, a constraint in the control signal should be applied to ensure that $-5 \leq u(k) \leq 5$. However, a limitation in the increment of the control signal could be also considered, for instance, the increment of the voltage at the input of the system must be limited to ± 0.5 V. In this case the formulation of the constraint is $-0.5 \leq \Delta u(k) \leq 0.5$. It should be noted that the constraints are usually be applied to the first value of the control, or incremental control signal. However, they can also be applied to several values of the vector U or ΔU. For instance, if some constraints in the first 3 values of the incremental control signal are necessary, then the following inequalities can be written:

$$\begin{aligned} \Delta u_{\min} &\leq \Delta u(k_i) \leq \Delta u_{\max} \\ \Delta u_{\min} &\leq \Delta u(k_i + 1) \leq \Delta u_{\max} \\ \Delta u_{\min} &\leq \Delta u(k_i + 2) \leq \Delta u_{\max} \end{aligned} \tag{5.64}$$

If the constraints are applied to the entire incremental control signal vector, one has

$$\begin{aligned} \Delta u_{\min} &\leq \Delta u(k_i) \leq \Delta u_{\max} \\ \Delta u_{\min} &\leq \Delta u(k_i + 1) \leq \Delta u_{\max} \\ \Delta u_{\min} &\leq \Delta u(k_i + 2) \leq \Delta u_{\max} \\ &\quad\vdots \\ \Delta u_{\min} &\leq \Delta u(k_i + N_c) \leq \Delta u_{\max} \end{aligned} \tag{5.65}$$

that in matrix form will be:

$$\Delta U_{\min} \leq \Delta U \leq \Delta U_{\max} \tag{5.66}$$

being $\Delta U_{\min}$ and $\Delta U_{\max}$ two vectors of dimension N_c, which can be expressed as:

$$\Delta U_{\min} = \Delta u_{\min} \begin{pmatrix} 1 \\ 1 \\ 1 \\ \vdots \\ 1 \end{pmatrix} \quad \Delta U_{\max} = \Delta u_{\max} \begin{pmatrix} 1 \\ 1 \\ 1 \\ \vdots \\ 1 \end{pmatrix}$$

The constraint expressed in (5.66) can be expressed as a set of two constraints as follows:

$$\begin{pmatrix} -I \\ I \end{pmatrix} \Delta U \leq \begin{pmatrix} -\Delta U_{\min} \\ \Delta U_{\max} \end{pmatrix} \tag{5.67}$$

where I is the identity matrix.

As stated in previous sections, a receding horizon is usually considered. Then, only a constraint in the first element of the vector ΔU is applied. In this situation, the aforementioned set of constraints can be reformulated taking in consideration the receding horizon principle:

$$\begin{pmatrix} -1 & 0 & 0 & \cdots & 0 \\ 1 & 0 & 0 & \cdots & 0 \end{pmatrix} \Delta U \leq \begin{pmatrix} -\Delta u_{\min} \\ \Delta u_{\max} \end{pmatrix} \tag{5.68}$$

or equivalently

$$\begin{pmatrix} -P_1 \\ P_1 \end{pmatrix} \Delta U \leq \begin{pmatrix} -\Delta u_{\min} \\ \Delta u_{\max} \end{pmatrix} \tag{5.69}$$

Being $P_1 = (1 \quad 0 \quad \cdots \quad 0)$, a vector which dimension is the horizon control N_c.

Example 5.5 Consider a system that the incremental control signal must be limited to $-1 \leq \Delta u \leq 1$. The constraints are applied to the first 2 elements of the incremental control vector ΔU. Considering a control horizon of $N_c = 4$, the mathematical formulation of the set of constraints are as follows:

The incremental control vector is defined:

$$\Delta U = [\Delta u(k_i) \quad \Delta u(k_i + 1) \quad \Delta u(k_i + 2) \quad \Delta u(k_i + 3) \quad \Delta u(k_i + 4)]$$

The constraints can be formulated in the following manner:

$$- \Delta u(k_i) \leq 1$$
$$\Delta u(k_i) \leq 1$$
$$- \Delta u(k_i + 1) \leq 1$$
$$\Delta u(k_i + 1) \leq 1$$

According to (5.67), this set of constraints can be expressed in a matrix form as follows:

$$
\begin{pmatrix}
-1 & 0 & 0 & 0 \\
1 & 0 & 0 & 0 \\
0 & -1 & 0 & 0 \\
0 & 1 & 0 & 0
\end{pmatrix}
\Delta U \le
\begin{pmatrix}
1 \\
1 \\
1 \\
1
\end{pmatrix}
$$

Example 5.6 In this example, the constraints presented in the last example will be applied to the control signal. That is $-1 \le u \le 1$. We need to reformulate the problem as follows:

$$
\begin{aligned}
-u(k_i) &= -u(k_i - 1) - \Delta u(k_i) \le 1 \\
u(k_i) &= u(k_i - 1) + \Delta u(k_i) \le 1 \\
-u(k_i + 1) &= -u(k_i) - \Delta u(k_i + 1) = -u(k_i - 1) - \Delta u(k_i) - \Delta u(k_i + 1) \le 1 \\
u(k_i + 1) &= u(k_i) + \Delta u(k_i + 1) = u(k_i - 1) + \Delta u(k_i) + \Delta u(k_i + 1) \le 1
\end{aligned}
$$

$$
\begin{pmatrix}
-1 \\
1 \\
-1 \\
1
\end{pmatrix}
u(k_i - 1) +
\begin{pmatrix}
1 & 0 & 0 & 0 \\
1 & 0 & 0 & 0 \\
-1 & -1 & 0 & 0 \\
1 & 1 & 0 & 0
\end{pmatrix}
\Delta U \le
\begin{pmatrix}
1 \\
1 \\
1 \\
1
\end{pmatrix}
$$

or equivalently

$$
\begin{pmatrix}
1 & 0 & 0 & 0 \\
1 & 0 & 0 & 0 \\
-1 & -1 & 0 & 0 \\
1 & 1 & 0 & 0
\end{pmatrix}
\Delta U \le
\begin{pmatrix}
1 \\
1 \\
1 \\
1
\end{pmatrix}
-
\begin{pmatrix}
-1 \\
1 \\
-1 \\
1
\end{pmatrix}
u(k_i - 1)
$$

The constraints can also be applied to the output variable y. This case of constraint is usually expressed as:

$$
Y_{\min} \le Y \le Y_{\max} \tag{5.70}
$$

Then, according to the predictive model described in (5.35), where the disturbance term has not been considered, the constraints on the output variable can be expressed as a function of the terms of the predictive model as follows:

$$
Y_{\min} \le Fx(k_i) + G_b \Delta U \le Y_{\max} \tag{5.71}
$$

With this notation, the constraints applied to the output variable are expressed as constraints on the incremental control signal vector. Now this problem is equivalent to the problem presented in the last examples. In fact, rearranging terms in (5.71), one has

$$Y_{\min} - Fx(k_i) \leq G_b \Delta U \leq Y_{\max} - Fx(k_i) \tag{5.72}$$

$$\begin{pmatrix} G_b \\ -G_b \end{pmatrix} \Delta U \leq \begin{pmatrix} Y_{\max} - Fx(k_i) \\ -Y_{\min} + Fx(k_i) \end{pmatrix} \tag{5.73}$$

Note that in all cases the constraints can be expressed as a function of the incremental control signal vector ΔU, as follows:

$$M \Delta U \leq \gamma \tag{5.74}$$

where M and γ are matrices or vectors generated according to the desired constraints, as shown in the aforementioned examples. In fact, the number of the constraints matches with the number of rows of M, while the number of columns is equal to the dimension of ΔU.

5.4.9 Optimization with Equality Constraints

As shown in previous sections, the minimization of the cost function (5.40) does not require any online procedure and it can be done off-line. In fact, the method has similitude with the well-known state-feedback controller. In this section, the case of equality constraints will be analyzed. If some equality constraints are considered, the cost function should be minimized considering the use of Lagrange multipliers. Considering (5.43), the cost function can be expressed using only the two last terms:

$$J = \Delta U^T \left(G_b^T G_b + R \right) \Delta U - 2 \Delta U^T G_b^T \left(Y_{ref}^T - Fx(k_i) \right) \tag{5.75}$$

where the first term of (5.43) can be omitted since it does not depend on ΔU and as a consequence, it does not affect the minimization procedure. In addition, the cost function can be divided by two, and the minimum will not change, then J can be expressed as

$$J = \frac{1}{2} \Delta U^T \left(G_b^T G_b + R \right) \Delta U - \Delta U^T G_b^T \left(Y_{ref}^T - Fx(k_i) \right) \tag{5.76}$$

and the constrained problem can be expressed as follows:

$$J = \frac{1}{2} \Delta U^T \Phi \Delta U + \Delta U^T P$$
$$M \Delta U = \gamma \tag{5.77}$$

where $\Phi = G_b^T G_b + R$ is a matrix assumed symmetric and defined positive, and $P = -G_b^T \left(Y_{ref}^T - Fx(k_i) \right)$.

Considering the Lagrange multiplier, the cost function can be expressed as follows:

$$J = \frac{1}{2}\Delta U^T \Phi \Delta U + \Delta U^T P + \lambda^T (M\Delta U - \gamma) \tag{5.78}$$

The original cost function now depends on the Lagrange multipliers. It is important to remark that the number of equality constraints should be less or equal to the number of variables that are minimized. It means that if $\Delta U = [\Delta u_1 \ \Delta u_2]^T$, the number of equality constraints must be less or equal to 2. However, the problem can be solved easily if only one constraint is considered. For a better understanding, the following example is presented.

Example 5.7 Minimize the cost function $J = (\Delta u_1 + 1)^2 + (\Delta u_2 - 1)^2$ subject to $2\Delta u_1 + \Delta u_2 = 4$.

Applying (5.78), the minimization problem can be written as follows:

$$J = (\Delta u_1 + 1)^2 + (\Delta u_2 - 1)^2 + \lambda(4 - 2\Delta u_1 - \Delta u_2)$$

If the constraint is not considered, it is $\lambda = 0$, the minimum of J is clearly $\Delta u_1 = -1$ and $\Delta u_2 = 1$, which is the global minimum of the function. Conversely, considering the constraint one has

$$J = (\Delta u_1 + 1)^2 + (4 - 2\Delta u_1 - 1)^2 + \lambda(4 - 2\Delta u_1 - 4 + 2\Delta u_1)$$
$$J = (\Delta u_1 + 1)^2 + (3 - 2\Delta u_1)^2$$

Now, by taking the derivative of J with respect u_1 and equalizing to zero,

$$\frac{\partial J}{\partial u_1} = 2(\Delta u_1 + 1) - 4(3 - 2\Delta u_1) = 10\Delta u_1 - 10 = 0$$

Obtaining $\Delta u_1 = 1$ and $\Delta u_2 = 4 - 2\Delta u_1 = 2$.

Note that in this case, the use of Lagrange multipliers has not been necessary.

Now, reconsidering (5.78), the minimization problem is usually expressed in terms of the Lagrange multipliers by taking the partial derivatives with respect to ΔU and λ:

$$\frac{\partial J}{\partial \Delta U} = \Phi \Delta U + P + M^T \lambda = 0 \tag{5.79}$$

$$\frac{\partial J}{\partial \lambda} = M\Delta U - \gamma = 0 \tag{5.80}$$

which solving for ΔU and λ, leads to the solution:

$$\lambda = -\left(M\Phi^{-1}M^T\right)^{-1}\left(\gamma + M\Phi^{-1}P\right) \tag{5.81}$$

$$\Delta U = -\Phi^{-1}(P + M^T\lambda) = -\Phi^{-1}P - \Phi^{-1}M^T\lambda \qquad (5.82)$$

As can be observed from (5.82), the optimal solution has two terms. The first term is the optimal solution without considering constraints, while the second term accounts for the constraints.

Example 5.8   In the last example, the cost function $J = (\Delta u_1 + 1)^2 + (\Delta u_2 - 1)^2$ can be expressed as:

$$J = \frac{1}{2}\Delta U^T \Phi \Delta U + \Delta U^T P$$

where $\Delta U = [\Delta u_1 \ \ \Delta u_2]^T$, $\Phi = \begin{pmatrix} 2 & 0 \\ 0 & 2 \end{pmatrix}$ and $P = [2 \ \ -2]^T$.

The constraint $2\Delta u_1 + \Delta u_2 = 4$ can be expressed as $M\Delta U = \gamma$, where $M = [2 \ 1]$ and $\gamma = 4$. Applying (5.81), the Lagrange multiplier can be found. Note that the number of Lagrange multipliers matches with the number of constraints:

$$\lambda = -\left(M\Phi^{-1}M^T\right)^{-1}(\gamma + M\Phi^{-1}P) = -2$$

Once the Lagrange multiplier is obtained, the optimum variable ΔU is obtained according to (5.82):

$$\Delta U = -\Phi^{-1}P - \Phi^{-1}M^T\lambda = -\frac{1}{2}P - \frac{1}{2}M^T\lambda = -\frac{1}{2}[2 \ \ -2] + \begin{bmatrix} 2 \\ 1 \end{bmatrix} = \begin{bmatrix} 1 \\ 2 \end{bmatrix}$$

which coincides with the result obtained in Example 5.8.

It must be noticed that with equality constraints, the minimization procedure can be done off-line, since the restriction has known parameters and the Lagrange multiplier can be obtained using (5.81). This kind of constraint is known as active constraint.

5.4.10   Optimization with Inequality Constraints

The minimization with inequality constraints is a more difficult task. In this case, the minimization procedure is done online, usually by means of a quadratic programming algorithm. The problem is to minimize J subject to several inequality constraints:

$$J = \frac{1}{2}\Delta U^T \Phi \Delta U + \Delta U^T P$$
$$M\Delta U \leq \gamma \qquad (5.83)$$

In this case, the number of inequality constraints can be larger than the number of variables. In this case, the constraints can be separated into active and inactive constraints. The active constraints are the constraints that are satisfied with the equality, $M\Delta U = \gamma$, while the inactive constraints are expressed as $M\Delta U < \gamma$. For active constraints, the Lagrange multipliers are positive or zero. The necessary conditions for this minimization problem are given by the Karush–Kuhn–Tucker (KKT) conditions:

$$\Phi\Delta U + P + M^T\lambda = 0$$
$$M\Delta U - \gamma \leq 0$$
$$\lambda^T(M\Delta U - \gamma) = 0 \qquad\qquad (5.84)$$
$$\lambda \geq 0$$

Example 5.9 Consider the cost function:

$$J = \frac{1}{2}\Delta U^T\Phi\Delta U + \Delta U^T P$$

where $\Delta U = [\Delta u_1 \ \Delta u_2]^T$, $\Phi = \begin{pmatrix} 2 & 0 \\ 0 & 2 \end{pmatrix}$ and $P = [-2 \ \ -3]^T$.

The inequality constraints are expressed as

$$\Delta u_1 + \Delta u_2 \leq 1$$
$$2\Delta u_1 - \Delta u_2 \leq 1$$

or equivalently:

$$M\Delta U \leq \gamma$$

with $M = \begin{pmatrix} 1 & 1 \\ 1 & -2 \end{pmatrix}$ and $\gamma = \begin{pmatrix} 1 \\ 1 \end{pmatrix}$.

The Lagrange multipliers considering equality constraints can be obtained from

$$\lambda = -\left(M\Phi^{-1}M^T\right)^{-1}\left(\gamma + M\Phi^{-1}P\right) = [1.78 \ \ -0.55]$$

As it can be seen, the second Lagrange multiplier is negative, which means that the second inequality constraint is inactive, while the first one is active. If the second constraint is removed, second row of M and γ are removed, yielding:

$$M = (1 \ 1) \ \text{ and } \ \gamma = 1$$

Now, the Lagrange multipliers are recalculated, obtaining

$$\lambda = -\left(M\Phi^{-1}M^T\right)^{-1}\left(\gamma + M\Phi^{-1}P\right) = 1.5$$

The Lagrange multiplier is positive, then the constraint is active.

From the set of positive Lagrange multipliers, the optimum of the cost function is obtained as

$$\Delta U = -\Phi^{-1}P - \Phi^{-1}M^T\lambda = (0.25 \ \ 0.75)^T$$

Note that the optimum values of ΔU satisfies both constraints.

Then, the solution of a minimization with inequality constraints is similar to the minimization problem with equality constraints, but only using the active constraints. It reduces the problem to use only the positive Lagrange multipliers. The solution to the optimization problem can be written as follows:

$$\lambda_a = -\left(M_a \Phi^{-1} M_a^T\right)^{-1}\left(\gamma_a + M_a \Phi^{-1} P\right) \tag{5.85}$$

$$\Delta U = -\Phi^{-1}\left(P + M_a^T \lambda_a\right) \tag{5.86}$$

where λ_a are the Lagrange multipliers associated with the active constraints, M_a and γ_a are matrices or vectors, which only contains the active constraints, which correspond with the corresponding rows of M and γ that contain the active constraints.

Usually, the aforementioned problem is simplified by means of the dual-primal method. The method presented in (5.83) is known as the primal method, which can be expressed also as follows:

$$\max_{\lambda \geq 0} \min_{\Delta U} \left[\frac{1}{2}\Delta U^T \Phi \Delta U + \Delta U^T P + \lambda^T (M\Delta U - \gamma)\right] \tag{5.87}$$

Considering the optimal solution in (5.82), and using it in (5.87), the optimization problem becomes

$$\max_{\lambda \geq 0} \left[-\frac{1}{2}\lambda^T H \lambda - \lambda^T K - \frac{1}{2}P^T \Phi^{-1} P\right] \tag{5.88}$$

where the matrices H and K are expressed as

$$H = M\Phi^{-1} M^T \tag{5.89}$$

$$K = \gamma_a + M_a \Phi^{-1} P \tag{5.90}$$

Note that by changing the sign of all terms in (5.88), the maximization problem becomes into a minimization problem as follows:

$$\min_{\lambda \geq 0} \left[\frac{1}{2}\lambda^T H \lambda + \lambda^T K + \frac{1}{2}P^T \Phi^{-1} P\right] \tag{5.91}$$

which is known as the dual problem of (5.83), and it is equivalent to minimize the following cost function:

$$J = \frac{1}{2}\lambda^T H \lambda + \lambda^T K + \frac{1}{2}P^T \Phi^{-1} P \tag{5.92}$$

subject to $\lambda \geq 0$, or equivalently, the set of active Lagrange multipliers. With this idea, when the positive Lagrange multipliers are known, the problem is solved. Note that the procedure to obtain the positive Lagrange multipliers is done online.

Usually, a quadratic programing algorithm is considered for this task [15]. In [9], the use of output constraints in a microgrid controller is presented. This is an interesting case for the reader to better understand the use of constraints in an MPC controller.

5.4.11 MPC for Multi-Input Multi-Output Systems

The same procedure explained for SISO systems can be applied to multi-input multi-output (MIMO) systems. First, the incremental model for MIMO systems can be derived in a similar way as explained in Section 5.3.1.

Let's consider a MIMO system defined as a space-state model that contains p inputs and q outputs:

$$x_m(k + 1) = A_m x_m(k) + B_m u(k) \tag{5.93}$$

$$y(k) = C_m x_m(k) \tag{5.94}$$

where the input of the system is defined by the vector u, containing p components, and the output is expressed by y, which contains q components. Consider also that the state vector x_m has n state variables. The incremental model for the MIMO system can be expressed as follows:

$$x(k + 1) = A x(k) + B \Delta u(k) \tag{5.95}$$

$$y(k) = C x(k) \tag{5.96}$$

where

$$A = \begin{pmatrix} A_m & 0^T_{q \times n} \\ C_m A_m & I_{q \times q} \end{pmatrix} \tag{5.97}$$

$$B = \begin{pmatrix} B_m \\ C_m B_m \end{pmatrix} \tag{5.98}$$

$$C = \begin{bmatrix} 0_{q \times n} I_{q \times q} \end{bmatrix} \tag{5.99}$$

being $0_{q \times n}$ a matrix of zeros of $q \times n$, and $I_{q \times q}$ is the identity matrix of dimension q.

According to the incremental control vector defined in (5.45), the incremental control vector for an MIMO system can be written as follows:

$$\Delta U = \left(G_b^T G_b + R \right)^{-1} G_b^T \left(R_r y^*(k_i) - F \begin{bmatrix} \Delta x_m(k_i) \\ y(k_i) \end{bmatrix} \right) \tag{5.100}$$

where

$$R_r y^*(k_i) = \underbrace{\begin{bmatrix} I_{q \times q} & I_{q \times q} & \cdots & I_{q \times q} \end{bmatrix}}_{N_p} y^*(k_i) \tag{5.101}$$

Note that the vector containing the outputs is expressed by

$$\boldsymbol{y}^* = \begin{bmatrix} y_1^* & y_2^* & \cdots & y_q^* \end{bmatrix} \tag{5.102}$$

Similar to that was explained for the SISO systems, the incremental control vector defined in (5.100) can be separated into two terms:

$$\Delta \boldsymbol{U} = -\boldsymbol{K}_r(\boldsymbol{y}(k_i) - \boldsymbol{y}^*(k_i)) - \boldsymbol{K}_c \Delta \boldsymbol{x}_m(k_i) \tag{5.103}$$

where the matrix gain $\boldsymbol{K}_r$ coincides with the last q columns of the matrix $\left(\boldsymbol{G}_b^T \boldsymbol{G}_b + \boldsymbol{R}\right)^{-1} \boldsymbol{G}_b^T \boldsymbol{F}$, while $\boldsymbol{K}_c$ are the remaining columns of $\left(\boldsymbol{G}_b^T \boldsymbol{G}_b + \boldsymbol{R}\right)^{-1} \boldsymbol{G}_b^T \boldsymbol{F}$.

5.4.12 Tutorial 2: MPC Design For a Grid-Connected VSI in *dq* Frame

In this example, an MPC controller will be designed for a three-phase grid-connected voltage source inverter. This is example of an MPC design for an MIMO system. The electrical circuit is depicted in Figure 5.9. Electrical circuit of a grid-connected voltage source inverter.

Where the inductor $L = 5$ mH, the grid voltages are a three-phase sinusoidal signals with a peak voltage $V_p = 110\sqrt{2}\,$V and $f = 60$ Hz. The DC-link voltage is $V_{dc} = 400$ V and the selected sampling frequency is $f_s = 10$ kHz.

First, we will find the space-state model of the system. The circuit equations in the dq-frame are defined as follows:

$$L\frac{di_d}{dt} = \frac{V_{dc}}{2}m_d - v_d - L\omega i_q \tag{5.104}$$

$$L\frac{di_q}{dt} = \frac{V_{dc}}{2}m_q - v_q + L\omega i_d \tag{5.105}$$

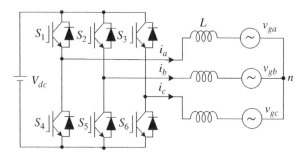

Figure 5.9 Electrical circuit of a grid-connected voltage source inverter.

The aforementioned equations can be expressed as a space-sate model considering that the system contains two inputs m_d and m_q and two outputs i_d and i_q:

$$\begin{pmatrix} \dfrac{di_d}{dt} \\ \dfrac{di_q}{dt} \end{pmatrix} = \begin{pmatrix} 0 & -\omega \\ \omega & 0 \end{pmatrix} \begin{pmatrix} i_d \\ i_q \end{pmatrix} + \begin{pmatrix} \dfrac{V_{dc}}{2L} & 0 \\ 0 & \dfrac{V_{dc}}{2L} \end{pmatrix} \begin{pmatrix} m_d \\ m_q \end{pmatrix} + \begin{pmatrix} -\dfrac{1}{L} & 0 \\ 0 & \dfrac{1}{L} \end{pmatrix} \begin{pmatrix} v_d \\ v_q \end{pmatrix}$$

(5.106)

That after a system discretization with a sampling period T_s, it is obtained:

$$x_m(k + 1) = A_m x_m(k) + B_m u(k) + D_m q(k) \tag{5.107}$$

$$y(k) = C_m x_m(k) \tag{5.108}$$

where $A_m = \begin{pmatrix} 1 & -\omega T_s \\ \omega T_s & 1 \end{pmatrix}$, $B_m = \begin{pmatrix} \dfrac{T_s \cdot V_{dc}}{2L} & 0 \\ 0 & \dfrac{T_s \cdot V_{dc}}{2L} \end{pmatrix}$, $D_m = \begin{pmatrix} -\dfrac{T_s}{L} & 0 \\ 0 & \dfrac{T_s}{L} \end{pmatrix}$,

$C_m = \begin{pmatrix} 1 & 0 \\ 0 & 1 \end{pmatrix}$

and

$$x_m(k) = [i_d \; i_q]^T, \; u(k) = [m_d \; m_q]^T \; \text{and} \; q(k) = [v_d \; v_q]^T$$

Note that the vector $q(k)$ can be considered as a disturbance.

Now to obtain the incremental model in the form as shown in (5.95) and (5.96), Eqs. (5.97), (5.98), and (5.99) are used, yielding:

$$A = \begin{pmatrix} 1 & -\omega T_s & 0 & 0 \\ \omega T_s & 1 & 0 & 0 \\ 1 & -\omega T_s & 1 & 0 \\ \omega T_s & 1 & 0 & 1 \end{pmatrix}, \; B = \begin{pmatrix} \dfrac{T_s \cdot V_{dc}}{2L} & 0 \\ 0 & \dfrac{T_s \cdot V_{dc}}{2L} \\ \dfrac{T_s \cdot V_{dc}}{2L} & 0 \\ 0 & \dfrac{T_s \cdot V_{dc}}{2L} \end{pmatrix}, \; D = \begin{pmatrix} -\dfrac{T_s}{L} & 0 \\ 0 & -\dfrac{T_s}{L} \\ -\dfrac{T_s}{L} & 0 \\ 0 & \dfrac{T_s}{L} \end{pmatrix},$$

$$C = \begin{pmatrix} 0 & 0 & 1 & 0 \\ 0 & 0 & 0 & 1 \end{pmatrix}$$

where the aforementioned system parameters can be substituted to obtain numerical matrices. Anyway, the incremental space-state model is defined by the following equations:

$$x(k + 1) = Ax(k) + B\Delta u(k) + D\Delta q(k)$$

$$y(k) = Cx(k)$$

In the dq-frame, v_d and v_q are constants, then the incremental signal of the perturbation, $\Delta q(k) \cong 0$, and it can be removed from the system equations, yielding

$$x(k + 1) = Ax(k) + B\Delta u(k)$$
$$y(k) = Cx(k)$$

The next step is to obtain the prediction model defined in (5.35). In this case, $\Delta q(k) \cong 0$ and the predictive model can be described as follows:

$$Y = Fx(k_i) + G_b\Delta U \tag{5.109}$$

where the matrices F and G_b can be obtained from the matrices of the systems A, B, and C using the expressions defined in (5.36) and (5.37), respectively, with the desired prediction and control horizons N_p and N_q. Note that in a real implementation, the prediction model is not necessary to be obtained. However, in this chapter and as an example, it is shown how the prediction model can be obtained for the grid connected VSI.

The last step is to use the matrices F and G_b in the expression of the optimal control signal (5.44) to obtain the control signals m_d and m_q. Note that in case of using constraints, the control signal can be obtained using (5.85) and (5.86) with a quadratic programming algorithm to find the active Lagrange multipliers, as shown in [15].

Figure 5.10 shows the control block diagram for the MPC system, and as shown in the figure, the grid voltages are measured and used in a phase-locked loop (PLL)

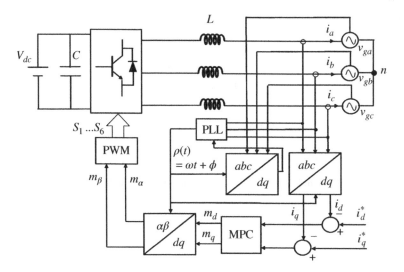

Figure 5.10 Block diagram of the control system.

to obtain the instantaneous phase $\rho(t)$ of the voltage. Once the phase is obtained, and by means of the Park transformation, the dq components of the voltages and currents are calculated. Since an incremental model is used, the dq voltages are not used in the controller since as shown before $\Delta \boldsymbol{q}(k) \cong 0$. The MPC block minimizes a cost function in order to minimize the error between the dq currents and their references. For the controller implementation in a digital signal processor (DSP), a complete pseudocode is presented in [3].

5.5 Design and Implementation Issues

5.5.1 Cost Function Selection

To meet the operational requirements in various power conversion applications, the definition of the cost function is an important design step. It is generally easy and flexible to design cost functions for multivariable systems with different physical natures, magnitudes, frequencies, and phase angles. A general classification of control objectives is provided in this section to explain the flexibility of the cost function. It is possible to divide the objectives of the cost function into two main categories as seen in Figure 5.11. Among the primary objectives, there are those related to the basic operation (current control, voltage control, etc.), and the secondary goals pertain to the technical requirements, constraints, and nonlinearities [13].

An error between the predicted value and its reference can be expressed in general form as follows:

$$g(k) = \|\hat{x}^*(k+1) - x^p(k+1)\| \tag{5.110}$$

where $\hat{x}^*$ is the extrapolated reference command, and x^p is the predicted value of the control variable; both of them correspond to the $(k+1)$ sampling instant.

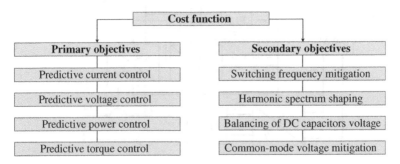

Figure 5.11 Cost function in MPC with some primary and secondary control objectives.

A norm $\|...\|$ measures the difference between reference and predicted values. The cost function can be realized as either an absolute value, a quadratic value, an exponential value, or an integral value.

$$
\left.\begin{aligned}
g(k) &= |\hat{x}^*(k+1) - x^p(k+1)| \\
g(k) &= [\hat{x}^*(k+1) - x^p(k+1)]^2 \\
g(k) &= \left| \int_k^{k+1} (x^*(t) - x^p(t))dt \right|
\end{aligned}\right\}
\tag{5.111}
$$

As long as only one error term is considered in the cost function, the absolute and the quadratic cost functions yield similar results. There can, however, be differences in results if the cost function has two or more different terms. When the cost function includes additional terms, the quadratic cost function performs better. Furthermore, the integral cost function takes into account the trajectory of error values between the $(k)^{\text{th}}$ and $(k+1)^{\text{th}}$ sampling instants by minimizing the average error between these sampling instants. In this way, reference tracking is improved, and steady-state errors are reduced. Despite this, the approach requires more calculations [12].

5.5.1.1 Examples for Primary Control Objectives

As mentioned before, current control is a primary objective of the cost function. The inner control loop of a power converter employs current control in a wide range of applications, no matter the topology of the converter. In order to implement the current control, there are three different reference frames available: a natural reference frame, a stationary reference frame, and a synchronous reference frame. In a similar manner, we can define the cost function in these reference frames as follows:

$$
\left.\begin{aligned}
g_i(k) &= \left|\hat{i}_a^* - i_a^p\right| + \left|\hat{i}_b^* - i_b^p\right| + \left|\hat{i}_c^* - i_c^p\right| \\
g_i(k) &= \left|\hat{i}_\alpha^* - i_\alpha^p\right| + \left|\hat{i}_\beta^* - i_\beta^p\right| \\
g_i(k) &= \left|\hat{i}_d^* - i_d^p\right| + \left|\hat{i}_q^* - i_q^p\right|
\end{aligned}\right\}
\tag{5.112}
$$

Various applications require voltage control, such as uninterruptible power supplies and voltage-forming inverters in islanded microgrids. No matter whether the load is linear, nonlinear, balanced or unbalanced, or single- or three-phase, the consumer demands high-quality sinusoidal voltages. This type of application requires predictive voltage control with the following cost function.

$$
g_v(k) = \left|\hat{v}_a^* - v_a^p\right| + \left|\hat{v}_b^* - v_b^p\right| + \left|\hat{v}_c^* - v_c^p\right|
\tag{5.113}
$$

Predictive power control is used to control active and reactive power in grid-connected applications such as PV inverters and wind power conversion system.

As a result of this control scheme, the error in the active and reactive powers is minimized by the following cost function.

$$g_{PQ}(k) = \left| \hat{P}^* - P^p \right| + \left| \hat{Q}^* - Q^p \right| \tag{5.114}$$

5.5.1.2 Examples for Secondary Control Objectives

Switching frequency is one of the major measures of control effort in power converters. In many applications, controlling or limiting the number of commutations of power switches is important. FCS-MPC reduces the converter's switching frequency by minimizing the number of commutations between sampling instants. The cost function of a three-phase power converter with Ns switching devices per phase is as follows [12]:

$$g_{sw}(k) = \sum_{j=1}^{N_a} \sum_{x=a,b,c} \left| s_{xj}^p(k) - s_{xj}^p(k-1) \right| \tag{5.115}$$

where $s_{xj}^p(k)$ and $s_{xj}^p(k-1)$ are the predicted and past sample optimal switching states of *j*th device, respectively.

To ensure low semiconductor device stress and reliable operation, DC-link capacitor voltages should be maintained at equal values in NPC and T-type converters. This can be achieved by defining sub-cost functions like these:

$$g_{PN}(k) = \left| V_{C1}^p - V_{C2}^p \right| \tag{5.116}$$

where V_{C1}^p and V_{C2}^p are the predicted DC-link capacitor voltages in NPC and T-type converters.

The control scheme must mitigate the common-mode voltage (CMV) in the converter to reduce electromagnetic interference and voltage stress on the generator/motor winding insulation. An even-level voltage source and matrix converter can be minimized by the following cost function [13].

$$g_{cm}(k) = \frac{v_{aN}^p + v_{bN}^p + v_{cN}^p}{3} \tag{5.117}$$

where v_{aN}^p, v_{aN}^p, and v_{aN}^p are the predicted converter output voltages.

5.5.2 Weighting Factor Design

From the previous section, we note that it is possible to control multiple control targets, variables, and constraints simultaneously with MPC. In other words, it is possible to control typical variables such as current, voltage, torque, or flux while also reducing switching frequency, reducing common-mode voltage, and reducing the harmonic spectrum. A cost function can be modified to include additional control targets simply by adding them to the cost function to be evaluated. On the other hand, it is not straightforward to combine two or more variables in a single cost function when their natures are different (different units, different magnitudes). Adding an additional term to the cost function has a specific weighting factor that determines how important or costly it is compared to other control targets. In linear control, PI regulator parameters must be tuned. The weighting factors in FCS-MPC, however, need to be optimized for optimal control performance. As a result, selecting the best weighting factor value is one of the most important steps in the FCS-MPC design process. A number of popular approaches for weighting factors design are discussed in terms of the available solutions/methodologies [13].

5.5.2.1 Empirical Selection Method

The tuning of the PI or hysteresis bandwidth limits of classical control techniques has been documented using a variety of methodologies. Despite this, FCS-MPC commonly uses an empirical procedure for selecting the weighting factors due to limited knowledge.

To analyze this approach, a quasi-Z source inverter with the following global cost function [16] is considered:

$$g(k + 1) = g_{v_o}(k + 1) + \lambda_i g_{i_L}(k + 1) + \lambda_v g_{v_C}(k + 1) \tag{5.118}$$

where λ_i and λ_v are the weighting factors to regulate the inductor current and capacitor voltage, respectively. The weighting factors tuning is conducted based on minimizing the error on the output voltage (e_{vo}) and minimizing the double-line frequency ($2*f_o$) ripple on the input current (Δi_{L1}) and the capacitor voltage (ΔV_{C2}). To observe the effect of λ_i and λ_v on the performance and selecting the optimum value for the best control results, e_{vo}, Δi_{L1}, and ΔV_{C2} (these variables are selected as performance indicators for the selection of the optimum weighting factors) were measured for different values of λ_i and λ_v. To simplify the analysis, λ_v is chosen equal to $0.1*\lambda_i$. To observe the effect of the $2f_o$ on i_{L1} and V_{C2}, this test was performed with unbalanced load condition. Figure 5.12 shows the variation of e_{vo}, Δi_{L1}, and ΔV_{C2} as a function of λ_i. The results have been obtained by performing several simulations, starting with $\lambda_i = 0$ and gradually increasing these values. It can be observed (see Figure 5.12) that a reduction of the ripples on the inductor current and the capacitor voltages introduces higher reference tracking error on the output voltage, which affects the quality of the output current as well. This tradeoff is very clear in Figure 5.12 [16]. Notice that the optimum performance is obtained for λ_i between 0.6 and 0.75 since the voltage error is still below 5% of the rated voltage and a significant reduction is achieved for the inductor current, and the capacitor voltage ripples. Thus, $\lambda_i = 0.75$ and $\lambda_v = 0.075$ are considered as the optimum weighting factors and are used for the simulation and experimental tests [16].

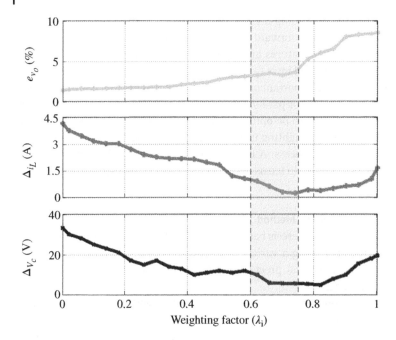

Figure 5.12 Weighting factors influence on the output voltage, the input current, and the capacitor voltage. *Source:* Bayhan et al. [16]/IEEE.

The control variables in this example are the primary control objectives since output voltage, input currents, and capacitor voltage should be regulated. In addition to primary control objectives, the power converter performance can be improved by using the secondary control objectives such as reducing common-mode voltage, minimizing switching frequency, and reducing switching loss, but these attributes are not required to operate the power converter. Therefore, the trial-and-error approach for this type of variable starts with a small weighting factor value.

5.5.2.2 Equal-Weighted Cost-Function-Based Selection Method

As mentioned before, the regulation of different variable types (voltages and currents) may be needed in controlling advanced power converters. The total cost term of an FCS-MPC can be defined as follows [17]

$$g = \lambda_1 \left| x_1^*(k+1) - x_1(k+1) \right| + \lambda_2 \left| x_2^*(k+1) - x_2(k+1) \right|$$
$$+ \ldots + \lambda_n \left| x_n^*(k+1) - x_n(k+1) \right| \tag{5.119}$$

The main difference between each sub-term is determined by the reference values used in the corresponding sub-term. The weighting factors (λ_1 to λ_n) in the total cost function should be tuned automatically to guarantee the desired performance. The idea behind this approach is to regulate all control variables that have the same importance in controlling the power converters. This method is described in more detail in [17]. To obtain such control structure, the weights of sub-terms in the total cost function should be equalized. According to this idea, each sub-term is normalized by dividing the sub-term with its reference value as follows:

$$g = \lambda_1 \frac{\left|x_1^*(k+1) - x_1(k+1)\right|}{X_1^*} + \lambda_2 \frac{\left|x_2^*(k+1) - x_2(k+1)\right|}{X_2^*}$$
$$+ \dots + \lambda_n \frac{\left|x_n^*(k+1) - x_n(k+1)\right|}{X_n^*} \tag{5.120}$$

where X_1^* to X_n^* denote the reference values used for each sub-term. It is worth noting that since AC control inputs are time-varying signals, their *rms* values are used for normalization. However, the measured value in each sampling interval is used for control input in nonlinear control methods. Therefore, the normalized value of the maximum control error is the amplitude of the AC signals. In (5.120), while the amplitude of AC references is used through normalization, the reference of DC signals can be used directly. Since the importance of each cost term is equalized by normalization, there is no need to use additional weighting factors (λ_1 to λ_n). Consequently, the equal-weighted cost function can be written as follows:

$$g = \frac{\left|x_1^*(k+1) - x_1(k+1)\right|}{X_1^*} + \frac{\left|x_2^*(k+1) - x_2(k+1)\right|}{X_2^*}$$
$$+ \dots + \frac{\left|x_n^*(k+1) - x_n(k+1)\right|}{X_n^*} \tag{5.121}$$

More information about this method can be found in 1.

5.5.2.3 Lookup Table-Based Selection Method

There are some specific applications where the magnitude of error terms or control variables changes based on operating conditions. In such cases, optimal weighting factors should be determined empirically for each operating condition and stored in lookup tables for use with FCS-MPC.

Based on the reference control variables, the optimal weighting factors are determined. Consequently, FCS-MPC provides excellent performance regardless of operating conditions. Lookup tables can also be used to select weighting factors for torque and flux controls, CMV mitigation, etc.

In [18], weighting factor optimization is performed by straightforward lookup table method for induction motor torque control. Furthermore, the switching frequency is regulated between 750 and 850 Hz with the help of a lookup-table-based weighting factor in [19].

References

1 J. W. Eaton and J. B. Rawlings, "Model-predictive control of chemical processes," *Chem. Eng. Sci.*, vol. 47, no. 4, pp. 705–720, June 1992.

2 H. Han and J. Qiao, "Nonlinear model-predictive control for industrial processes: an application to wastewater treatment process," *IEEE Trans. Ind. Electron.*, vol. 61, no. 4, pp. 1970–1982, Apr. 2014.

3 R. Guzman, L. G. de Vicuña, A. Camacho, J. Miret, and J. M. Rey, "Receding-horizon model-predictive control for a three-phase VSI With an LCL Filter," *IEEE Trans. Ind. Electron.*, vol. 66, no. 9, pp. 6671–6680, Sept. 2019.

4 A. Saleh, A. Deihimi, and R. Iravani, "Model predictive control of distributed generations with feed-forward output currents," *IEEE Trans. Smart Grid*, vol. 10, no. 2, pp. 1488–1500, Mar. 2019.

5 C. Alfaro, R. Guzman, L. G. de Vicuña, J. Miret, and M. Castilla, "Dual-loop continuous control set model-predictive control for a three-phase unity power factor rectifier," *IEEE Trans. Power Electron.*, vol. 37, no. 2, pp. 1447–1460, Feb. 2022.

6 J. Sawma, F. Khatounian, E. Monmasson, L. Idkhajine, and R. Ghosn, "Cascaded dual-model-predictive control of an active front-end rectifier," *IEEE Trans. Ind. Electron.*, vol. 63, no. 7, pp. 4604–4614, Jul. 2016, doi: 10.1109/TIE.2016.2547874.

7 R. Guzmán, L. García de Vicuña, M. Castilla, J. Miret, and A. Camacho, "Finite control set model predictive control for a three-phase shunt active power filter with a Kalman filter-based estimation," *Energies*, vol. 10, no. 10, p. 1553, Mar. 2017.

8 H. Komurcugil, S. Bayhan, N. Guler, and F. Blaabjerg, "An effective model predictive control method with self-balanced capacitor voltages for single-phase three-level shunt active filters," *IEEE Access*, vol. 9, pp. 103811–103821, Oct. 2021.

9 C. A. Aragon, R. Guzman, L. G. de Vicuña, J. Miret, and M. Castilla, "Constrained predictive control based on a large-signal model for a three-phase inverter connected to a microgrid," *IEEE Trans. Ind. Electron.*, vol. 69, no. 7, pp. 6497–6507, Jul. 2022, doi: 10.1109/TIE.2021.3097608.

10 Z. Yi, Y. Xu, W. Gu, and Z. Fei, "Distributed model predictive control based secondary frequency regulation for a microgrid with massive distributed resources," *IEEE Trans. Sustainable Energy*, vol. 12, no. 2, pp. 1078–1089, Apr. 2021.

11 S. Bayhan and H. Abu-Rub, "Chapter 40 - Predictive control of power electronic converters," in *Power Electronics Handbook*, M. Rashid, Ed., 4th ed. UK: Elsevier, 2018. doi: 10.1016/b978-0-12-811407-0.00044-1.

12 J. Rodriguez and P. Cortes, *Predictive Control of Power Converters and Electrical Drives*. USA: IEEE-Wiley, 2012.

13 W. Yaramasu and B. Wu, *Model Predictive Control of Wind Energy Conversion Systems*. USA: Wiley-IEEE Press, 2017.

14 W. H. Chen, "Stability analysis of classic finite horizon model predictive control," *Int. J. Control Autom. Syst.*, vol. 8, pp. 187–197, Apr. 2010.

15 L. Wang, *Model Predictive Control System Design and Implementation Using MATLAB*. New York, NY, USA: Springer, 2009.

16 S. Bayhan, M. Trabelsi, H. Abu-Rub, and M. Malinowski, "Finite-control-set model-predictive control for a Quasi-Z-source four-leg inverter under unbalanced load condition," *IEEE Tran. Ind. Elect.*, vol. 64, no. 4, Apr. 2017.

17 N. Guler, S. Bayhan, and H. Komurcugil, "Equal weighted cost function-based weighting factor tuning method for model predictive control in power converters," *IET Power Electron.*, vol. 15, pp. 203–215, June 2022.

18 S. A. Davari, D. A. Khaburi and R. Kennel, "Using a weighting factor table for FCS-MPC of induction motors with extended prediction horizon," *IECON 2012 - 38th Ann. Conf. IEEE Industrial Electronics Society*, Aug. 2012, pp. 2086–2091, doi: 10.1109/IECON.2012.6388737.

19 V. Yaramasu, B. Wu, and J. Chen, "Model-predictive control of grid-tied four-level diode-clamped inverters for high-power wind energy conversion systems," *IEEE Trans. Power Electron.*, vol. 29, no. 6, pp. 2861–2873, Jun. 2014, doi: 10.1109/TPEL.2013.2276120.

6

MATLAB/Simulink Tutorial on Physical Modeling and Experimental Setup

6.1 Introduction

In the previous chapters, the theoretical background of advanced control techniques was presented along with the implementation steps. We believe that the best learning strategy is to practice what we learned. In this chapter, it will be shown how to build the physical models of power converters using MATLAB/Simulink. It will also be described how to build the real-time model of the power converters in the OPAL-RT simulator. Following that, we'll explore experimentally how power electronics systems can be controlled using the developed control techniques. For those who are not familiar with Simulink simulations of power converters, the tutorials are written in a step-by-step manner to guide the learner through them.

6.2 Building Simulation Model for Power Converters

Simulation is a computer program that tests a controller's or system's performance under defined conditions to accelerate the product design process. This allows for a faster launch of products at a lower development cost. When compared to manual calculations, simulation software offers considerable advantages. The following are some of the advantages of simulation studies in engineering.

- Using simulation software, you can analyze the behavior of a system without having to build a prototype. Engineers can simulate various designs and come up with alternative solutions. As a result, simulation during product design can ensure that the most appropriate prototyping solution is obtained before the product is built. Reducing the number of iterations in the design process is one of its benefits as well.

Advanced Control of Power Converters: Techniques and MATLAB/Simulink Implementation,
First Edition. Hasan Komurcugil, Sertac Bayhan, Ramon Guzman, Mariusz Malinowski, and Haitham Abu-Rub.
© 2023 The Institute of Electrical and Electronics Engineers, Inc.
Published 2023 by John Wiley & Sons, Inc.
Companion website: www.wiley.com/go/komurcugil/advancedcontrolofpowerconverters

- Engineers can rapidly iterate and test designs using simulation software. By doing so, one can design well-functioning products for the first time. This results in significant cost reductions in manufacturing and testing.
- Engineers can export simulation results as graphical representations with the help of simulation software. By using this method, engineers can better analyze the behavior of systems and the performance of products.

Although many simulations software have the capability to simulate power electronic systems, we will focus here on MATLAB/Simulink. MATLAB is a programming platform that offers a wide range of tools for engineers and scientists, such as the ability to analyze and design systems and products. A key component of MATLAB is the matrix-based MATLAB language, which allows the most natural expression of computations. On the other hand, The Simulink environment uses block diagrams to model and simulate multidomain systems. It is used to design embedded systems at the system level, simulate them, generate automated code, and test them continuously. In Simulink, dynamic systems are modeled and simulated with a graphical editor, customizable block libraries, and solvers. The program is integrated with MATLAB, which allows you to integrate MATLAB algorithms into models and export simulation results to MATLAB for further analysis [1], [2].

In the next subsections, it will be explained step by step how to model various power converters and their control algorithms in MATLAB/Simulink.

6.2.1 Building Simulation Model for Single-Phase Grid-Connected Inverter Based on Sliding Mode Control

The objective of this tutorial is to produce a Simulink model to simulate the dynamics of a single-phase grid-connected inverter based on sliding mode control (SMC). The mathematical model and simulation results of this model are given in Section 7.1. Before the focus on modeling in this subsection, we recommend the reader to review the system's mathematical model.

The simulated model consists of two parts as can be seen in Figure 6.1: the physical single-phase grid-connected inverter including DC voltage supply, voltage and current sensors, and the AC voltage supply, the controller including the subsystem for Error Variables and, embedded m-file.

Step by Step

1) Create a Simulink file called *SVSI_SMC.slx*.
2) Defining the power electronics simulation environment is the first step in building the model-based simulation. Perform the following tasks:
   a) Inside the directory of Specialized Power Systems, find *"powergui"* and move it to *SVSI_SMC.slx*.

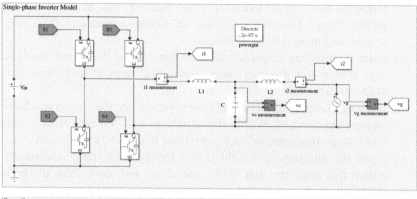

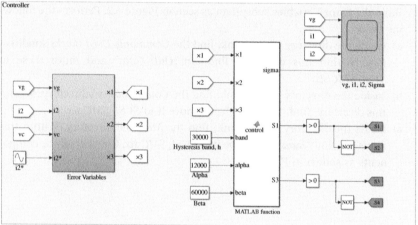

Figure 6.1 Simulink diagram for simulation of single-phase grid-connected inverter based on SMC.

    b) Click on *"powergui"* to find *"configure parameters."* Under the item of *"Simulation type,"* choose *"discrete,"* and under the item of sample time, enter *"Ts,"* which is the sampling interval for power electronic model. Please note that sampling time is 0.2 μs in this model.

3) After the simulation environment is defined, a physical model is to be created for the single-phase grid-connected inverter as seen in Figure 6.1. Perform the following tasks:

    a) Inside the directory of *Specialized Power Systems,* find the *"Power Electronics"* subdirectory. Within this directory, find *"IGBT"* and move four pieces to *SVSI_SMC.slx.*

b) Inside the directory of "*Specialized Power Systems*," find the Sources subdirectory. Within this directory, find "*AC Voltage Source and DC Voltage Source*" and move it to *SVSI_SMC.slx*.

c) Inside the directory of *Specialized Power Systems*, find the *Passives* subdirectory. Within this directory, find "*Series RLC Branch*" and move it to *SVSI_SMC.slx*.

d) Inside the directory of *Specialized Power Systems*, find the *Sensors and Measurement* subdirectory. Within this directory, find "*Current Measurement and Voltage Measurement*" and move these blocks to *SVSI_SMC.slx*.

e) Inside the directory of Simulink, find the *Signal Routing* subdirectory. Within this directory, find "*From and Goto*" and move these blocks to *SVSI_SMC.slx*. Connect all the components as shown in Figure 6.1.

4) Build the Error Variables subsystem as seen in Figure 6.2. Perform the following tasks:

a) Inside the directory of Simulink, find the *Commonly Used Blocks* subdirectory. Within this directory, find "*In, Out, Gain*" and move these to *SVSI_SMC.slx*.

b) Inside the directory of Simulink, find the *Continuous* subdirectory. Within this directory, find "*Derivative*" and move it to *SVSI_SMC.slx*.

c) Inside the directory of Simulink, find the *Math* subdirectory. Within this directory, find "*Sum*" and move it to *SVSI_SMC.slx*. Connect all the components as shown in Figure 6.2.

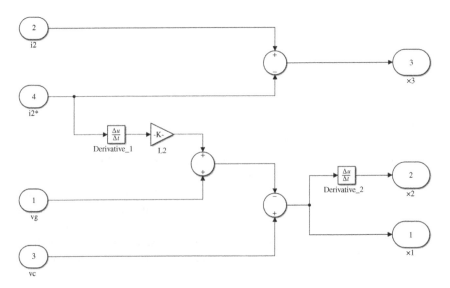

Figure 6.2 Subsystem for the Error Variables.

5) Build the subsystem for implementation of the SMC algorithm. The following tasks need to be completed:

a) Inside the directory of Simulink, find the *User-Defined Functions* subdirectory. Within this directory, find "*MATLAB Function*," and move this to *SVSI_SMC.slx*.

b) Double click on the "*MATLAB Function*" and copy the below m-codes inside the *MATLAB Function*.

```
function [Sigma,S1,S3]= control(x1,x2,x3,band,
    alpha,beta)
%%%%%%%%%%%%%%%%%CONTROL%%%%%%%%%%%%%%%%%%%%%%%%%%%%%

Sigma=x1*alpha+x2+x3*beta;

%%%%%%%%%%%%%SWITCHING SIGNAL%%%%%%%%%%%%%%%%%%%%%%%%
persistent s1_old s3_old;
if isempty(s1_old)
    s1_old=0;
    s3_old=0;
end
% band=4000;
if (Sigma>band)
    S3=1;
elseif (Sigma<0)
    S3=0;
else
    S3=s3_old;
end
if (-Sigma>band)
    S1=1;
elseif(Sigma>0)
    S1=0;
else
    S1=s1_old;
end
s1_old=S1;
s3_old=S3;
```

6) Figure 6.1 shows how to connect all the components. Finally, save and run the model to monitor the results on the scopes.

6.2.2 Building Simulation Model for Three-Phase Rectifier Based on Lyapunov-Function-Based Control

The objective of this tutorial is to produce a Simulink model to simulate the dynamics of a three-phase rectifier based on Lyapunov-function-based control. The mathematical model and simulation results of this model is given in Section 8.2. We recommend reviewing the system's mathematical model.

The simulated model consists of two parts as seen in Figure 6.3: the physical three-phase rectifier including, voltage and current sensors, and the AC voltage supplies, the controller including the subsystem for Park Transformation, Main Controller, and PWM Signals subsystems.

Step by Step

1) Create a Simulink file called *Rectifier_Lyapunov.slx*.
2) To begin building the physical model-based simulation, we need to define the simulation environment for the power electronics. Perform the following tasks:
   a) Inside the directory of Specialized Power Systems, find "powergui" and move it to *Rectifier_Lyapunov.slx*.
   b) Click on "powergui" to find "configure parameters." Under the item of "Simulation type," choose "discrete," and under the item of sample time, enter "Ts," which is the sampling interval for power electronic model. Please note that sampling time is 10 µs in this model.
3) Once the simulation environment is defined, a physical model is to be created for three-phase rectifier as seen in Figure 6.3. Perform the following tasks:
   a) Inside the directory of Specialized Power Systems, find the Power Electronics subdirectory. Within this directory, find "*IGBT*" and move six pieces to *Rectifier_Lyapunov.slx*.
   b) Inside the directory of Specialized Power Systems, find the Source subdirectory. Within this directory, find "*AC Voltage Source*" and move three pieces to *Rectifier_Lyapunov.slx*.
   c) Inside the directory of Specialized Power Systems, find the Passives subdirectory. Within this directory, find "Three-Phase *Series RLC Branch and Series RLC Branch*" and move the set to *Rectifier_Lyapunov.slx*.
   d) Inside the directory of Specialized Power Systems, find the Sensors and Measurement subdirectory. Within this directory, find "Three-phase *Current Measurement and Voltage Measurement*" and move these blocks to *Rectifier_Lyapunov.slx*.
   e) Inside the directory of Specialized Power Systems, find the Power Grid Elements subdirectory. Within this directory, find "Three-phase Breaker" and move this blocks to *Rectifier_Lyapunov.slx*.

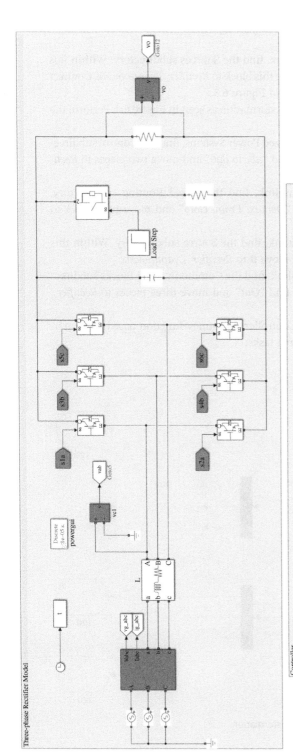

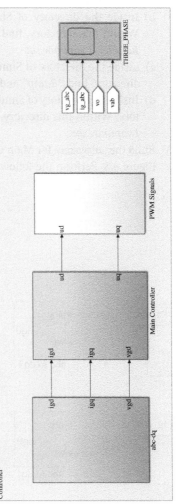

Figure 6.3 Simulink diagram for simulation of three-phase Rectifier based on Lyapunov-function-based control.

f) Inside the directory of Simulink, find the Sources subdirectory. Within this directory, find "*Step*" and move this block to *Rectifier_Lyapunov.slx*. Connect all the components as shown in Figure 6.3.

4) Build the subsystem for Park Transformation as seen in Figure 6.4. Perform the following tasks:

a) Inside the directory of Specialized Power Systems, find the Control subdirectory. Within this directory, find "*abc to dq0*" and move two pieces to *Rectifier_Lyapunov.slx*.

b) Inside the directory of Simulink, find the Signal Routing subdirectory. Within this directory, find "*Demux, From, Goto*" and move two pieces to *Rectifier_Lyapunov.slx*.

c) Inside the directory of Simulink, find the Source subdirectory. Within this directory, find "*Ramp*" and move it to *Rectifier_Lyapunov.slx*.

d) Inside the directory of Simulink, find the Commonly Used Blocks subdirectory. Within this directory, find "*Out*" and move three pieces to *Rectifier_ Lyapunov.slx*.

5) Build the subsystem for Main Controller (Lyapunov-function-based) as seen in Figure 6.5. Perform the following tasks:

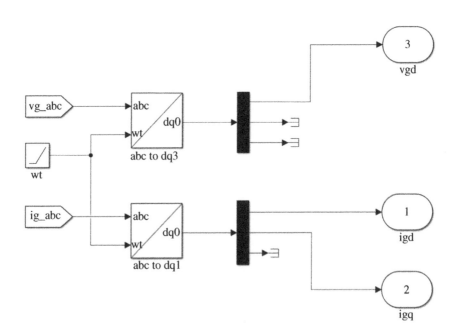

Figure 6.4 Subsystem for the Park transformation.

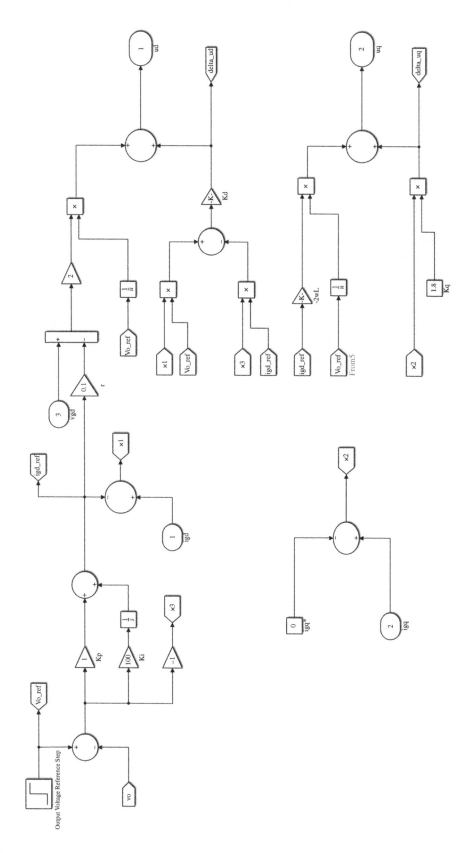

Figure 6.5 Subsystem for Main Controller.

   a) Inside the directory of Simulink, find the Commonly Used Blocks subdirectory. Within this directory, find "Constant, In, *Out, Gain, Integrator*" and move these blocks to *Rectifier_Lyapunov.slx*.
   b) Inside the directory of Simulink, find the Math Operations subdirectory. Within this directory, find "Subtract, Product, Math Function" and move these blocks to *Rectifier_Lyapunov.slx*.
   c) Inside the directory of Simulink, find the Sources subdirectory. Within this directory, find "Step" and move this block to *Rectifier_Lyapunov.slx*.
   d) Inside the directory of Simulink, find the Signal Routing subdirectory. Within this directory, find "From, Goto" and move this block to *Rectifier_Lyapunov.slx*.
6) Build the subsystem for PWM Generation as seen in Figure 6.6. Perform the following tasks:
   a) Inside the directory of Simulink, find the Commonly Used Blocks subdirectory. Within this directory, find "Constant, In, Mux, Demux, Data Type Conversion" and move these blocks to *Rectifier_Lyapunov.slx*.

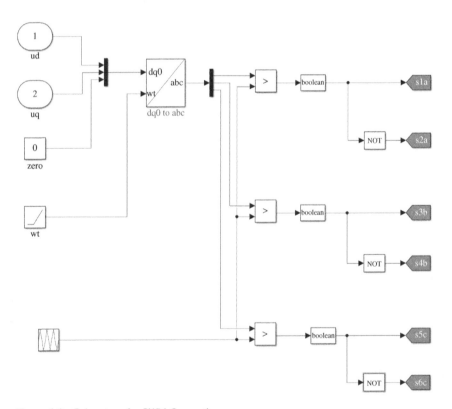

Figure 6.6 Subsystem for PWM Generation.

b) Inside the directory of Simulink, find the Logic and Bit Operations subdirectory. Within this directory, find "Relational Operator" and move this block to *Rectifier_Lyapunov.slx*.

c) Inside the directory of Simulink, find the Math Operations subdirectory. Within this directory, find "Subtract, Product, Math Function" and move these blocks to *Rectifier_Lyapunov.slx*.

d) Inside the directory of Simulink, find the Sources subdirectory. Within this directory, find "Ramp, Repeating Sequence" and move these blocks to *Rectifier_Lyapunov.slx*.

e) Inside the directory of Simulink, find the Signal Routing subdirectory. Within this directory, find "Goto" and move this block to *Rectifier_Lyapunov.slx*.

f) Inside the directory of Specialized Power Systems, find the Control subdirectory. Within this directory, find "dq0 to *abc*" and move two pieces to *Rectifier_Lyapunov.slx*.

7) Connect all the components as shown in Figure 6.3. Run the model to monitor the results on the scope.

6.2.3 Building Simulation Model for Quasi-Z Source Three-Phase Four-Leg Inverter Based on Model Predictive Control

The objective of this tutorial is to produce a Simulink model to simulate the dynamics of a three-phase four-leg quasi-Z source inverter (qZSI). Please note that creating the simulation model is explained step by step here. The mathematical model and simulation results of three-phase four-leg qZSI is given in Section 9.3. We recommend reviewing the system's mathematical model.

The simulated model consists of three parts as can be seen in Figure 6.7: the physical three-phase four-leg qZSI including DC voltage supply and voltage and current sensors, the controller including the subsystem for reference current generation and, embedded m-file, the monitoring part consist of scopes to monitor voltage and current signals of the four-leg qZS inverter.

Step by Step

1) Create a Simulink file called *qZSI_MPC.slx*.

2) For begin building the physical model-based simulation, we need to define the simulation environment for the power electronics. Perform the following tasks:

a) Inside the directory of Specialized Power Systems, find "powergui" and move it to *qZSI_MPC.slx*.

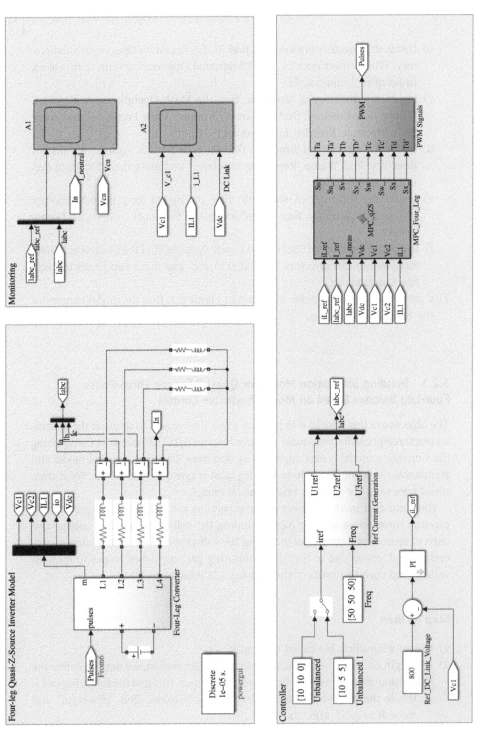

Figure 6.7 Simulink diagram for simulation of three-phase four-leg qZSI.

b) Click on "powergui" to find "configure parameters." Under the item of "Simulation type," choose "discrete," and under the item of sample time, enter "Ts," which is the sampling interval for power electronic model. Please note that sampling time is 10 μs in this model.

3) Once the simulation environment is defined, a physical model is to be created for the three-phase four-leg qZSI as seen in Figure 6.8. Perform the following tasks:

   a) Inside the directory of Specialized Power Systems, find the Power Electronics subdirectory. Within this directory, find *"IGBT"* and move it to *qZSI_MPC.slx*.

   b) Inside the directory of Specialized Power Systems, find the Passives subdirectory. Within this directory, find *"Series RLC Branch"* and move it to *qZSI_MPC.slx*.

   c) Inside the directory of Specialized Power Systems, find the Sensors and Measurement subdirectory. Within this directory, find *"Current Measurement and Voltage Measurement"* and move it to *qZSI_MPC.slx*.

   d) Inside the directory of Simulink, find the Signal Routing subdirectory. Within this directory, find *"From, Goto, Connection Port, Mux, Demux"* and move it to *qZSI_MPC.slx*.

4) Build the subsystem for the reference current generation as seen in Figure 6.9. Perform the following tasks:

   a) Inside the directory of Simulink, find the *"Commonly used blocks"* subdirectory. Within this directory, find *"Constant, Demux, Gain, In, Out"* and move these to *qZSI_MPC.slx*.

   b) Inside the directory of Simulink, find the *"Math Operations"* subdirectory. Within this directory, find *"Add, Product,"* and move these to *qZSI_MPC.slx*.

   c) Inside the directory of Simulink, find the *"User-Defined Functions"* subdirectory. Within this directory, find *"MATLAB Function,"* and move this to *qZSI_MPC.slx*.

   d) Inside the directory of Simulink, find the *"Logic and Bit Operations"* subdirectory. Within this directory, find *"Compare To Constant,"* and move this to *qZSI_MPC.slx*.

   e) Inside the directory of Simulink, find the *"Discrete"* subdirectory. Within this directory, find *"Discrete-Time Integrator and Unit Delay,"* and move these to *qZSI_MPC.slx*.

5) Build the subsystem for implementation of the model predictive control algorithm. Perform the following tasks:

   a) Inside the directory of Simulink, find the *"User-Defined Functions"* subdirectory. Within this directory, find *"MATLAB Function,"* and move this to *qZSI_MPC.slx*.

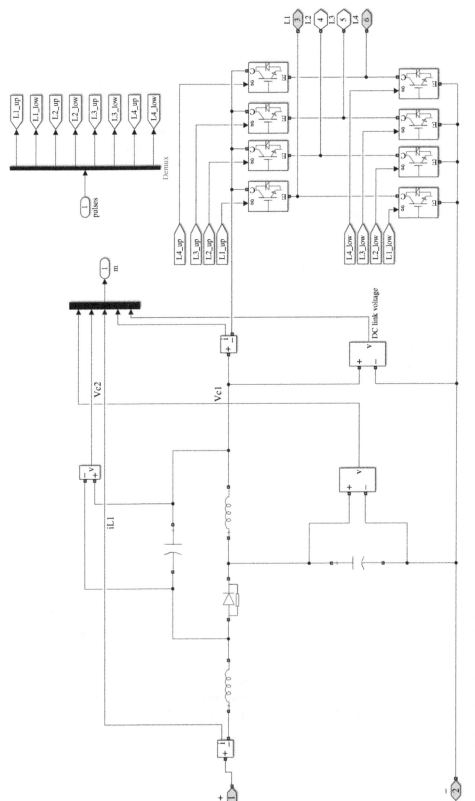

Figure 6.8 A three-phase four-leg qZSI model.

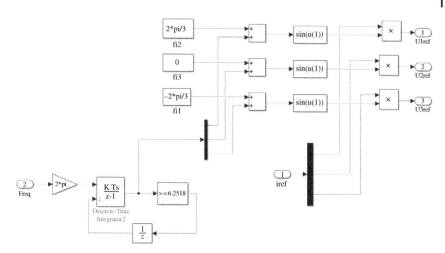

Figure 6.9 Subsystem for the reference current generation.

b) Double click on the "*MATLAB Function*" and copy the below m-codes inside the *MATLAB Function*.

```
function  [Su,Su_,Sv,Sv_,Sw,Sw_,Sx,Sx_]  =  MPC_qZS
   (iLref,I_ref,I_meas, Vdc, Vc1, Vc2, iL1)
% Variables defined in the parameters file
%global Ts R L v states
Ts = 10e-6;
R = 10; % Resistance [Ohm]
L = 10e-3; % Inductance [H]
Lx= 20e-3;
%iLref=7;
Vin=500;
C=1000e-6;
Lamda1=100.01;
% Switching states
S=[1 0 0 0;1 1 0 0;0 1 0 0;0 1 1 0;0 0 1 0;1 0 1 0;1 1
1 0;0 0 0 0;1 0 0 1;1 1 0 1;0 1 0 1;0 1 1 1;0 0 1 1;1
0 1 1;1 1 1 1;0 0 0 1;1 0 0 0;0 1 0 0;0 0 1 0;0 0 0
1;1 1 0 0;1 0 1 0;1 0 0 1;0 1 1 0;0 1 0 1;0 0 1 1;1 1
1 0;1 1 0 1;1 0 1 1;0 1 1 1;1 1 1 1];
S1=[0 1 1 1;0 0 1 1;1 0 1 1;1 0 0 1;1 1 0 1;0 1 0 1;0
0 0 1;1 1 1 1;0 1 1 0;0 0 1 0;1 0 1 0;1 0 0 0;1 1 0
0;0 1 0 0;0 0 0 0;1 1 1 0;1 0 0 0;0 1 0 0;0 0 1 0;0 0
0 1;1 1 0 0;1 0 1 0;1 0 0 1;0 1 1 0;0 1 0 1;0 0 1 1;1
```

```
1 1 0;1 1 0 1;1 0 1 1;0 1 1 1;1 1 1 1];
%%%%%%%%%%%%%%%%%%%%%%%%%%%%%%%%%%%%%%%%%%%%%%%%%%%%%%
%%%%%%%%%%%%%%%%%
% Optimum vector and measured current at instant k-1
 persistent vux_opt vvx_opt vwx_opt
 % Initialize values
if isempty(vux_opt),vux_opt = 1; end
if isempty(vvx_opt),vvx_opt = 1; end
if isempty(vwx_opt),vwx_opt = 1; end
iu_ref=I_ref(1);
iv_ref=I_ref(2);
iw_ref=I_ref(3);
iu=I_meas(1);
iv=I_meas(2);
iw=I_meas(3);
Cv=Ts/(L+R*Ts);
Ci=L/(L+R*Ts);
%Current Prediction in k+1
iuk1=Cv*vux_opt+Ci*iu;
ivk1=Cv*vvx_opt+Ci*iv;
iwk1=Cv*vwx_opt+Ci*iw;

%Prediction Algorithm
index_s_min=1;
g_min=2e20;
for i = 1:18
vux=(S(i,1)-S(i,4))*Vdc;
vvx=(S(i,2)-S(i,4))*Vdc;
vwx=(S(i,3)-S(i,4))*Vdc;
%Current Prediction in k+2
iuk2=Cv*vux+Ci*iuk1;
ivk2=Cv*vvx+Ci*ivk1;
iwk2=Cv*vwx+Ci*iwk1;

%input current prediction according to Nonshoot-
through and Shoot-through states
if i<=16;
    iLk1=iL1+(Ts/Lx)*(Vin-Vc1); %Nonshoot-through
state
else
   iLk1=iL1+(Ts/Lx)*(Vin+Vc2); %Shoot-through state
end
```

```
% Cost function
aux1=(iu_ref-iuk2)*(iu_ref-iuk2);
aux2=(iv_ref-ivk2)*(iv_ref-ivk2);
aux3=(iw_ref-iwk2)*(iw_ref-iwk2);
aiL=abs(iLref-iLk1);
g = aux1+aux2+aux3+Lamda1*aiL;
% Selection of the optimal value
 if (g<g_min)
      index_s_min=i;
      g_min = g;
end
end
% Output switching states
Su = S(index_s_min,1);
Sv = S(index_s_min,2);
Sw = S(index_s_min,3);
Sx = S(index_s_min,4);
Su_= S1(index_s_min,1);
Sv_= S1(index_s_min,2);
Sw_= S1(index_s_min,3);
Sx_= S1(index_s_min,4);
End
```

6) Build the subsystem for the PWM as seen in Figure 6.10. Perform the following tasks:
   a) Inside the directory of Simulink, find the "*Commonly Used Blocks*" subdirectory. Within this directory, find "*Data Type Conversion, In, Out, Mux,*" and move this to *qZSI_MPC.slx*.
7) Connect all the components as shown in Figure 6.7. Run the model to monitor the results on the scope.

6.2.4 Building Simulation Model for Distributed Generations in Islanded AC Microgrid

The objective of this tutorial is to produce a Simulink model to simulate two distributed generations (DGs) in islanded AC microgrid. Please note that creating the simulation model is explained step by step here. The mathematical model and simulation results of this model are given in Section 9.5. First, we recommend the reader to review the system's mathematical model.

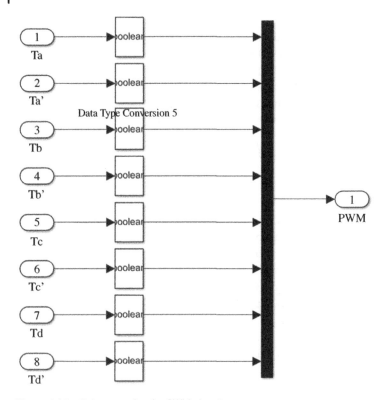

Figure 6.10 Subsystem for the PWM signals.

The simulated model consists of three parts as can be seen in Figure 6.11. The islanded AC microgrid model consists of two distributed inverters that include DC supplies, three-phase inverters, LCL filters, and controllers, load banks, and a monitoring the monitoring part consist of scopes to monitor voltage, current, and power signals of the AC microgrid.

Step by Step

1) Create a Simulink file called *AC_Microgrid.slx.*
2) To begin building the physical model-based simulation, we need to define the simulation environment for the power electronics. Perform the following tasks:
   a) Inside the directory of Specialized Power Systems, find "powergui" and move it to *AC_Microgrid.slx.*
   b) Click on "powergui" to find "configure parameters." Under the item of "Simulation type," choose "discrete," and under the item of sample time, enter "*Ts*," which is the sampling interval for power electronic model. Please note that sampling time is 10 μs in this model.

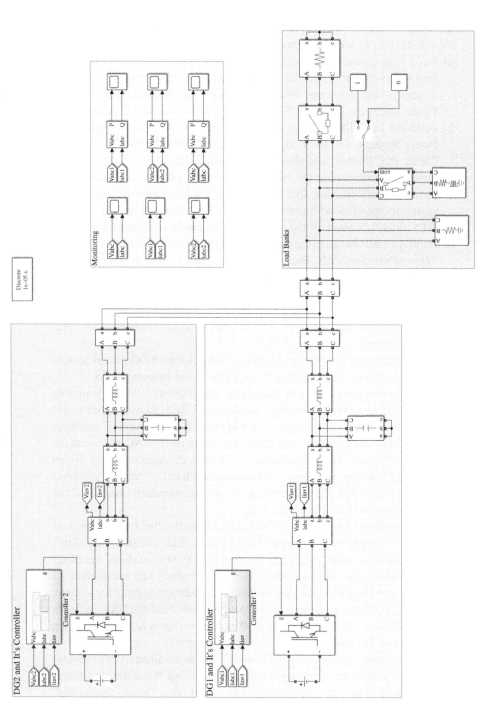

Figure 6.11 Simulink diagram for simulation of distributed generation inverters in islanded AC microgrid.

3) Once the simulation environment is defined, a physical model is to be created for the DG unit as seen in Figure 6.11 Perform the following tasks:

a) Build three-phase inverter in each DG unit. Inside the directory of Specialized Power Systems, find the Power Electronics subdirectory. Within this directory, find *"Universal Bridge"* and move it to *AC_Microgrid.slx*. Select Power Electronic Device to IGBT/Diodes and number of bridge arms 3.

b) Build the LCL filter in each DG unit. Inside the directory of Specialized Power Systems, find the Passives subdirectory. Within this directory, find *"Series RLC Branch"* and move it to *AC_Microgrid.slx*.

c) Build measurements for the three phase voltages and currents. At the *"Measurements"* subdirectory, find *"Three-Phase V-I Measurement"* and move it to *AC_Microgrid.slx* with position between the "Three phase VSI" and the "LCL filter." Connect all the components as shown in Figure 6.11.

d) Build the Load Banks in the model. At the "Passives" subdirectory, find the *"Three-Phase Series RLC Branch"* and copy it to *AC_Microgrid.slx*. Select "R and RL" and define its resistance (R_s) and inductance (L_s) accordingly. At the *"Power Grid Elements"* subdirectory, find the *"Three-Phase Breaker"* and copy it to *AC_Microgrid.slx*. At *the "Signal Routing"* subdirectory, find the *"Manual Switch"* and copy it to *AC_Microgrid.slx*. Connect all the components as shown in Figure 6.11.

4) Build the subsystem for a Droop Based Controller for each DG unit as seen in Figure 6.12. Perform the following tasks: Perform the following tasks:

a) Build the Droop equations in the model. Inside the directory of Simulink, find the *"Commonly used blocks"* subdirectory, find *"Constant and Gain"* and move these to *AC_Microgrid.slx*. Find the *"Math Operations"* subdirectory. Within this directory, find *"Sum"* and move these to *AC_Microgrid.slx*. Find the *"Signal Routing"* subdirectory. Within this directory, find *"From and Goto"* and move these to *AC_Microgrid.slx*. Find the *"Discrete"* subdirectory, find *"Unit Delay"* and move these to *AC_Microgrid.slx*. Connect all the components as shown in Figure 6.12.

b) Build a Voltage Control Loop in the model. Inside the directory of Simulink, find the *"Commonly used blocks"* subdirectory, find *"Constant and Gain"* and move these to *AC_Microgrid.slx*. Find the *"Math Operations"* subdirectory. Within this directory, find *"Sum and Product"* and move these to *AC_Microgrid.slx*. Find the *"Signal Routing"* subdirectory, find *"From, Goto, Mux, Demux"* and move these to *AC_Microgrid.slx*. Find the *"Discrete"* subdirectory, find *"Discrete PI Controller"* and move this to AC_Microgrid.slx. Connect all the components as shown in Figure 6.12.

c) Build a Current Control Loop in the model. Inside the directory of Simulink, find the *"Commonly used blocks"* subdirectory, find *"Constant and Gain"* and move these to *AC_Microgrid.slx*. Find the *"Math Operations"*

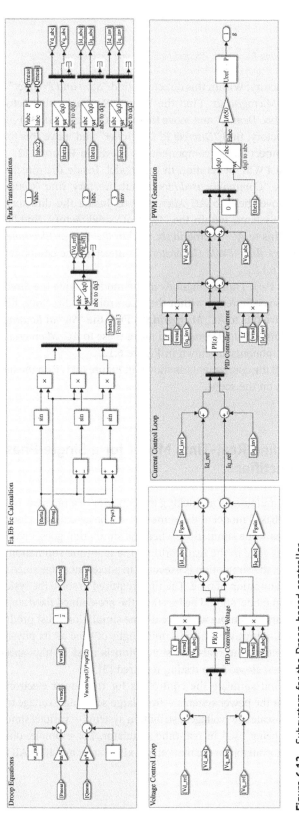

Figure 6.12 Subsystem for the Droop-based controller.

subdirectory. Within this directory, find "*Sum and Product*" and move these to *AC_Microgrid.slx*. Find the "*Signal Routing*" subdirectory, find "*From, Goto, Mux, Demux*" and move these to *AC_Microgrid.slx*. Find the "*Discrete*" subdirectory, find "*Discrete PI Controller*" and move this to AC_Microgrid. slx. Connect all the components as shown in Figure 6.12.

d) Build a PWM Generation in the model. Inside the directory of Simulink, find the "*Commonly used blocks*" subdirectory, find "*Constant, Gain, Out*" and move these to *AC_Microgrid.slx*. Inside the directory of Specialized Power Systems, within the "Control" subdirectory, find "*dq0 to abc*" and move this to *AC_Microgrid.slx. Within the "Power Electronics Control" sub-directory, find "PWM Generator."* Connect all the components as shown in Figure 6.12.

e) Build a Park Transformations in the model. Inside the directory of Specialized Power Systems, within the "Control" subdirectory, find "abc to *dq0*" and move this to *AC_Microgrid.slx*. Find the "*Signal Routing*" subdirectory, find "*From, Goto, Demux*" and move these to *AC_Microgrid.slx*. Connect all the components as shown in Figure 6.12.

5) Connect all the components as shown in Figure 6.11. Run the model to monitor the results on the scope.

6.3 Building Real-Time Model for a Single-Phase T-Type Rectifier

A real-time simulation is mimicking the behavior of a physical system by running its computer-based model at the same rate as the actual wall clock time. It means that, in the real-time simulation, when the simulation goes a particular time, the same time has gone in the real world. In the real-time simulation, instead of variable step size, the simulation is executed in a fixed step size since time moves forward in equal duration of time. The time required to solve the system equations in the simulation model should be less than the pre-defined fixed step. Otherwise, an over run occurs. In simple words, real-time simulation must produce the internal variables and output within the same length of time as its physical counterpart would. In general, the real-time simulation is used in high-speed simulations, especially when closed-loop testing required [3].

To model and simulate the controllers for the power electronics in different applications or the power systems – from large-scale high-voltage transmission systems to small-scale low-voltage distribution systems – various simulation tools are available for being used in real-time simulator. For example, different real-time simulators are available in the market like xPC Target and RT-LAB for mechatronic systems and using MATLAB/Simulink, eFPGASIM and eDRIVESIM for power electronic simulation and eMEGASIM, HYPERSIM and RTDS for power grid real-time (RTS) simulation [3].

The fundamental concept of real-time simulation is illustrated in Figure 6.13. The first step is to build the MATLAB/Simulink model of the considered power electronics system by using Simulink and OPAL-RT libraries. After that, the OPAL-RT software is used to compile and build the developed model for the OPAL-RT simulator. Then, finally load the real-time model to the OPAL-RT simulator and execute the model.

Step by Step

1) Double-click the RT-LAB shortcut icon on PC's Desktop to launch the RT-LAB interface. The Workspace Launcher window prompts you to select a workspace (the workspace is the directory where RT-LAB will store all the files required for your simulations) as you can see in Figure 6.14 [4]. Click the OK button and wait for the Welcome page to appear (see Figure 6.15). The Welcome page provides quick access to tutorials and documentation. Click Go to the workbench to open the main RT-LAB window (you can access the Welcome page at any time from the Help menu). RT-LAB is now ready to create and run real-time simulations.

2) Before using RT-LAB, the target needs to be configured. This step describes the basic tools and steps to setup RT-LAB and runs a sample model. In the *Project Explorer*, right-click to *Targets* and then click *Discover targets*; this process may take some time (see Figure 6.16). Once RT-LAB detects targets, the *Detected RT-LAB Targets* window appears. Select the target you want to use and click *Finish*. Your simulator is now available in the RT-LAB interface [4].

3) In the RT-LAB Project Explorer (see Figure 6.17), double-click "*Create a new project*" or click the File and then click to New→RT-LAB Project.

4) Insert the project name (e.g. 002_MPC_TType) in New RT-LAB Project Menu then click **Next** (see Figure 6.18).

5) The next window that appears allows you to select the model for the project (see Figure 6.19). Select **empty** for this project. Then, click Finish. Then, the created project (002_MPC_TType) will appear in the Project Explorer (as seen in Figure 6.20).

6) To create a MATLAB/Simulink model in this project, right click on the Models→New→Model. Then, New RT-LAB Model window will appear as seen in Figure 6.21. Name your Model (e.g. T-Type) and select your Model T_type (in this cased MATLAB/Simulink (.slx)) then click Next. The next window that appears allows you to select the model for the project. Select **empty** for this project (see Figure 6.22). Then, click **Finish**.

Matlab/Simulink Model with OPAL-RT Libraries

RT-LAB Software

Real-time simulation in OPAL-RT

Figure 6.13 The fundamental concepts of real-time simulation.

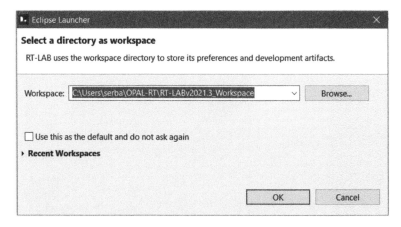

Figure 6.14 The screenshot of OPAL-RT launcher.

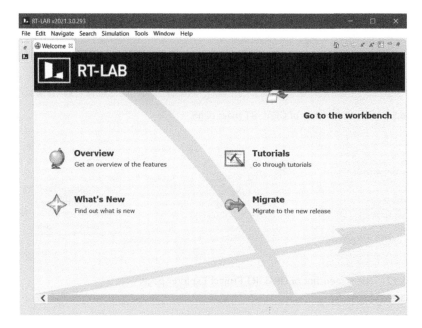

Figure 6.15 The screenshot of OPAL-RT welcome page.

7) Your MATLAB/Simulink model (T_type) is now available in the Project Explorer (see Figure 6.23).

8) To build/modify the MATLAB/Simulink model, right-click to Simulink Model (e.g. T_type.slx). Then, select **Edit with** as seen in Figure 6.24. After

Figure 6.16 The screenshot of OPAL-RT main page.

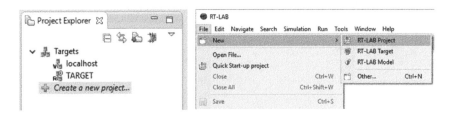

Figure 6.17 The screenshot of OPAL-RT Project Explorer page.

that, choose your MATLAB version and click on it. MATLAB software and empty Simulink model (T-Type.slx) will be opened automatically.

9) The simulation model of T-type rectifier now can be built by using model in Figure 6.25. On the other hand, the simulation model needs to be modified to run on the OPAL-RT RTS. In the root layer of the Simulink model, you need to

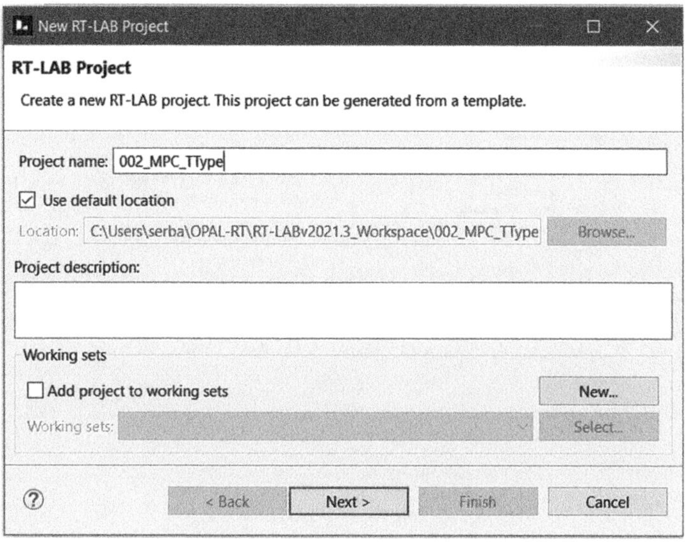

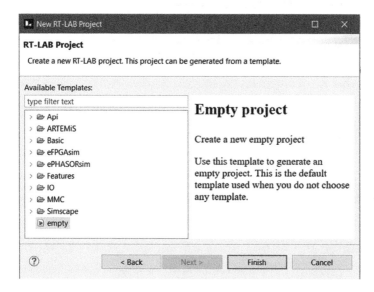

Figure 6.18 The screenshot of New RT-LAB Project Menu.

Figure 6.19 The screenshot of RT-LAB Project Menu.

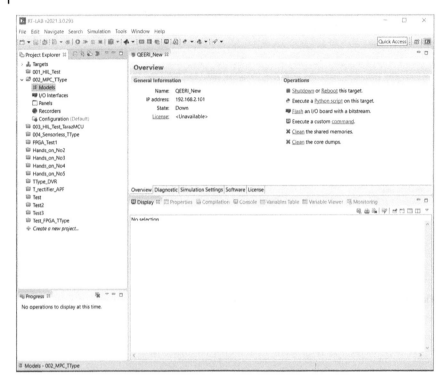

Figure 6.20 The screenshot of RT-LAB Project Explorer.

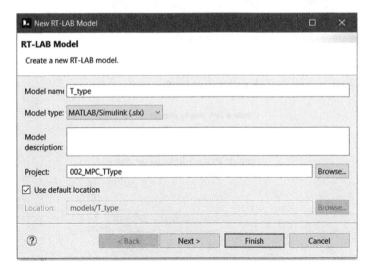

Figure 6.21 The screenshot of RT-LAB Model menu.

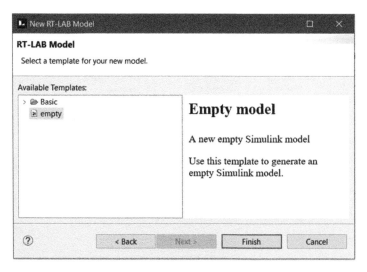

Figure 6.22 The screenshot of RT-LAB Model selection menu.

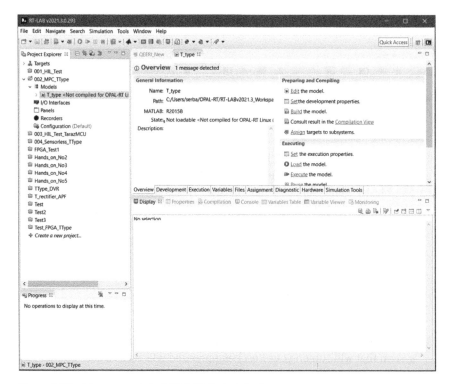

Figure 6.23 The screenshot of RT-LAB Project Explorer.

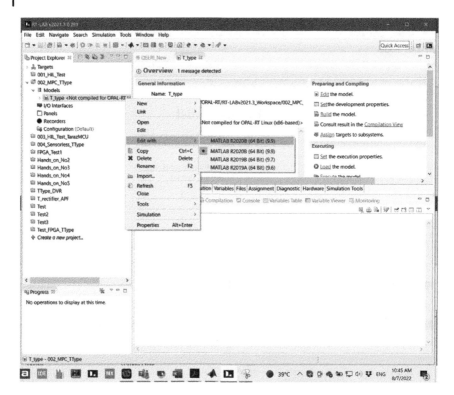

Figure 6.24 The screenshot of RT-LAB Project Explorer – Editing Simulink Model.

create two subsystems: SM_[name] and SC_[name] as seen in Figure 6.26. In this example, subsystem names are SM_Master and SC_Console.

10) SM_Master stands for subsystem master. This is where all the real-time simulation will occur. This is the only subsystem that will be run on the simulator and that contains I/Os. The SM_Master subsystem is a block from the OPAL-RT library that controls I/Os as seen in Figure 6.26. Depending on your system, you may see the following blocks [4]:
    - Analog input (Ain) and output blocks (Aout),
    - Static digital input (Din) and output blocks (Dout),
    - PWM input (PWMin) and output blocks (PWMout), and,
    - Event detector (TSDin) and Event generator blocks (TSDout).

11) SC_Console stands for subsystem console. This will be an asynchronous subsystem that will run on your host computer and will act as a user interface as

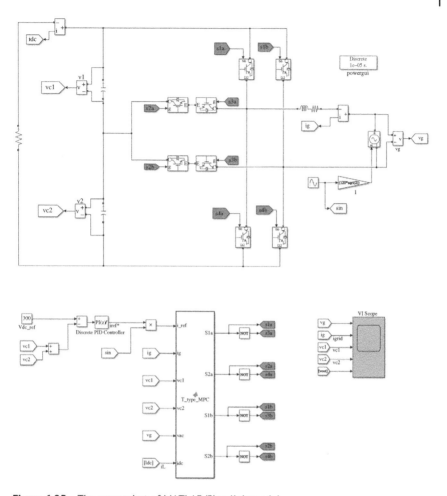

Figure 6.25 The screenshot of MATLAB/Simulink model.

seen in Figure 6.26. No critical mathematical logic should be included in this subsystem.

12) Save and run the simulation model and make sure the model runs without any issue. After that Return the RT-LAB software for compiling and building the RT model.

13) The build process allows RT-LAB to transform the Simulink model into a full real-time simulation. This process must be repeated each time the Simulink model is modified. In the Project Explorer window, expand the project you created to find the T-type model. Drag the model onto your target. This will

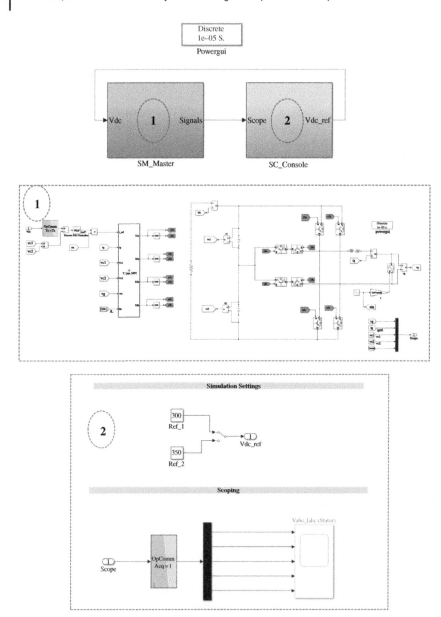

Figure 6.26 The structure of MATLAB/Simulink model for RT-LAB implementation.

automatically configure your model to run on the OPAL-RT target. Right-click on the model, then select **Simulation > Build Configurations** (see Figure 6.27). This opens the build configurations window. Verify that your target is set as the Development Node. The Development Node is the target that RT-LAB will use to perform the build. (To set as Development node, right-click the target and select Set as development node.) Click OK then wait for the build process to complete. You can view the progress of the build in the Compilation View at the bottom of the RT-LAB interface [4].

14) You need to load your model into the target. The load process prepares the real-time target to perform the simulation. Click the **Load** toolbar button. It may take a few moments for the model to load. When it has loaded, the **T_type_sc_console** Simulink console window appears as seen in Figure 6.28.

15) The final step is to run the model on the target. To do that, click the Execute toolbar button. Executing the model starts the real-time simulation on the target.

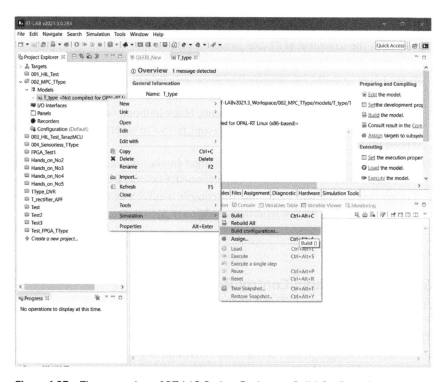

Figure 6.27 The screenshot of RT-LAB Project Explorer – Build Configuration.

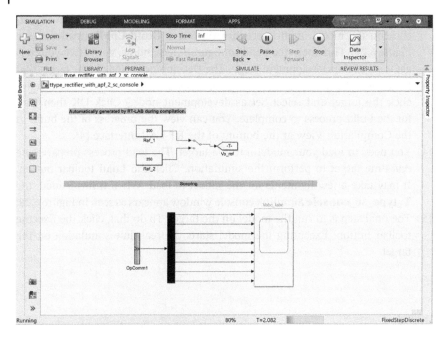

Figure 6.28 The screenshot of RT-LAB Simulink Console.

16) The user console (Figure 6.28) is now receiving and sending data to the simulation. Double-click the adjust reference block to modify the set point of the DC-link voltage and double-click the scope blocks to observe signals received from the simulator.
17) Stopping the simulation releases the target (makes it available for use) and allows for another simulation to be performed. Click the **Reset** toolbar button to stop the simulation. Verify that the console is automatically closed. You are now ready to test your integration model.

6.4 Building Rapid Control Prototyping for a Single-Phase T-Type Rectifier

The testbed shown in Figure 6.29 has been developed for this book to validate the control system designs in the real hardware environment. As shown in this figure, a single-phase three-level T-type rectifier in the laboratory set-up consists of a programmable regenerative grid simulator, which is used to emulate the grid voltage.

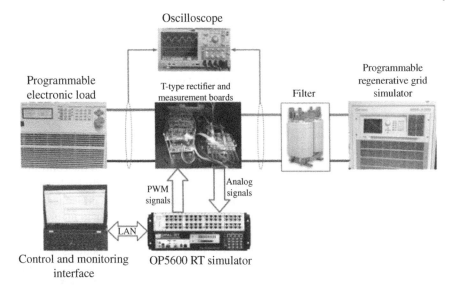

Figure 6.29 The block diagram of the experimental setup.

From the grid simulator, there is a line inductor connected between the rectifier and the grid emulator. The T-type rectifier is made up of mainly two components: a measurement board, and power module (power switches and gate drivers). The measurement board includes AC current, AC voltage, and DC-link voltage sensors. The power module consists of 12 IGBT devices including freewheeling diode and gate drivers. A programmable electronic load is connected to the system to verify the controller's performance under various load conditions. The developed control algorithm is run on OPAL-RT OP5707 controller.

6.4.1 Components in the Experimental Testbed

6.4.1.1 Grid Simulator
The Grid Simulator is used to simulate a three-phase symmetric or asymmetric grid. The grid simulator used in this testbed is Chroma 61800 regenerative grid simulator. The 61800 is a full four quadrant, fully regenerative, AC power supply designed to simulate worldwide power line conditions and disturbances for testing various power converters. In addition to supplying clean, precise, and stable AC voltage for regular applications, the 61800 is capable of simulating various types of distorted voltage waveforms and transient conditions required by product validation testing. These are accomplished using built-in programmable waveform functions such as LIST, STEP and PULSE modes. The STEP and PULSE functions

allow users to perform single or continuous step changes of output voltage. LIST mode is a more versatile function and allows users to compose complex waveforms of up to 100 sequences. Voltage waveforms required by immunity specifications such as IEC 61000-4-11 (short interruption and voltage dropout) can easily be achieved by the 61800 Regenerative Grid Simulator [5].

6.4.1.2 A Single-Phase T-Type Rectifier Prototype

The T-type rectifier used in this testbed is based on three-phase two-level voltage-source converter and middle leg switches. The rectifier module has two layers. The first layer is a drive board, which reads the control commands from the controller to generate PWM signals for the IGBTs, the second layer contains the high voltage side of the IGBTs, which has the DC-link bus capacitors and the protections, the IGBT modules are connected to a piece of heat-sink material as seen in Figure 6.30.

The gate driver used in this test bed is GDC-2A6S1. The GDC Series include high performance fully isolated SiC/IGBT/MOSFET gate driver modules for six Switches. It is specifically designed for fastest inverter prototyping in research and educational projects. These drives use Texas Instrument's ISO5852 smart and high-performance gate driver IC, and feature's dead time generation logic, fault latch logic, input and output indication LEDs, test points and built in 5 V regulator which could be used to power up external control circuitry.

The most notable feature of these modules is that they identify short circuit condition using desaturation detection. Consequently, it safely turns off the switch and gives the controller an isolated fault feedback signal. This series also has a very high common mode immunity of 100 kV/μs. In other words, it utilizes the high switching speed of SiC FETs to its fullest.

Figure 6.30 The photograph of T-type rectifier prototype.

The power module used in this test bed is SPM-VFD-KIT. The SPM Series of Power Modules are simple and modular blocks that could be used for fast prototyping and validation of popular power converter circuits such as, Single & Multi-Phase Inverters, Buck/Boost Converters, Single & Multi-Phase Active Rectifiers and Modular Multi Level Converters etc. It can cover wide range of applications such as Variable Frequency Drives (VFDs), BLDC Motor Drives, PV inverters and converters in research and educational environments. User can connect Input and output terminals using pluggable terminal blocks or banana connectors, providing ease of use in labs. Test points are also available for pain free testing.

6.4.1.3 Measurement Board

The measurement board used in this testbed is shown in Figure 6.31. The USM Series of Measurement Modules are High Performance, Fully Isolated and multi-channel voltage and current sensing modules. They are designed to be used in wide range of applications such as Feedback Block of Power Electronics Converters & Inverters, 3 Phase Systems Monitoring, Motor Drives and PV Monitoring applications. The Module have both unipolar and bipolar output IDCs, with configurable 3 V/5 V unipolar output voltage, which makes it compatible with all type of Controllers such as dSPACE and OPAL-RT controllers. BNC output can be used for direct interface with Oscilloscope as well. The Current Sensors are non-intrusive, so they can be used with bus bars. Voltage range can be selected with 10X or 100X attenuation as well [6].

Figure 6.31 The photograph of measurement board.

6.4.1.4 Programmable Load

As shown in Figure 6.29, on the left hand wide of the test bench, a programmable electronic load is connected to the output of the T-type rectifier. The programmable load used in this testbed is 63800 series from Chroma. This programmable load can simulate load conditions under high crest factor and varying power factors with real time compensation even when the voltage waveform is distorted. This special feature provides real world simulation capability and prevents overstressing resulting in reliable and unbiased test results. Chroma's 63800 Series AC/DC Electronic Loads also include built-in 16-bits precision measurement circuits to measure the steady-state and transient responses for true RMS voltage, true RMS current, true power(P), apparent power(S), reactive power(Q), crest factor, power factor, THDv and peak repetitive current. In addition to these discrete measurements, two analog outputs, one for voltage and one for current, are provided as a convenient means of monitoring these signals via an external oscilloscope [7].

6.4.1.5 Controller

The tasks of controller algorithm design, editing and simulation are accomplished using MATLAB/Simulink in the host computer. This host computer is connected to the OP5707 simulator via a LAN cable. OPAL-RT simulator can be used not only for the plant simulation (power grid, smart grids, microgrids, etc.), but also in a rapid control prototyping mode (RCP) as controller, where the researcher can design any control algorithm on Simulink, adds the necessary I/Os (PWM Outputs, Analog Inputs, Encoder and resolver inputs), and quickly upload it to the real-time computer. Such readily operational controller that can be used to control any equipment in the lab (power converter or others), all is without any additional cost. Once the control program is designed, the host computer will compile and upload the program to the OP5707, which will operate the control experiment in real time mode.

6.4.2 Building Control Structure on OP-5707

Step by Step

1) Section 6.3 explains how to implement the real-time simulation model of a single-phase T-type rectifier. In this section, we will use the same model to build the control structure for the real plant that is the single-phase T-type rectifier in our case.
2) The modified **SM_Master** subsystem is shown in Figure 6.32. First, the physical component blocks (IGBTs, Power Source, Passive Components, etc.) should be removed from the model.

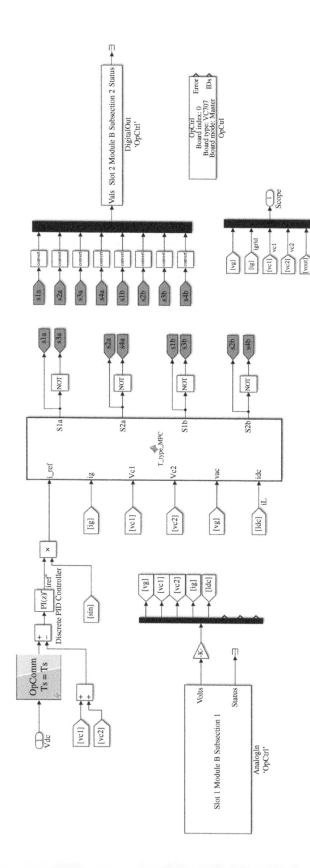

Figure 6.32 The MATLAB/Simulink model for RCP.

3) Then, the **OpCtrl** block should be added in the model. The **OpCtrl** provides an interface to different Opal-RT card via the PCIe bus. The card holds a Xilinx FPGA chip and can control up to 256 I/O lines. The Figure 6.33 shows the **OpCtrl** configured to control a VC707 board. The **OpCtrl** block allows the user to specify the card type and which bitstream is required by the model. Programming of the flash memory of the card is performed accordingly before model loading.

4) To read voltage and current signals from the measurement board, use **AnalogIn** block that is used to return analog input acquisition data obtained from the analog input channels of an analog module installed on a different type of carrier compatible with the selected FPGA board. The data values are transferred from the FPGA to the RT-LAB model through one DataOut port of the bitstream, via the PCIe bus of the target computer.

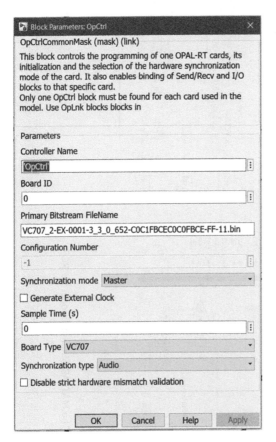

Figure 6.33 The screenshot of the configuration menu of OpCtrl block.

5) The **Digital Out** block is used to transfer 0/1 values to be applied to the digital output channels module installed a different type of carrier compatible with the selected FPGA board. These digital signals are applied to the gate-drivers to control T-type rectifier prototype.

6) Run the model and make sure the revised model is error-free. Then, save the model and return RT-LAB to build the model.

7) Since you have already configured the build process, simply click the **Build** toolbar button, and wait a few seconds while the model is compiled.

8) Since this model uses I/Os, some additional steps are required. First, go to the Assignation tab and ensure that the XHP box is checked (on) as seen in Figure 6.34. Then, go to the Execution tab and ensure that Real-time simulation mode is Hardware synchronized (see Figure 6.35).

9) Finally, click the Load toolbar button and wait for the load process to be completed. Then, click the Execute button. A new console window appears. Your I/Os are up and running! You can now change reference value to regulate DC-link voltage. You can observe analog signals using the scope block in RT-LAB Simulink Console.

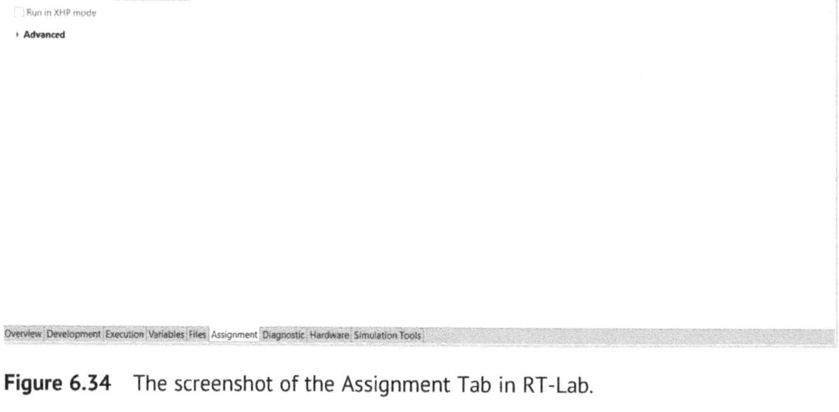

Figure 6.34 The screenshot of the Assignment Tab in RT-Lab.

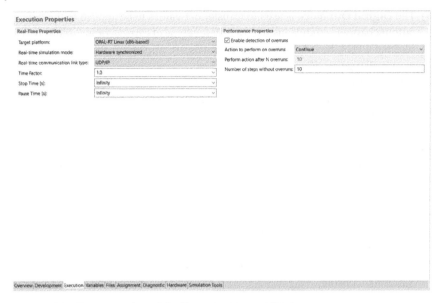

Figure 6.35 The screenshot of the Execution Tab in RT-Lab.

References

1 L. Wang, S. Chai, D. Yoo, G. Lu, and K. Ng, *PID and Predictive Control of Electrical Drives and Power Converters Using MATLAB®/Simulink*. UK: Wiley, 2015.

2 Matlab and Simulink Tutorials. Available: https://www.mathworks.com/support/learn-with-matlab-tutorials.html

3 S. S. Refaat, O. Ellabban, S. Bayhan, H. Abu-Rub, F. Blaabjerg, and M. Begovic, *Smart Grid and Enabling Technologies*. UK: Wiley/IEEE Press, 2021.

4 OPAL-RT User Documentation Hub. Available: https://wiki.opal-rt.com

5 Regenerative Grid Emulator. Available: https://www.chromausa.com/product/regenerative-grid-simulator-61800

6 *Isolated Voltage & Current Sensor Module*. Available: https://www.taraztechnologies.com/product/power-electronics-modules/sensors-measurement/usm3iv-isolated-voltage-current-sensor

7 Bidirectional and regenerative automated test equipment. Available: https://www.chromausa.com

7

Sliding Mode Control of Various Power Converters

7.1 Introduction

This chapter presents the sliding mode control of various power converters that include single-phase grid-connected inverter with LCL filter, three-phase grid-connected inverter with LCL filter, three-phase AC–DC rectifier, three-phase transformerless dynamic voltage restorer, and three-phase shunt active power filter. The design of sliding mode control for each converter is presented in detail. The MATLAB/Simulink model of each converter together with the sliding mode control is made available to help the practicing engineers, researchers, as well as graduate students who are interested in sliding mode control.

7.2 Single-Phase Grid-Connected Inverter with LCL Filter

In this section, the sliding mode control of a single-phase voltage-source inverter (VSI), which is introduced in [1], is described. Since the inverter is connected to the grid via an LCL filter, the complexity of controller design is increased due to the resonance issue of the LCL filter. On the other hand, the difficulty in generating the required filter capacitor voltage reference is identified. It is pointed out that the sliding mode control is able to tackle the resonance issue without using any damping method. Moreover, the filter capacitor voltage reference can be generated by employing a proportional-resonant controller, which leads to almost zero steady-state error in the grid current.

Advanced Control of Power Converters: Techniques and MATLAB/Simulink Implementation,
First Edition. Hasan Komurcugil, Sertac Bayhan, Ramon Guzman, Mariusz Malinowski, and Haitham Abu-Rub.
© 2023 The Institute of Electrical and Electronics Engineers, Inc.
Published 2023 by John Wiley & Sons, Inc.
Companion website: www.wiley.com/go/komurcugil/advancedcontrolofpowerconverters

7.2.1 Mathematical Modeling of Grid-Connected Inverter with LCL Filter

The circuit diagram of a single-phase grid-connected VSI with *LCL* filter is shown in Figure 7.1. The main function of the VSI is to inject the required AC current into the grid utility. This injection is achieved through appropriate switching of the four switching devices (S_1–S_4). On the other hand, the main function of the *LCL* filter is to attenuate the high-frequency ripple occurring due to the switching in the grid current. As can be seen in Figure 7.1, the *LCL* filter has an inverter-side inductor (L_1), grid-side inductor (L_2), and a parallel connected capacitor (C). The differential equations of the system can be written as [1]

$$L_1 \frac{di_1}{dt} = uV_{in} - v_C \tag{7.1}$$

$$L_2 \frac{di_2}{dt} = v_C - v_g \tag{7.2}$$

$$C \frac{dv_C}{dt} = i_1 - i_2 \tag{7.3}$$

where $v_g = V_g \sin(\omega t)$ is the grid voltage, i_1 is the inverter current, i_2 is the grid current, v_C is the capacitor voltage, V_{in} is the DC input voltage, and u is the control input, which takes the values in the finite set $\{-1, 0, 1\}$. The error variables can be defined as

$$x_1 = v_C - v_C^* \tag{7.4}$$

$$x_2 = \frac{dv_C}{dt} - \frac{dv_C^*}{dt} \tag{7.5}$$

$$x_3 = i_2 - i_2^* \tag{7.6}$$

where v_C^* and i_2^* are the references for v_C and i_2, respectively. In grid-connected inverter systems, both active and reactive power control is required. In the case

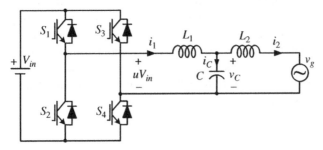

Figure 7.1 Single-phase grid-connected VSI with an LCL filter. *Source:* Adapted from Komurcugil et al. [1].

of active power control, the current injected into the grid is required to be sinusoidal and mostly in phase with the grid voltage. In order to meet this requirement, the reference of i_2 should be defined as

$$i_2^* = I_2^* \sin(\omega t) \tag{7.7}$$

By making use of (7.2), v_C^* can be generated as

$$v_C^* = L_2 \frac{di_2^*}{dt} + v_g \tag{7.8}$$

7.2.2 Sliding Mode Control

Let the sliding surface function be defined as follows [1]:

$$\sigma = \alpha x_1 + x_2 + \beta x_3 \tag{7.9}$$

where α and β are coefficients that should be positive to guarantee the stability of closed-loop system. Substituting the equations of i_2 and i_2^* obtained from (7.2) and (7.8) into (7.6) yields x_3 as follows:

$$x_3 = \frac{1}{L_2} \int x_1 dt \tag{7.10}$$

Note that the integral in (7.10) eliminates the steady-state errors in i_2 and v_C. During the sliding mode, the sliding surface function should be zero ($\sigma = 0$). In this case, the error variables move on the sliding surface toward the origin ($x_1 = 0$, $x_2 = 0$, and $x_3 = 0$). However, the state trajectory reaches the sliding surface provided that the following reaching condition is satisfied [2]:

$$\sigma \frac{d\sigma}{dt} < 0 \tag{7.11}$$

The time derivative of (7.9) can be written as

$$\frac{d\sigma}{dt} = \alpha \frac{dx_1}{dt} + \frac{dx_2}{dt} + \beta \frac{dx_3}{dt} \tag{7.12}$$

Taking time derivative of the error variables results in

$$\frac{dx_1}{dt} = x_2 \tag{7.13}$$

$$\frac{dx_2}{dt} = \omega_1^2 u V_{in} - \omega_r^2 x_1 + D(t) \tag{7.14}$$

$$\frac{dx_3}{dt} = \frac{x_1}{L_2} \tag{7.15}$$

where $\omega_1 = 1/\sqrt{L_1 C}$ and $\omega_2 = 1/\sqrt{L_2 C}$. The disturbance term, $D(t)$, and resonant angular frequency of the *LCL* filter, ω_r, are given by

$$D(t) = -\omega_r^2 v_C^* - \frac{d^2 v_C^*}{dt^2} + \omega_2^2 v_g \tag{7.16}$$

$$\omega_r = \sqrt{\frac{L_1 + L_2}{L_1 L_2 C}} \tag{7.17}$$

Since $\sigma = 0$ in the sliding mode, then using (7.12), (7.13), and (7.15), one can obtain

$$\frac{d^2 x_1}{dt^2} + \alpha \frac{dx_1}{dt} + \frac{\beta}{L_2} x_1 = 0 \tag{7.18}$$

It is evident that (7.18) is a second-order differential equation whose solution is given by

$$x_1(t) = A_1 e^{s_1 t} + A_2 e^{s_2 t} \tag{7.19}$$

where A_1 and A_2 are arbitrary constants, s_1 and s_2 are the closed-loop poles defined as

$$s_{1,2} = \frac{-\alpha L_2 \pm \sqrt{\alpha^2 L_2^2 - 4 L_2 \beta}}{2 L_2} \tag{7.20}$$

Obviously, the poles are real and can be placed on the left-half of s-plane when $\alpha^2 L_2^2 > 4 L_2 \beta$. In this case, $x_1(t) \to 0$. Hence, according to Eqs. (7.13)–(7.15), $x_2(t) \to 0$ and $x_3(t) \to 0$. This implies that the origin is stable.

Now, let us define the control input u as [3]

$$u = -sign(\sigma) \tag{7.21}$$

Then, according to (7.11) the existence of the sliding mode is guaranteed for the following cases:

i) $\sigma < 0 \implies u = 1$;

$$l_1 = \frac{d\sigma}{dt} = -\left(\omega_r^2 - \frac{\beta}{L_2}\right) x_1 + \alpha x_2 + d_1(t) > 0 \tag{7.22}$$

ii) $\sigma > 0 \implies u = -1$;

$$l_2 = \frac{d\sigma}{dt} = -\left(\omega_r^2 - \frac{\beta}{L_2}\right) x_1 + \alpha x_2 + d_2(t) < 0 \tag{7.23}$$

where

$$d_1(t) = \omega_1^2 V_{in} + D(t) \tag{7.24}$$

$$d_2(t) = -\omega_1^2 V_{in} + D(t) \tag{7.25}$$

Equations (7.22) and (7.23) represent two parallel lines that constitute the boundaries of the stability region of the reaching mode. This region, determined by l_1 and l_2, is depicted in Figure 7.2. It is evident that the entire region has two distinct sub-regions. While one is above the sliding line ($\sigma = 0$), the second region is below $\sigma = 0$ line. The sliding mode exists when the state trajectory hits the sliding line segment that is common in both regions. From Figure 7.2, one can easily see that the common line segment is the line between points σ_1 and σ_2. The x_1 intercepts of lines l_1 and l_2 can be written as

$$E_1 = \frac{L_2 d_1(t)}{\left(L_2 \omega_r^2 - \beta\right)}, \quad E_2 = \frac{L_2 d_2(t)}{\left(L_2 \omega_r^2 - \beta\right)} \tag{7.26}$$

It is obvious from (7.22) and (7.23) that l_1 and l_2 are parallel to each other, which implies that their slopes are equal as follows:

$$m_1 = m_2 = \frac{1}{\alpha}\left(\omega_r^2 - \frac{\beta}{L_2}\right) \tag{7.27}$$

The slopes (m_1 and m_2) become positive for $\beta < L_2 \omega_r^2$. However, it should be noted that m_1 and m_2 are no longer positive for $\beta > L_2 \omega_r^2$ and $\beta = L_2 \omega_r^2$, which change the borders of the existence region. When the borders of the existence region are changed, the stability of the entire system is affected.

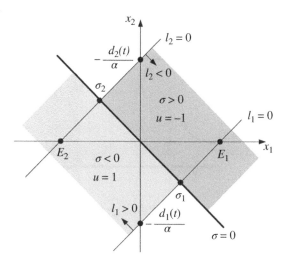

Figure 7.2 Stability regions of the reaching mode for $\beta < L_2 \omega_r^2$. *Source:* Adapted from Komurcugil et al. [1].

7.2.3 PWM Signal Generation Using Hysteresis Modulation

It is worth noting that when the inverter is controlled using (7.21), the chattering occurs which makes the switching frequency uncontrollable due to the *sign* function. As a remedy to this, the *sign* function is usually replaced by a hysteresis function that has a hysteresis band resulting in a controllable switching frequency as reported in [3]. The hysteresis function can be implemented using a single band as well as double bands. The following sub-sections that are studied extensively in [1] focus on the single-band, double-band, and switching frequency computation of both hysteresis function implementations.

7.2.3.1 Single-Band Hysteresis Function

The traditional single-band hysteresis switching is depicted in Figure 7.3. It is obvious that the sliding surface function varies between the upper and lower boundaries of the hysteresis band. In this case, inverter output voltage has only two levels: $+V_{in}$ and $-V_{in}$. In the single-band hysteresis modulation, S_1 and S_4 are turned on/off simultaneously. Similarly, S_2 and S_3 are turned on/off simultaneously. While S_1 and S_4 are turned on, S_2 and S_3 are turned off. Such switching causes high switching frequency since the zero level of inverter output voltage is not accessed.

7.2.3.2 Double-Band Hysteresis Function

The switching frequency in the single-band hysteresis modulation can be reduced if the zero level of inverter output voltage is enabled. In other words, the inverter output voltage should have three levels. In order to generate such three-level voltage, a double-band hysteresis modulation scheme shown in Figure 7.4 can be used.

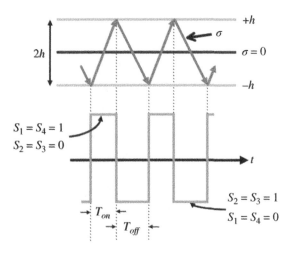

Figure 7.3 Single-band hysteresis switching in one-cycle. *Source:* Adapted from Komurcugil et al. [1].

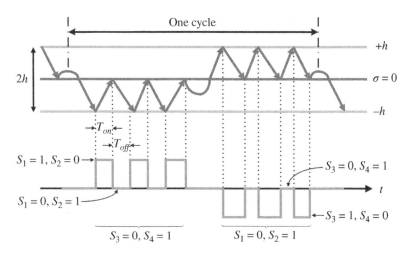

Figure 7.4 Double-band hysteresis switching in one-cycle. *Source:* Adapted from Komurcugil et al. [1].

Table 7.1 Driving Signals in Double-Band Hysteresis Scheme.

	Condition	S_1	S_2	S_3	S_4
Negative half-cycle of σ	$\sigma < -h$	1	0	0	1
	$\sigma > 0$	0	1	0	1
Positive half-cycle of σ	$\sigma > +h$	0	1	1	0
	$\sigma < 0$	0	1	0	1

The switching logic of this scheme is given in Table 7.1 where h is the width of this hysteresis band. When $\sigma < 0$ in the half-cycle, the switching occurs for S_1 and S_2 on one leg of the inverter. Similarly, when $\sigma > 0$ in the other half-cycle, the switching is done between S_3 and S_4 on the second leg of the inverter. In the negative half-cycle of σ, when σ moves from 0 toward $-h$, it hits $-h$ boundary (at the start of T_{on}), and immediately after it goes beyond $-h$ ($\sigma < -h$), S_1 is turned on (S_2 is turned off). As soon as S_1 is turned on, σ changes its direction and moves from $-h$ toward 0 and hits 0 boundary (at the start of T_{off}). When it goes beyond 0 ($\sigma > 0$), S_2 is turned on (S_1 is turned off). The switching of S_3 and S_4 in the positive half-cycle of σ occurs in the same way. An important consequence of using double-band hysteresis scheme is that all switches have equal switching losses during one-cycle. On the other hand, while S_2 and S_4 have equal conduction losses in one-cycle, S_1 and S_3 have no conduction losses, except in the half-cycle where they are switched on and off.

7.2.4 Switching Frequency Computation

The relationship between the switching frequency and the hysteresis band together with other parameters is very crucial and should be investigated for practical applications. Here, the switching frequency computation is based on the assumption that the state variables x_1 and x_2 are negligibly small in the steady state. Substituting (7.13), (7.14), and (7.15) into (7.12) and assuming that $x_1 \approx 0$ and $x_2 \approx 0$, one can obtain the following approximate equation [1]:

$$\frac{d\sigma}{dt} \cong \omega_1^2 u V_{in} - K_d \sin(\omega t + \phi) \tag{7.28}$$

where

$$K_d = \sqrt{V_g^2(\omega_1^2 - \omega^2)^2 + (\omega L_2 I_2^*)^2 (\omega_r^2 - \omega^2)^2} \tag{7.29}$$

$$\phi = \tan^{-1}\left(\frac{\omega L_2 I_2^* (\omega_r^2 - \omega^2)}{V_g(\omega_1^2 - \omega^2)}\right) \tag{7.30}$$

7.2.4.1 Switching Frequency Computation with Single-Band Hysteresis Modulation

The expressions of T_{on} and T_{off} can be written by using Figure 7.3 as follows:

$$T_{on} = \frac{2h}{\left.\dfrac{d\sigma}{dt}\right|_{u=1}} = \frac{2h}{\omega_1^2 V_{in} - K_d \sin(\omega t + \phi)} \tag{7.31}$$

$$T_{off} = \frac{-2h}{\left.\dfrac{d\sigma}{dt}\right|_{u=-1}} = \frac{2h}{\omega_1^2 V_{in} + K_d \sin(\omega t + \phi)} \tag{7.32}$$

Since $\omega L_2 I_2^* (\omega_r^2 - \omega^2) \ll V_g(\omega_1^2 - \omega^2)$, the phase shift ϕ can be neglected. Hence, the equation of instantaneous switching frequency can be written as

$$f_{sw}^{SB} = \frac{1}{T_{on} + T_{off}} = \frac{\omega_1^2 V_{in}}{4h} - \frac{K_d^2}{8h\omega_1^2 V_{in}} + \frac{K_d^2}{8h\omega_1^2 V_{in}} \cos(\omega t) \tag{7.33}$$

Considering (7.33), the expression of average switching frequency can be written as

$$f_{sw,av}^{SB} = \frac{\omega_1^2 V_{in}}{4h} - \frac{K_d^2}{8h\omega_1^2 V_{in}} \tag{7.34}$$

7.2.4.2 Switching Frequency Computation with Double-Band Hysteresis Modulation

Similarly, the expressions of T_{on} and T_{off} can be written by using Figure 7.4 as follows:

$$T_{on} = \frac{h}{\left.\dfrac{d\sigma}{dt}\right|_{u=1}} = \frac{h}{\omega_1^2 V_{in} - K_d \sin(\omega t + \phi)} \tag{7.35}$$

$$T_{off} = \frac{-h}{\left.\dfrac{d\sigma}{dt}\right|_{u=0}} = \frac{h}{K_d \sin(\omega t + \phi)} \tag{7.36}$$

Similarly, the phase shift ϕ is negligible since $\omega L_2 I_2^* \left(\omega_r^2 - \omega^2\right) \ll V_g \left(\omega_1^2 - \omega^2\right)$. It should be noted that the sum of T_{on} and T_{off} gives the switching period for S_1 and S_2 in the half-cycle while S_3 is off and S_4 is continuously on. Therefore, the switching period of the inverter in one cycle is twice of $T_{sw} = T_{on} + T_{off}$. Hence, the equation of instantaneous switching frequency during one cycle can be written as

$$f_{sw}^{DB} = \frac{1}{2\left(T_{on} + T_{off}\right)} = \frac{K_d}{2h\omega_1^2 V_{in}} \left[\omega_1^2 V_{in} \sin(\omega t) - K_d \sin^2(\omega t)\right] \tag{7.37}$$

The expression of average switching frequency in one cycle can be written as

$$f_{sw,av}^{DB} = \frac{K_d}{h\pi} - \frac{K_d^2}{4h\omega_1^2 V_{in}} \tag{7.38}$$

The influence of variations in h and L_1 on the average switching frequency is depicted in Figure 7.5. The average switching frequency for each modulation method

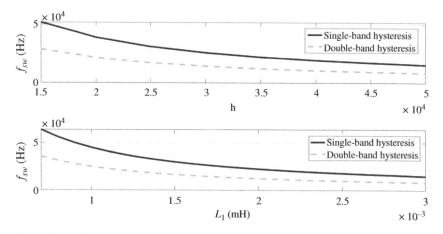

Figure 7.5 Influence of variations in h and L_1 on the average switching frequency.

Table 7.2 System Parameters.

Symbol	Value
DC-link voltage, V_{in}	400 V
Inverter-side inductance, L_1	1.74 mH
Filter capacitance, C	50 μF
Grid-side inductance, L_2	0.68 mH
Grid voltage amplitude, V_g	$230\sqrt{2}$V
Grid frequency, f_g	50 Hz

is computed by substituting the required constants (see Table 7.2) into (7.34) and (7.38). Comparing the average switching frequencies, one can see that the double-band hysteresis modulation leads to smaller switching frequency in both cases.

7.2.5 Selection of Control Gains

When the system is in sliding mode, the sliding surface function should satisfy $\sigma = 0$. In this case, Eq. (7.9) can be written as

$$\frac{dx_1}{dt} = -\alpha x_1 - \frac{\beta}{L_2}\int x_1 dt \tag{7.39}$$

Taking Laplace transform of both sides and solving for the quadratic equation, one can obtain the closed-loop poles as

$$s_{1,2} = \frac{-\alpha L_2 \pm \sqrt{\alpha^2 L_2^2 - 4L_2\beta}}{2L_2} \tag{7.40}$$

Equation (7.40) gives approximated values of the closed-loop poles since the state variables are not ideally zero in the sliding mode. However, the values of α and β can be selected to place the poles at desired positions.

Figure 7.6 shows the root locus of the closed-loop poles plotted when α is varied from 500 to 20000 using $L_2 = 0.68$ mH and fixed β values (25000 and 60000), respectively. It should be noted that the initial α value ($\alpha = 500$) yields complex conjugate poles that are located close to the imaginary axis for both β values. However, when α is increased, both poles move away from the imaginary axis until they reach to the break-in point where they become real with equal magnitudes. Thereafter, while one pole moves away from the imaginary axis, the other pole moves toward the imaginary axis. Comparing these root locus plots, one can see that the break-in point moves away from the imaginary axis when β is increased.

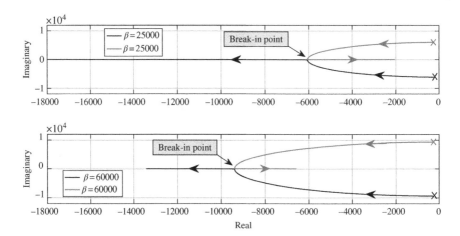

Figure 7.6 Root locus of closed-loop poles when α is varied from 500 to 20000.

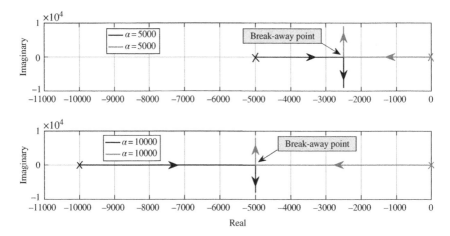

Figure 7.7 Root locus of closed-loop poles when β is varied from 0 to 60000.

Figure 7.7 shows the root locus of the closed-loop poles plotted when β is varied from 0 to 60000 using $L_2 = 0.68$ mH and fixed α values (5000 and 10000), respectively. It is worth noting that the initial β value ($\beta = 0$) yields real poles for both α values. While one pole is located very close to zero, the other pole is located away from the imaginary axis. However, when β is increased, while one pole moves toward the imaginary axis, the other pole moves away from the imaginary axis. It can be easily observed that that the break-away point shifts to the left when α is increased.

7.2.6 Simulation Study

The block diagram of single-phase grid-connected VSI with the SMC method is shown in Figure 7.8. The sliding surface function calculation is based on Eq. (7.9). The switching signals are produced using the double-band hysteresis switching logic shown in Table 7.1. A discussion on building the Simulink model of this diagram is presented in Chapter 6. The system parameters used in the simulation studies are given in Table 7.2.

Figure 7.9 shows the steady-state responses of sliding surface function (σ), inverter current (i_1), grid current (i_2), and grid voltage (v_g) for $I_2^* = 10\,\text{A}$, $h = 30000$, $\alpha = 12 \times 10^3$, and $\beta = 60 \times 10^3$. Substituting the values of α, β, and L_2 into Eq. (7.40), one can compute the closed-loop poles as $s_{1,2} = -6000 \pm j7227.4$. Obviously, the poles are complex–conjugate pair and are located away from the imaginary axis. In this case, the poles cause fast dynamic response (see Chapter 1). On the other hand, it can be seen that the positive cycle of sliding surface function varies between 0 and $+h$ while the negative cycle varies between 0 and $-h$. It can be noticed that the zero crossing of sliding surface function causes distortions in the inverter current. However, these distortions and switching harmonics in the inverter current are suppressed by *LCL* filter before they interact with the grid current. It can be seen that the grid current is in phase with the grid voltage. In addition, the grid current has reasonably low total harmonic distortion (THD = 1.26%) as shown in Figure 7.10.

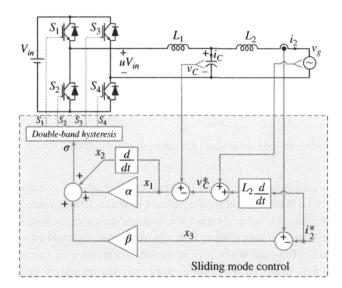

Figure 7.8 The block diagram of single-phase grid-connected VSI with the SMC method. *Source:* Adapted from Komurcugil et al. [1].

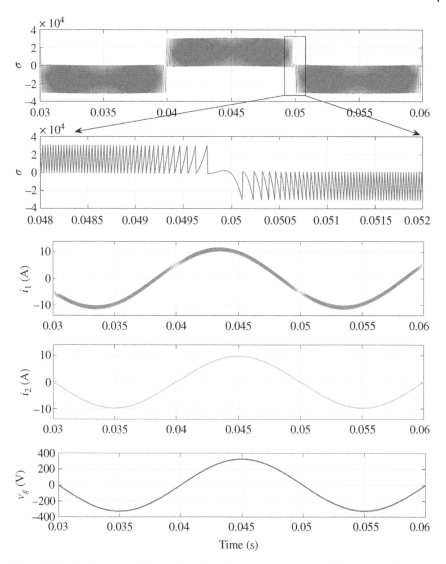

Figure 7.9 Steady state sliding surface function, inverter current, grid current and grid voltage for $I_2^* = 10$ A.

Figure 7.11 shows the response of grid current, grid current reference, and capacitor voltage on a sudden change in I_2^* from 10 to 20 A obtained by various α values, while the other control gains are the same as in Figure 7.9. The first result shown in Figure 7.11 is obtained for $\alpha = 12 \times 10^3$, while the other one is obtained for $\alpha = 2 \times 10^3$. Comparing these results, one can see that i_2 tracks its reference very

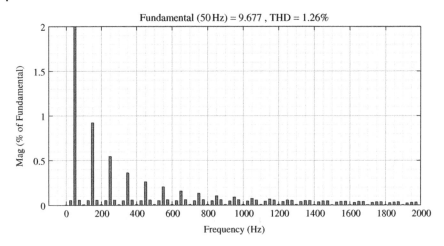

Figure 7.10 Spectrum of grid current for $I_2^* = 10$ A.

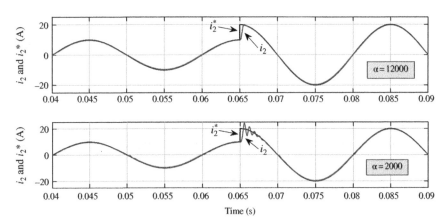

Figure 7.11 Dynamic response of grid current and grid current reference for step change in I_2^* from 10 to 20 A obtained by various α values.

fast for the large α value. On the other hand, undesired oscillations occur during step change transient for the small α value.

As explained before, the main goal of adding βx_3 term into the sliding function was to eliminate the steady-state error in the grid current and capacitor voltage. The effect of various β values on the grid-current steady-state error has been investigated. Figure 7.12 shows the grid current and its reference obtained by various β values for $I_2^* = 10$ A. It can be noticed that steady-state error and distortion exists

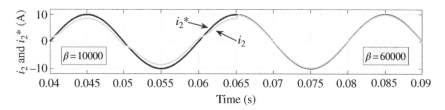

Figure 7.12 Grid current and its reference obtained by various β values for $I_2^* = 10$ A.

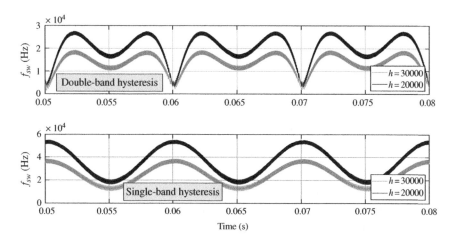

Figure 7.13 Instantaneous switching frequencies obtained by double-band and single-band hysteresis modulation schemes using various band values for $I_2^* = 10$ A.

in the grid current when $\beta = 10000$. However, both steady-state error and distortion are eliminated when β is changed to 60000.

Figure 7.13 shows the instantaneous switching frequencies obtained by double-band and single-band hysteresis modulation schemes using various band values for $I_2^* = 10$ A. It can be observed that the double-band hysteresis scheme results in smaller switching frequency than the single-band hysteresis scheme for the same band value. In addition, it is clear that the switching frequency in both schemes increases when the hysteresis band is decreased.

7.2.7 Experimental Study

The double-band hysteresis function-based SMC method was implemented by using a TMS320F28335 floating point digital signal processor. The sensed signals are applied to the analog-to-digital converter of the digital signal processor.

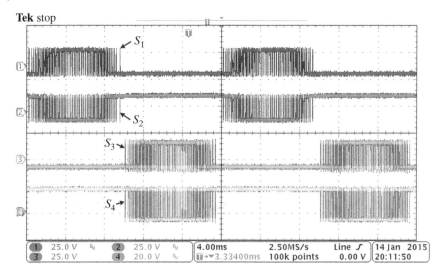

Figure 7.14 Driving signals for the switching devices. *Source:* Komurcugil et al. [1]/IEEE.

The average switching frequency was around 18.8 kHz. The driving signals of the switching devices S_1–S_4 are shown in Figure 7.14. Because of double-band hysteresis scheme, the control signals are generated in such a way that the switching devices S_1 and S_2 are switched on and off, while S_4 is always on and S_3 is always off during the positive half-cycle. Similarly, S_3 and S_4 are switched on and off, while S_2 is always on and S_1 is always off during the negative half-cycle. In this way, the double-band hysteresis switching leads to a lower switching frequency than that obtained by the conventional single-band hysteresis scheme.

Figure 7.15 shows the dynamic response of the grid (v_g) and current (i_2) for a step change in the reference grid current amplitude (I_2^*) from 10 to 20 A and back to 10 A. It is obvious that i_2 exhibits a fast response in tracking i_2^* due to the change in I_2^*. Based on the disturbance term $D(t)$ in (7.16) and reference capacitor voltage (v_C^*) in (7.8) which contains i_2^*, any change in i_2^* can be considered as a new disturbance applied to the closed-loop system. The results in Figure 7.15 clearly demonstrate that the SMC method is not affected by the disturbances in the system.

Figure 7.16 shows the measured grid current spectrum for $I_2^* = 20$ A. The THD is measured to be 1.4%, which is reasonable. It is apparent that the harmonic components in the grid current are negligibly small.

Figure 7.17 shows the dynamic response of the grid voltage and current for a step change in β from 15000 to 1000. It is clear that the closed-loop system achieves a sinusoidal grid current in phase with the grid voltage with a large β value.

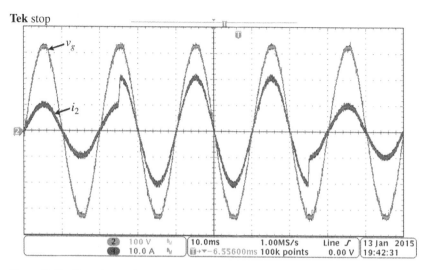

Figure 7.15 Experimental responses of v_g and i_2 for a step change in I_2^* from 10 to 20 A and back to 10 A. (CH2 (v_g): 100 V/div, CH4 (i_2):10 A/div). *Source:* Komurcugil et al. [1]/IEEE.

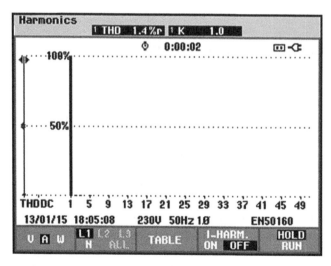

Figure 7.16 Measured grid current spectrum for $I_2^* = 20$ A. *Source:* Komurcugil et al. [1]/IEEE.

However, after the step change in β, there exists a considerable steady-state error in the grid current (approximately 2 A). In addition, the grid current is distorted after the step change in β. These results show the necessity of having βx_3 term in the sliding surface function.

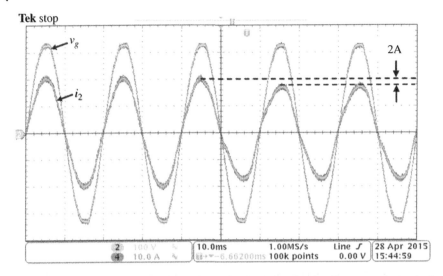

Figure 7.17 Experimental responses of v_g and i_2 in the steady-state for a step change in β from 15000 to 1000. (CH2 (v_g): 100 V/div, CH4 (i_2): 10 A/div). *Source:* Komurcugil et al. [1]/IEEE.

7.3 Three-Phase Grid-Connected Inverter with LCL Filter

This section presents a new concept in active damping techniques combined to sliding mode control (SMC) using a reduced model of an LCL filtered grid-connected voltage source inverter (VSI). This technique was previously presented in [4]. It is well known that the presence of the LCL filter complicates the design of the inverter control scheme, particularly when uncertainties in the system parameters, especially in the grid inductance, are considered. The control algorithm is addressed to overcome such difficulties using a reduced model of the inverter in a state observer. In this method, two of the three-state variables of the system are obviated from the physical inverter model, and only the inverter-side current is considered. Therefore, the inverter-side current can be estimated emulating the case of an inverter with only one inductor, thus eliminating the resonance problem produced by the LCL filter. Besides, in the case of a distorted grid, the method allows us to estimate the voltages at the point of common coupling free of noise and distortion without using any phase-locked loop-based synchronization algorithm. This method provides the following features to the closed-loop system: first, robust and simple active damping control under system parameters deviation; second, robustness against grid voltage unbalance and distortion; and third, an important reduction in the computational load of the control algorithm, which allows us to increase the switching frequency. To complete the control scheme, a theoretical stability analysis is developed considering the effect of the observer, the system discretization, and the system parameters deviation. Experimental and comparative evaluation results are presented to validate the effectiveness of the control scheme.

7.3.1 Physical Model Equations for a Three-Phase Grid-Connected VSI with an LCL Filter

Figure 7.18 shows a three-phase grid-connected VSI with an *LCL* filter, where the grid impedance is considered pure inductive. The VSI equations can be obtained from the circuit as follows:

$$L_1 \frac{d\mathbf{i}_1}{dt} = \frac{V_{dc}}{2}\mathbf{u} - \mathbf{v}_c - v_n\mathbf{1} \tag{7.41}$$

$$C \frac{d\mathbf{v}_c}{dt} = \mathbf{i}_1 - \mathbf{i}_2 \tag{7.42}$$

$$\left(L_2 + L_g\right)\frac{d\mathbf{i}_2}{dt} = \mathbf{v}_c - \mathbf{v}_g \tag{7.43}$$

where $\mathbf{i}_1 = \begin{bmatrix} i_{1a} & i_{1b} & i_{1c} \end{bmatrix}^T$ is the inverter-side current vector, $\mathbf{i}_2 = \begin{bmatrix} i_{2a} & i_{2b} & i_{2c} \end{bmatrix}^T$ is the grid-side current vector, $\mathbf{v}_c = \begin{bmatrix} v_{ca} & v_{cb} & v_{cc} \end{bmatrix}^T$ is the capacitor voltage vector, $\mathbf{v}_g = \begin{bmatrix} v_{ga} & v_{gb} & v_{gc} \end{bmatrix}^T$ is the grid voltage vector, $\mathbf{u} = \begin{bmatrix} u_a & u_b & u_c \end{bmatrix}^T$ is the input control signals vector and v_n is the voltage at neutral point expressed as:

$$v_n = \frac{V_{dc}}{6}\left(u_a + u_b + u_c\right) \tag{7.44}$$

and $\mathbf{1}$ is a column vector defined as $\begin{bmatrix} 1 & 1 & 1 \end{bmatrix}^T$.

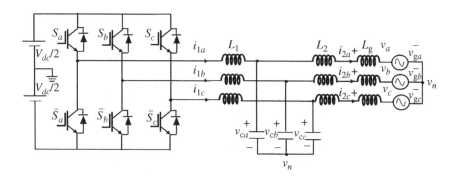

Figure 7.18 Circuit diagram of three-phase grid-connected inverter with an *LCL* filter.

The above differential equations can be rewritten as a discrete state-space model where the process and measurement discrete equations for each phase-leg i, where $i \in \{a, b, c\}$, are expressed as follows:

$$\mathbf{x}_i(k + 1) = \mathbf{A}\mathbf{x}_i(k) + \mathbf{B}u_i(k) + \mathbf{D}v_{gi}(k) + \mathbf{E}v_n(k) \tag{7.45}$$

$$y_i(k) = \mathbf{C}\mathbf{x}_i(k) \tag{7.46}$$

where $\mathbf{x}_i = (i_{1i}, v_{ci}, i_{2i})$ is the space-state vector, $\mathbf{C}$ is the output matrix defined as

$$\mathbf{C} = (1 \quad 0 \quad 0) \tag{7.47}$$

and $\mathbf{A}$, $\mathbf{B}$, $\mathbf{D}$ and $\mathbf{E}$ are the discrete matrices of the system that can be computed using the first-order approximation as follows:

$$\mathbf{A} = \mathbf{I} + \mathbf{A}_c T_s = \begin{pmatrix} 1 & -T_s/L_1 & 0 \\ T_s/C & 1 & -T_s/C \\ 0 & T_s/(L_2 + L_g) & 1 \end{pmatrix} \tag{7.48}$$

$$\mathbf{B} = \mathbf{B}_c T_s = (V_{DC}T_s/2L_1 \quad 0 \quad 0)^T \tag{7.49}$$

$$\mathbf{D} = \mathbf{D}_c T_s = (0 \quad 0 \quad -T_s(L_2 + L_g))^T \tag{7.50}$$

$$\mathbf{E} = \mathbf{E}_c T_s = (-T_s/L_1 \quad 0 \quad 0)^T \tag{7.51}$$

with T_s the sampling time and $\mathbf{A}_c$, $\mathbf{B}_c$, $\mathbf{D}_c$ and $\mathbf{E}_c$ the continuous system matrices, which can be derived from the model defined in (7.41)–(7.43).

7.3.2 Control System

The control scheme is oriented to achieve a robust active damping algorithm using an SMC with a reduced model. Figure 7.19 depicts the control scheme for phase-leg a, which has the same structure for the phases b and c. The controller is based on a state observer, a Kalman Filter (KF), which uses a reduced VSI model in order to estimate the inverter-side current, the PCC voltage and its quadrature. As shown in the figure, the reference current is obtained according to the desired active power, and using the estimated PCC voltages.

Once the reference is computed, the reference current and the estimated inverter-side current are used in a sliding surface, and a switching algorithm obtains the control signal u_a. This algorithm was introduced in [5, 6] and it will be explained in detail in this chapter. Its main objective is to achieve a quasi-fixed switching frequency. Finally, the generated control signal is applied to control the phase-leg a of the VSI, generating a current without oscillations.

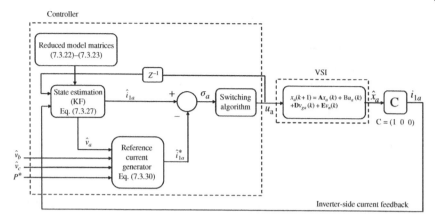

Figure 7.19 Control system for phase-leg a.

7.3.2.1 Reduced State-Space Model of the Converter

Figure 7.20 shows the filter equivalent circuits of an L and LCL structures from the point of view of its impedance. The input impedance of the LCL filter presented in Figure 7.20 can be obtained easily, yielding:

$$Z_{LCL} = \frac{L_1 C (L_2 + L_g) s^3 + (L_1 + L_2 + L_g) s}{C(L_2 + L_g) s^2 + 1} \tag{7.52}$$

While the input impedance for the L topology is expressed as

$$Z_L = (L_1 + L_2 + L_g) s \tag{7.53}$$

It can be easily found that $Z_{LCL} \cong Z_L$ when $\omega_o \ll \omega_{res}$, with ω_o the grid frequency. Note that the resonance frequency can be obtained by making equal to zero the denominator of (7.52) and solving for ω, which leads to

$$\omega_{res} = \frac{1}{\sqrt{(L_2 + L_g) C}} \tag{7.54}$$

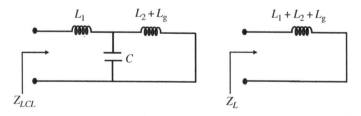

Figure 7.20 Input impedances of the LCL filter and for the L filter.

Usually, it is accomplished that $\omega_o \ll \omega_{res}$ and according to this approximation, the effect of the capacitor may be neglected in the input impedance computation. This fact leads to accomplish that $\mathbf{i}_1 \cong \mathbf{i}_2$. For a better understanding, a Bode plot of the filter impedance is presented in Figure 7.21, using the system parameters listed in Table 7.3. The behavior of both impedances is similar at the grid frequency ω_o.

Using the aforementioned approximation, and considering (7.43), the capacitor voltage vector can be rewritten as follows:

$$\mathbf{v}_c = \left(L_1 + L_g\right)\frac{d\mathbf{i}_1}{dt} + \mathbf{v}_g \qquad (7.55)$$

Now, if (7.55) is used in (7.41), the following differential equation can be found:

$$\left(L_1 + L_2 + L_g\right)\frac{d\mathbf{i}_1}{dt} = \frac{V_{dc}}{2}\mathbf{u} - \mathbf{v}_g - v_n\mathbf{1} \qquad (7.56)$$

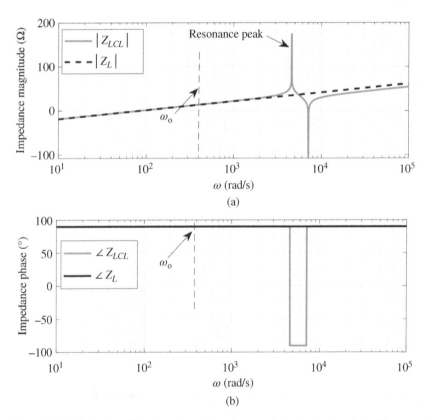

Figure 7.21 Bode plot of the input impedance in case of the physical model and in case of the reduced model. (a) Magnitude and (b) phase.

Table 7.3 System parameters.

Symbol	Description	Value
L_{1o}	Nominal filter input inductance	5 mH
C_o	Nominal filter capacitor	6.8 µF
L_{2o}	Nominal Filter output inductance	2 mH
L_g	Grid inductance	0.5 mH
V_{dco}	Nominal DC-link voltage	450 V
f_s	Sampling frequency	60 kHz
f_{sw}	Switching frequency	6 kHz
f_o	Grid frequency	60 Hz
V_{grid}	Peak grid voltage	110 V
P^*	Active power	1.5 kW
Q^*	Reactive power	0 kVAr
R_i	Single-phase system noise power	$0.26\ V^2$

The last equation describes the current dynamics of a VSI with only one inductor value of which is $L_1 + L_2 + L_g$. According to the Bode plots shown before, the impedance of both filters is similar at the grid frequency. Therefore, at the frequency ω_o, the dynamics of $\mathbf{i}_1$ is approximated to the dynamics of $\mathbf{i}_2$. This idea will be applied to derive the reduced model according to the following premises:

1) Since the sliding-mode controller is designed in natural frame, the three current controllers are cross coupled through the neutral point voltage (see (7.44)). For this reason, and in order to eliminate these interferences between controllers, the neutral point voltage expression is not considered in the reduced model [7].
2) In order to achieve active damping dynamics, the model only considers the inverter-side current state variable. The capacitor voltage and the grid-side current are obviated in this model. With this consideration, the inverter-side current is estimated as in the case of a VSI with only one inductor where the resonance does not exist.
3) The voltage at the PCC and its quadrature can be estimated instead of using the measured grid voltage. This fact leads to a version of the PCC voltage free of noise and distortion even in the case of a distorted grid. Besides, the model is independent of the grid inductance [5, 8].

According to the aforementioned conditions, the reduced model can be represented as follows:

$$(L_{1o} + L_{2o})\frac{di_1}{dt} = \frac{V_{dco}}{2}\mathbf{u} - \mathbf{v} \tag{7.57}$$

$$\frac{d\mathbf{v}}{dt} = \omega_o\mathbf{v}^\perp \tag{7.58}$$

$$\frac{d\mathbf{v}^\perp}{dt} = -\omega_o\mathbf{v} \tag{7.59}$$

where L_{1o} and L_{2o} are the inductors nominal values, ω_o the grid angular frequency, V_{dco} the nominal DC-link voltage, and $\mathbf{v}$ and $\mathbf{v}^\perp$ the voltage at the point of common coupling (PCC) and its quadrature, respectively.

Figure 7.22 shows the input impedance magnitude of the physical LCL filter for two different values of L_g, 0.5 and 5 mH, respectively. The input impedance magnitude of the reduced model, $|Z_L|$ regarding nominal values L_{1o} and L_{2o} is also depicted. The figure shows that at the grid frequency $\omega_o = 2\pi 60$ rad/s, the impedance magnitude of the reduced model matches with good accuracy with the magnitude impedance of the physical model, $|Z_{LCL}|$, for both values of the grid impedance. Note that for the reduced model, it is not necessary to know the value of the grid inductance, which makes the system robust against this parameter.

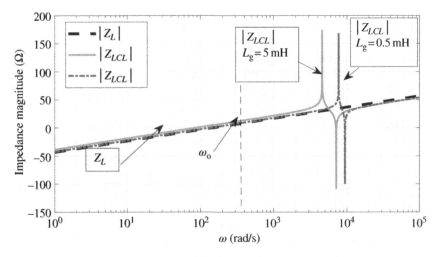

Figure 7.22 Bode plot of the input impedance magnitude in the case of the physical model for two different values of the grid inductance and in the case of the reduced model the magnitude of the physical model for both values of the grid impedance.

7.3.2.2 Model Discretization and KF Adaptive Equation

For the digital implementation of the control algorithm, this model is discretized in order to be used in a KF algorithm. The reduced discrete space-state model, where the symbol ^ denotes estimated variables, is represented as follows:

$$\hat{\mathbf{x}}_i(k+1) = \hat{\mathbf{A}}\hat{\mathbf{x}}_i(k) + \hat{\mathbf{B}}u_i(k) + \boldsymbol{\eta}_i(k) \tag{7.60}$$

$$y_i(k) = \mathbf{C}\mathbf{x}_i(k) + \mathbf{w}_i(k) \tag{7.61}$$

where

$$\hat{\mathbf{A}} = \begin{pmatrix} 1 & -T_s/(L_{1o} + L_{2o}) & 0 \\ 0 & 1 & T_s\omega_o \\ 0 & -T_s\omega_o & 1 \end{pmatrix} \tag{7.62}$$

$$\hat{\mathbf{B}} = \left(V_{dco}T_s/2(L_{1o} + L_{2o}) \quad 0 \quad 0 \right)^T \tag{7.63}$$

$$\hat{\mathbf{x}}_i(k) = \left(\hat{i}_{1i} \quad \hat{v}_i \quad \hat{v}_i^{\perp} \right)^T \tag{7.64}$$

and $\boldsymbol{\eta}_i$ and $\mathbf{w}_i$ the process and the measurement noise vector, respectively, which defines the following covariance matrices:

$$\mathbf{R}_i(k) = E\left\{\mathbf{w}_i(k)\mathbf{w}_i^T(k)\right\} \tag{7.65}$$

$$\mathbf{Q}_i(k) = E\left\{\boldsymbol{\eta}_i(k)\boldsymbol{\eta}_i^T(k)\right\} \tag{7.66}$$

To design the observer-based SMC, a KF can be used. According to the KF features, a KF-based observer is optimum in the presence of noise and improves the dynamics of the closed-loop system in noisy environments. For this reason, this method is focused on the state-space model (7.60)–(7.61) used in a KF algorithm. This model forces the estimated inverter-side currents to get similar dynamics as in the case of a VSI with only one inductor. Then, a virtual damping factor is introduced in the estimated currents. Besides, the reduced model has no dependence on the neutral point voltage, which makes the dynamics of the three-phase currents only dependent on its control signal [5]. Finally, if the estimated currents are used in an SMC, a fixed switching frequency can be achieved since the three sliding surfaces are decoupled as it will be shown later. The main properties and the implementation of the KF algorithm has been explained in [9], therefore, only the adaptive equation used for the estimation is given in this section. The equation for the estimation is expressed as follows:

$$\hat{\mathbf{x}}_i(k+1) = \hat{\mathbf{A}}\hat{\mathbf{x}}_i(k) + \hat{\mathbf{B}}u_i(k) + \mathbf{L}_i\left(i_{1i}(k) - \hat{i}_{1i}(k)\right) \tag{7.67}$$

where the measured inverter-side current i_{1i} is used to compute the estimation error, and the Kalman gain is computed using the following expression:

$$\mathbf{L}_i(k) = \mathbf{P}_i(k)\mathbf{C}^T\left(\mathbf{C}\mathbf{P}_i(k)\mathbf{C}^T + \mathbf{R}_i(k)\right)^{-1} \tag{7.68}$$

where $\mathbf{P}_i(k)$ is the error covariance matrix. Note that Kalman gain is computed in order to minimize the noise in the estimation whose algorithm is explained in [5, 9].

7.3.2.3 Sliding Surfaces with Active Damping Capability

As it was explained, in order to achieve a damped dynamic, the estimated inverter-side current can be used in a sliding-mode controller. Then, the following sliding surface for each phase-leg i can be written:

$$\sigma_i = \hat{i}_i^* - \hat{i}_{1i} \tag{7.69}$$

where $\hat{i}_i^*$ is the reference current, which can be computed according to the active and reactive power references:

$$\hat{i}_i^* = \frac{P^*}{|\hat{v}|^2}\hat{v}_i + \frac{Q^*}{\sqrt{3}|\hat{v}|^2}\hat{v}_i^\perp \tag{7.70}$$

From (7.69) and (7.70), the expression of the sliding surface can be rewritten as follows:

$$\sigma_i = \mathbf{G}\hat{\mathbf{x}}_i \tag{7.71}$$

being the vector $\mathbf{G}$ defined as:

$$\mathbf{G} = \left(1 \quad \frac{-P^*}{|\hat{v}^2|} \quad \frac{-Q^*}{\sqrt{3}|\hat{v}^2|}\right) \tag{7.72}$$

The use of the estimated PCC voltages in the reference currents computation must be noticed. This fact allows us to reduce the current distortion even in the case of a highly distorted grid [7, 8].

The existence of a sliding mode controller can be guaranteed if the following inequality holds:

$$\dot{\sigma}_i\sigma_i < 0 \tag{7.73}$$

then, to determine the switching action of the SMC, the following control variable is defined:

$$u_i = \begin{cases} u_i^+ & \text{if } \sigma_i > 0 \\ u_i^- & \text{if } \sigma_i < 0 \end{cases} \tag{7.74}$$

where $u_i \in \{1, -1\}$. Then, considering (7.57), (7.69), and (7.73), the following condition is obtained:

$$\frac{V_{dco}}{2(L_{1o} + L_{2o})}\left(u_i^+ - u_i^-\right) < 0 \tag{7.75}$$

From the last condition, the switching action must be defined as follows:

$$u_i = \begin{cases} u_i^+ = -1 & \text{if } \sigma_i > 0 \\ u_i^- = 1 & \text{if } \sigma_i < 0 \end{cases} \tag{7.76}$$

7.3.3 Stability Analysis

To conclude with the robust controller design, a stability analysis is performed in this section in order to validate the control method. The stability of the system will be also analyzed accounting for the system parameters deviation. The system stability is ensured if the eigenvalues of the closed-loop system are inside the unity circle in the z-plane.

The three-phase system has nine poles, each one related to one state variable. Due to the decoupling introduced by the KF, a decoupling between controllers is achieved and the three-phase system can be considered as three independent single-phase systems from the control viewpoint [5]. Therefore, only the stability of a single-phase system is analyzed, and consequently, only three poles due to the LCL filter are considered. However, the state observer will add three additional poles. Then, the system together with the observer will have a total of six poles (see Figure 7.23b).

While the system is in sliding regime, it is found that there is one pole fixed at the origin ($z = 0$) [10]. This is because the SMC forces one state variable to be equal to its reference, $\hat{i}_{i1} = \hat{i}_i^*$, and the order of the system is then reduced in one. The position of the remaining poles will depend on the model parameters and the KF gain. The following sections are dedicated to find the closed loop equations.

7.3.3.1 Discrete-Time Equivalent Control Deduction

The so-called equivalent control, $\hat{u}_{i,eq}$, can be obtained by assuming that the system is in sliding regime ($\sigma_i = \dot{\sigma}_i = 0$). In this case, the control signal u_i can be replaced by the equivalent control $\hat{u}_{i,eq}$. Note that the hat symbol in the equivalent control is due to the fact that the sliding surfaces are computed using estimated variables. Then, according to the equivalent control, the estimated variables can be computed using the following expression:

$$\hat{\mathbf{x}}_i(k+1) = \hat{\mathbf{A}}\hat{\mathbf{x}}_i(k) + \hat{\mathbf{B}}\hat{u}_{i,eq}(k) + \mathbf{L}_i(k)(\mathbf{C}\mathbf{x}_i(k) - \mathbf{C}\hat{\mathbf{x}}_i(k)) \tag{7.77}$$

The equivalent control can be obtained by using the aforementioned equation in (7.31), at time $k+1$, yielding:

$$\sigma_i(k+1) = \mathbf{G}^T\hat{\mathbf{x}}_i(k+1) = \mathbf{G}^T\left(\hat{\mathbf{A}}\hat{\mathbf{x}}_i(k) + \hat{\mathbf{B}}\hat{u}_{i,eq}(k) + \mathbf{L}_i(k)(\mathbf{C}\mathbf{x}_i(k) - \mathbf{C}\hat{\mathbf{x}}_i(k))\right) = 0 \tag{7.78}$$

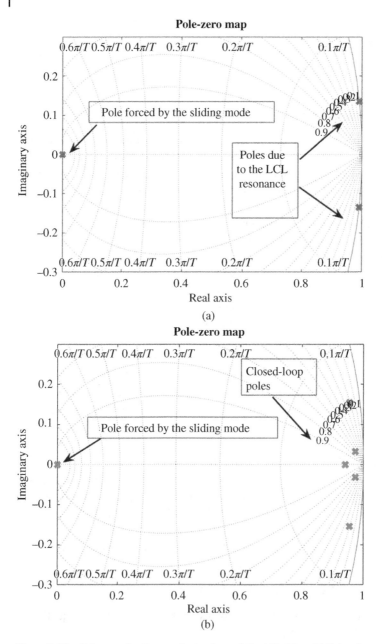

Figure 7.23 Pole map for the nominal values of the LCL filter. (a) Using the conventional SMC without observer and (b) Using the reduced model with a KF.

then, solving for the equivalent control $\hat{u}_{i,eq}(k)$

$$\hat{u}_{i,eq}(k) = \left(\mathbf{G}^T\hat{\mathbf{B}}\right)^{-1}\left(\sigma_i(k+1) - \mathbf{G}^T\hat{\mathbf{A}}\hat{\mathbf{x}}_i(k) - \mathbf{G}^T\mathbf{L}(k)e_i(k)\right) \tag{7.79}$$

where $e_i(k)$ is the error between the measured inverter-side current and the estimated one, and defined as:

$$e_i(k) = \mathbf{C}\mathbf{x}_i(k) - \mathbf{C}\hat{\mathbf{x}}_i(k) \tag{7.80}$$

In sliding regime, it is accomplished that $\sigma_i(k+1) = \sigma_i(k) = 0$. Then, the equivalent control can be rewritten as follows:

$$\hat{u}_{i,eq}(k) = \mathbf{K}_1\hat{\mathbf{x}}_i(k) + \mathbf{K}_2e_i(k) \tag{7.81}$$

where the gains $\mathbf{K}_1$ and $\mathbf{K}_2$ are expressed as a function of the system matrices as follows:

$$\mathbf{K}_1 = -\left(\mathbf{G}^T\hat{\mathbf{B}}\right)^{-1}\mathbf{G}^T\hat{\mathbf{A}} \tag{7.82}$$

$$\mathbf{K}_2 = -\left(\mathbf{G}^T\hat{\mathbf{B}}\right)^{-1}\mathbf{G}^T\mathbf{L}_i(k) \tag{7.83}$$

7.3.3.2 Closed-Loop System Equations

The closed-loop system equations can be defined by the vector $(\mathbf{x}_i(k)\ \ \hat{\mathbf{x}}_i(k))$. In order to find these equations, the grid voltages will be considered as disturbances. Thus, these voltages can be removed from the VSI system equations since they have no effect in the stability analysis. Since the control signal is obtained using estimated variables, the equivalent control signal is $\hat{u}_{i,eq}(k)$, and the discrete state-space equations for each phase-leg i can be defined as follows:

$$\mathbf{x}_i(k+1) = \mathbf{A}\mathbf{x}_i(k) + \mathbf{B}\hat{u}_{i,eq}(k) \tag{7.84}$$

Replacing (7.81) in (7.84) and in (7.77), and considering (7.80), the following state-space equations for the real system variables and for the estimated ones can be obtained:

$$\mathbf{x}_i(k+1) = (\mathbf{A} + \mathbf{B}\mathbf{K}_2\mathbf{C})\mathbf{x}_i(k) + \mathbf{B}(\mathbf{K}_1 - \mathbf{K}_2\mathbf{C})\hat{\mathbf{x}}_i(k) \tag{7.85}$$

$$\hat{\mathbf{x}}_i(k+1) = \left(\mathbf{L}_i + \hat{\mathbf{B}}\mathbf{K}_2\right)\mathbf{C}\mathbf{x}_i(k) + \left(\hat{\mathbf{A}} + \hat{\mathbf{B}}(\mathbf{K}_1 - \mathbf{K}_2\mathbf{C}) - \mathbf{L}_i\mathbf{C}\right)\hat{\mathbf{x}}_i(k) \tag{7.86}$$

Equations (7.85) and (7.86) define the closed-loop system. These equations can be expressed in a matrix form as follows:

$$\begin{pmatrix} \mathbf{x}_i(k+1) \\ \hat{\mathbf{x}}_i(k+1) \end{pmatrix} = \mathbf{F}\begin{pmatrix} \mathbf{x}_i(k) \\ \hat{\mathbf{x}}_i(k) \end{pmatrix} \tag{7.87}$$

where the matrix $\mathbf{F}$ contains the closed-loop eigenvalues and is defined as:

$$\mathbf{F} = \begin{pmatrix} \mathbf{A} + \mathbf{B}\mathbf{K}_2\mathbf{C} & \mathbf{B}(\mathbf{K}_1 - \mathbf{K}_2\mathbf{C}) \\ \left(\mathbf{L}_i + \hat{\mathbf{B}}\mathbf{K}_2\right)\mathbf{C} & \hat{\mathbf{A}} + \hat{\mathbf{B}}(\mathbf{K}_1 - \mathbf{K}_2\mathbf{C}) - \mathbf{L}_i\mathbf{C} \end{pmatrix} \tag{7.88}$$

Note that the parameters uncertainties are considered in the system matrices **A** and **B**, whereas the matrices of the reduced model, $\hat{\mathbf{A}}$ and $\hat{\mathbf{B}}$, contain the nominal values.

The stability of the closed-loop system will be given by the eigenvalues of (7.88), which can be obtained by solving the following equation with respect to λ:

$$(\mathbf{A} + \mathbf{B}\mathbf{K}_2\mathbf{C} - \lambda\mathbf{I})(\hat{\mathbf{A}} + \hat{\mathbf{B}}(\mathbf{K}_1 - \mathbf{K}_2\mathbf{C}) - \mathbf{L}_i\mathbf{C} - \lambda\mathbf{I}) - (\mathbf{B}(\mathbf{K}_1 - \mathbf{K}_2\mathbf{C}))((\mathbf{L}_i + \hat{\mathbf{B}}\mathbf{K}_2)\mathbf{C}) = 0$$

$$(7.89)$$

Figure 7.23 compares the position of the closed-loop poles in two different scenarios when the nominal *LCL* filter parameters presented in Table 7.3 are considered. In the first case, the VSI is controlled using the conventional SMC without any damping strategy. The position of the poles in this case is represented in Figure 7.23a. The figure shows one pole at the origin forced by the sliding mode and two more poles provided by the *LCL* filter on the unity circle. However, when the model is used in a KF, the *LCL* poles are attracted inside the unity circle in the z plane, as shown in Figure 7.23b. Note that three more poles are added due to the presence of the observer.

7.3.3.3 Test of Robustness Against Parameters Uncertainties

Although the system can be stable for the nominal values of the system parameters, deviations in the *LCL* filter and in the grid inductance should be studied. In this analysis, deviations of $\pm 30\%$ in L_1, C and L_2, and 900% in the grid inductance L_g are considered. Figure 7.24a–d depicts the root locus obtained in each case. As can be seen, stability is ensured for large variations of these parameters. Small variations in the position of the poles can be observed when the system parameters vary, even for a large variation of the grid inductance. This fact proves the high robustness of the control algorithm even in the case of system parameters deviations.

7.3.4 Experimental Study

In this section, experimental results are reported in order to validate the control algorithm. For the experimental results, a three-phase three-wire inverter prototype has been built using a 4.5-kVA SEMIKRON full-bridge as the power converter. The TMS320F28M36 floating-point digital signal processor (DSP) has been chosen as the control platform with a sampling frequency of 60 kHz. The grid and the DC-link voltages have been generated using a PACIFIC 360-AMX and an AMREL SPS1000-10-K0E3 sources, respectively. The system parameters are listed in Table 7.3.

7.3.4.1 Test of Robustness Against Grid Inductance Variations

Figure 7.25 shows the PCC voltages and the grid currents for three different values of the grid inductance $L_g = 0.5$, 2, and 5 mH, respectively. The controller

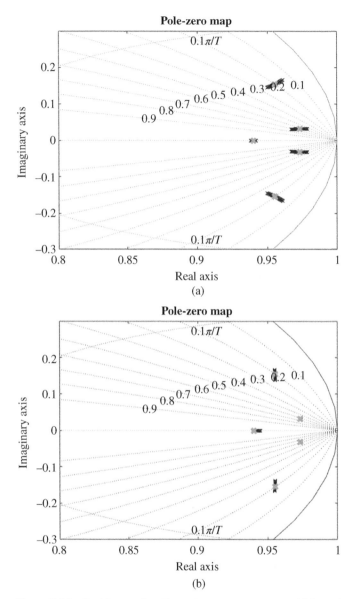

Figure 7.24 Root locus when the system parameters vary. (a) L_1 varies ±30%, (b) C varies ±30%, (c) L_2 varies ±30%, and (d) L_g varies ±900%.

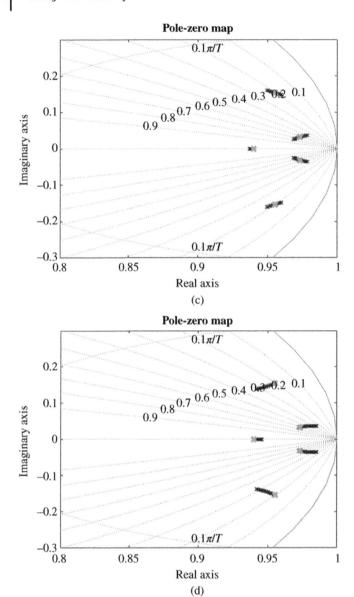

Figure 7.24 (Continued)

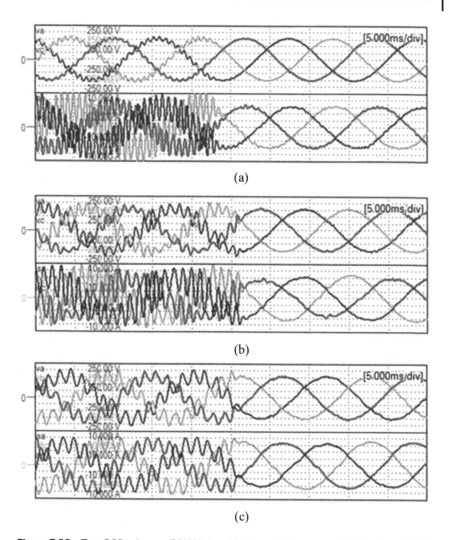

(a)

(b)

(c)

Figure 7.25 Top: PCC voltages (50 V/div) and bottom. Grid currents (2 A/div, 5 ms/div) for three different values of L_g. (a) L_g = 0.5 mH, (b) L_g = 2 mH, and (c) L_g = 5 mH.

performances have been tested for these three different values that represent a 900% of variation in the grid inductance nominal value. In this test, the active damping is enabled when the control method is used. Otherwise, the conventional SMC is applied without using any observer. The results obtained in this test are discussed as follows. Figure 7.25a shows the PCC voltages and the grid-side currents for a nominal grid inductance of $L_g = 0.5$ mH. As it can be seen, when the conventional SMC is used, the system oscillates since the controller (7.69) is working without a damping resistor. According to (7.14), the resonance frequency in this case is around $f_{res} = 1220$ Hz. Conversely, when the active damping is enabled, the oscillation disappears and all the roots are attracted inside the unity circle in the z-plane, as shown in Figure 7.23. Figure 7.25b shows the result of the same test but when the grid inductance is incremented to a value of $L_g = 2$ mH. Now, the resonance frequency is reduced by around 21%, which corresponds to a value of $f_{res} = 965$ Hz. However, the control algorithm works as expected, and the oscillation also disappears.

Finally, in Figure 7.25c, the grid inductance is changed to $L_g = 5$ mH. This variation represents a reduction of 40% of the nominal resonance frequency, being now $f_{res} = 729$ Hz. Once again, satisfactory results are obtained.

We can conclude that the effectiveness of the control algorithm is practically independent of the grid inductance variations. This is a superior feature of this controller in comparison with other filter-based active damping techniques that need autotuning procedures, where if the grid inductance exceeds from a critical value the system may become unstable [11, 12]. Note that according to Figure 7.24d, even in a wide variation of the grid inductance, only small variations in the system poles are observed. This fact proves the high robustness of the control algorithm against important grid inductance variations without the implementation of tuning algorithms.

7.3.4.2 Test of Stability in Case of Grid Harmonics Near the Resonance Frequency

The system is also tested when a grid harmonic is near to the resonance frequency. In this case, the system could become unstable depending on which kind of controller is used. Figure 7.26 shows the current behavior when the grid voltage contains one harmonic whose amplitude is 30 V near the resonance frequency. As shown in the figure, the currents are slightly oscillating due to the harmonic, but the system is maintained stable.

7.3.4.3 Test of the VSI Against Sudden Changes in the Reference Current

Figure 7.27 shows the transient behavior against changes in the reference active power. The figure shows the active and reactive powers in case of reference step change when two different controllers are used. A step change in the reference of the active power from 750 to 1500 W is applied, whereas the reactive power is set to 0.

As it can be seen, when the control algorithm is used, the active and reactive powers have a fast transient dynamic. Besides, a small ripple is achieved.

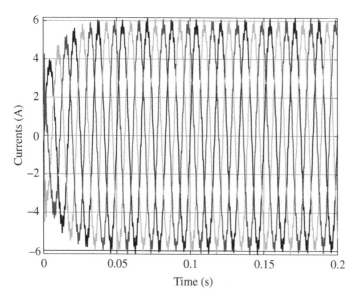

Figure 7.26 Three-phase currents with a grid harmonic near the resonance frequency.

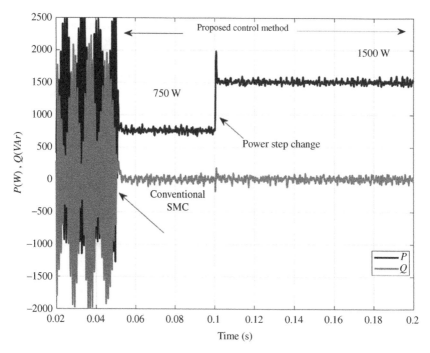

Figure 7.27 Active and reactive powers in case of a sudden step change.

As can be observed, the active and reactive powers do not oscillate with a good dynamic. However, when the conventional SMC is used, the powers are oscillating due to the lack of damping term.

7.3.4.4 Test of the VSI Under Distorted Grid

In Figure 7.28, a comparison between the conventional SMC and the presented control method is shown. The PCC voltages with a total harmonic distortion (THD) of 14% and the three-phase grid currents are shown in the case of a distorted grid. As shown in the figure, when the conventional SMC is used, the grid currents are distorted since the reference currents generation method uses distorted PCC voltages. In contrast, when the control method with a reduced model is used, the grid currents only have a slight distortion since the reference current is generated by using only the fundamental component of the PCC voltage (see (7.70)) obtained from the KF. Note that with this method, the use of extra filters for the grid voltages is not necessary.

7.3.4.5 Test of the VSI Under Voltage Sags

This controller can also operate in case of voltage sags. Figure 7.29 shows the VSI performance under a grid voltage sag characterized by a positive and negative sequence of $V^+ = 0.7$ p.u. and $V^- = 0.3$ p.u., respectively, and with a phase angle between sequences of $c\varphi = \pi/6$. In this case, the reference currents are obtained using the positive sequences of the PCC voltages as $\hat{i}^*_{a,b,c} = \dfrac{P^*}{|v^+|^2} \hat{v}^+_{a,b,c}$. The positives sequences can be obtained according to the expressions presented in [7]. Since the grid current tracks only the positive sequence of the PCC voltage, the current amplitude is balanced during the voltage sag. Note that the current amplitude is increased in order to maintain the desired active power due to the sag.

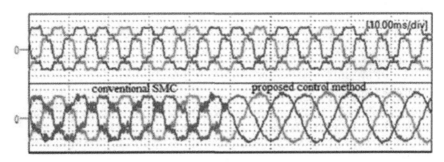

Figure 7.28 (Top) Distorted PCC voltages (50 V/div) with THD = 14%. (Bottom) Grid-side currents (2 A/div).

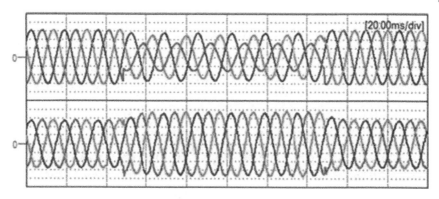

Figure 7.29 (Top) Distorted PCC voltages (50 V/div) with THD = 14%. (Bottom) Grid-side currents (2 A/div) under a voltage sag.

7.3.5 Computational Load and Performances of the Control Algorithm

Table 7.4 presents the execution time t_{ex} of the controller, computational load, memory usage in bytes, the maximum switching frequency, and the THD. The computational load can be computed according to

$$\text{Load}(\%) = f_s \times t_{ex} \times 100 \tag{7.90}$$

And the maximum switching frequency is considered $f_s/10$ with the objective of ensure enough samples for the correct operation of the hysteresis band comparator.

In order to obtain the execution time, one timer of the DSP is used to measure the time of the controller task. As can be seen, the total time used by the algorithm is noticeably smaller in this method. This fact allows us to increase the sampling frequency up to 60 kHz, which limits the maximum switching frequency to 6 kHz, as shown in Figure 7.30. Besides, a low THD of the grid current is achieved in both cases. However, it is slightly reduced when the algorithm with the reduced model is used. This algorithm could be implemented in other kinds of digital processors, such as field programmable gate arrays (FPGAs) or other floating-point multicore DSPs [13].

Table 7.4 Comparative analysis between models.

f_s	t_{ex}	Load (%)	Mem.	f_{sw}	THD
60 kHz	10 μs	60	8149	6 kHz	1.5%

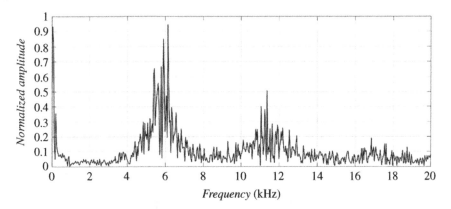

Figure 7.30 Switching spectrum using the reduced model.

7.4 Three-Phase AC–DC Rectifier

In this chapter, an improved variable hysteresis-band current-control in natural frame for a three-phase unity power factor rectifier (UPFR) will be explained [5]. The control algorithm uses a sliding model control (SMC) and is based on three decoupled sliding-mode controllers. These controllers will be combined with three independent Kalman filters. The use of Kalman filters instead of a standard-state observer improves the quality of the estimated signals in the presence of noise and allows a reduction of the number of sensors. This fact leads to an immunity increasing the control loop in noisy environments. To reduce drastically the computational load in the Kalman algorithm, a reduced bilinear model is derived, which allows to use a Kalman filter algorithm instead of an extended Kalman filter used in nonlinear systems. A fast output-voltage control is also presented, which avoids output voltage variations when a sudden change in the load or a voltage sag appears. Moreover, a fixed switching frequency algorithm is presented, which uses a variable hysteresis-band in combination with a switching decision algorithm, ensuring a switching spectrum concentrated around the desired switching frequency.

7.4.1 Nonlinear Model of the Unity Power Factor Rectifier

A circuit scheme of an UPFR is depicted in Figure 7.31. An equivalent circuit for each phase-leg i, with $i \in \{a, b, c\}$, is shown in Figure 7.32. From this circuit, the three-phase system equations can be written as

$$L\frac{d\mathbf{i}}{dt} = v - \frac{v_o}{2}\mathbf{u} + v_n\mathbf{1}^T \tag{7.91}$$

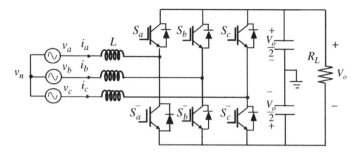

Figure 7.31 Circuit diagram of a three-phase unity power factor rectifier.

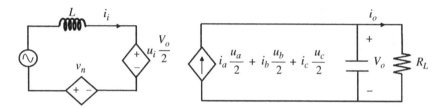

Figure 7.32 Equivalent per-phase circuit of the UPFR.

$$C\frac{dv_o}{dt} = i_a\frac{u_a}{2} + i_b\frac{u_b}{2} + i_c\frac{u_c}{2} - i_o \qquad (7.92)$$

$$v_n = \frac{v_o}{6}(u_a + u_b + u_c) \qquad (7.93)$$

where v_n is the neutral point voltage, $\mathbf{i} = \begin{bmatrix} i_a & i_b & i_c \end{bmatrix}^T$ is the inductor current vector, $\mathbf{v} = \begin{bmatrix} v_a & v_b & v_c \end{bmatrix}^T$ is the grid voltage vector, $\mathbf{u} = \begin{bmatrix} u_a & u_b & u_c \end{bmatrix}^T$ is the control variable vector with $u_{a,b,c} \in \{\pm 1\}$, v_o is the output voltage and $\mathbf{1}^T = \begin{bmatrix} 1 & 1 & 1 \end{bmatrix}^T$. The previous equations can be written as a nonlinear state-space model:

$$\frac{dx}{dt} = \mathbf{f}(x, u) + \mathbf{E}(v + v_n) \qquad (7.94)$$

$$y = \mathbf{h}(x, u) \qquad (7.95)$$

where

$$\mathbf{x} = \begin{bmatrix} i_a & i_b & i_c & v_o \end{bmatrix}^T \qquad (7.96)$$

$$
\mathbf{f} = \begin{pmatrix} 0 & 0 & 0 & -\dfrac{u_a}{2L} \\[2mm] 0 & 0 & 0 & -\dfrac{u_b}{2L} \\[2mm] 0 & 0 & 0 & -\dfrac{u_c}{2L} \\[2mm] \dfrac{u_a}{2C} & \dfrac{u_b}{2C} & \dfrac{u_c}{2C} & \dfrac{-1}{R_L} \end{pmatrix}
\tag{7.97}
$$

$$
\mathbf{E} = \frac{1}{L}[1 \quad 1 \quad 1 \quad 0]^T
\tag{7.98}
$$

And the output matrix is given by **h**, where

$$
\mathbf{H} = \frac{\partial \mathbf{h}}{\partial \mathbf{x}} = \begin{pmatrix} 1 & 0 & 0 & 0 \\ 0 & 1 & 0 & 0 \\ 0 & 0 & 1 & 0 \\ 0 & 0 & 0 & 1 \end{pmatrix}
\tag{7.99}
$$

7.4.2 Problem Formulation

In this section, the problem formulation is presented. Three different control objectives will be presented to design an appropriate SMC controller. The main control objective in a UPFR is to guarantee sinusoidal input currents in phase with the grid voltages (i.e. unity power factor). For purpose, a simple sliding surface can be used:

$$
\boldsymbol{\sigma} = \mathbf{i}^* - \mathbf{i}
\tag{7.100}
$$

where $\mathbf{i}^* = k\mathbf{v} = [i_a^* \quad i_b^* \quad i_c^*]$ is the reference current vector. Using (7.91)–(7.93) and (7.100), the dynamics for each sliding surface can be obtained as follows:

$$
\frac{d\sigma_a}{dt} = \frac{di_a^*}{dt} - \frac{1}{L}\left(v_a - \frac{V_o}{2}u_a + \frac{V_o}{2}\frac{u_a + u_b + u_c}{3}\right)
\tag{7.101}
$$

$$
\frac{d\sigma_b}{dt} = \frac{di_b^*}{dt} - \frac{1}{L}\left(v_b - \frac{V_o}{2}u_b + \frac{V_o}{2}\frac{u_a + u_b + u_c}{3}\right)
\tag{7.102}
$$

$$
\frac{d\sigma_c}{dt} = \frac{di_c^*}{dt} - \frac{1}{L}\left(v_c - \frac{V_o}{2}u_c + \frac{V_o}{2}\frac{u_a + u_b + u_c}{3}\right)
\tag{7.103}
$$

The aforementioned expressions exhibit a cross-coupling term introduced by the neutral point voltage (7.93). This term may produce changes on the sliding surfaces slope, depending on the discrete values of the control variables u_a, u_b and

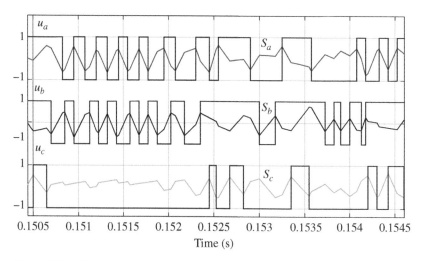

Figure 7.33 Simulation result of cross-coupled sliding mode controllers.

u_c. This phenomenon is depicted in Figure 7.33 where it can be seen that the slope of each sliding surface is affected by the three control variables. The procedure to achieve decoupled controllers is based on the utilization of three independent Kalman filters (KF), which use a model of the converter where the neutral point voltage term has been removed.

The second objective is to achieve a fixed switching frequency. It is well-known that the SMC technique operates a variable switching frequency, which is not desirable in most industry applications. Then, the control design can be modified to solve this problem. Finally, the third and last objective is to make the system robust against load changes and voltage sags. This will be achieved by an output-current feedforward control, which also linearizes the output voltage dynamics.

7.4.3 Axis-Decoupling Based on an Estimator

In this section, a linear simplified model of the UPFR is deduced to achieve a perfect axis-decoupling. This simplified model will be used in a KF estimator. The model is based on the axis-decoupling approach which has the first objective to remove the dependence of the neutral point voltage from the control dynamics (7.101)–(7.103). Then, the sliding surfaces σ_a, σ_b, and σ_c will only depend on its control variable u_a, u_b, and u_c, respectively, and whose dynamics are

$$\frac{d\sigma_a}{dt} = \frac{di_a^*}{dt} - \frac{1}{L}\left(v_a - \frac{V_o}{2}u_a\right) \tag{7.104}$$

$$\frac{d\sigma_b}{dt} = \frac{di_b^*}{dt} - \frac{1}{L}\left(v_b - \frac{V_o}{2}u_b\right) \tag{7.105}$$

$$\frac{d\sigma_c}{dt} = \frac{di_c^*}{dt} - \frac{1}{L}\left(v_c - \frac{V_o}{2}u_c\right) \tag{7.106}$$

Figure 7.34 shows the sliding surfaces and the three control actions. As shown in the figure, the switching surfaces are perfectly synchronized with their corresponding control variables. In order to derive the converter model, some assumptions can be considered:

1) The value of $v_n = 0$ is taken in the converter model that will be used in the KF. With this consideration the cross-coupling among controllers is removed.
2) The output capacitor is usually large, and the output voltage has a slow dynamic. Hence, the state variable v_o can be assumed constant between different sampling instants of the same switching period. Besides, since this variable is measured, it can be considered as a parameter and its dynamics can be neglected. With this consideration, the space-state model for each phase leg i could be considered linear in a switching period [14].

The second objective is to reduce the number of sensors. With this objective in mind, the state-space model be incremented with two additional states, the grid voltage v_i and its quadrature v_{iq} [15]. These estimated voltages will be used in the sliding surfaces and in the hysteresis band expressions as will be shown in the next sections.

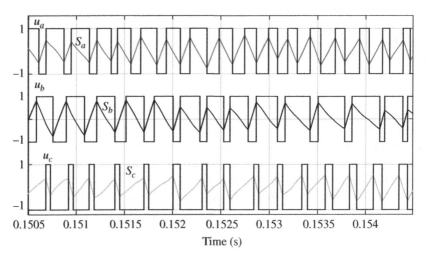

Figure 7.34 Simulation result of decoupled sliding mode controllers.

The new differential equations for each phase-leg i can be obtained from (7.91) to (7.93) but considering the aforementioned considerations. Then, one has

$$\frac{di_i}{dt} = \frac{1}{L}v_i - \frac{V_o}{2L}u_i \tag{7.107}$$

$$\frac{dv_i}{dt} = \omega_o v_{iq} \tag{7.108}$$

$$\frac{dv_{iq}}{dt} = -\omega_o v_i \tag{7.109}$$

where ω_o is the grid angular frequency.

With Eqs. (7.107)–(7.109), the new state-space model can be expressed:

$$\frac{d\mathbf{x}_i}{dt} = \mathbf{A}\mathbf{x}_i + \mathbf{B}V_o\mathbf{u}_i \tag{7.110}$$

$$y_i = \mathbf{C}\mathbf{x}_i \tag{7.111}$$

where $x_i = [i_i \quad v_i \quad v_{iq}]^T$, $A = \begin{pmatrix} 0 & 1/L & 0 \\ 0 & 0 & \omega_o \\ 0 & -\omega_o & 0 \end{pmatrix}$ and $B = [-1/2L \quad 0 \quad 0]^T$.

The control algorithm will not use grid voltage sensors. Hence, the value of vector $\mathbf{C}$ can be defined as $\mathbf{C} = [1 \quad 0 \quad 0]$. According to matrix $\mathbf{C}$ only the current i_i for each phase-leg i is measured. Then, it is necessary to prove if the system is observable and controllable using only the measured variable i_i. The observability matrix is defined by

$$\mathbf{O} = [\mathbf{C} \quad \mathbf{CA} \quad \mathbf{CA}^2]^T \tag{7.112}$$

and the controllability matrix is

$$\mathbf{\Gamma} = [\mathbf{B} \quad \mathbf{AB} \quad \mathbf{A}^2\mathbf{B}] \tag{7.113}$$

Matrices $\mathbf{O}$ and $\mathbf{\Gamma}$ are both of full rank, (i.e. rank$\{\mathbf{O}\}$ = rank $\{\mathbf{\Gamma}\}$ = 3), so that the system is controllable and can be observed using only the measured current i_i.

7.4.4 Control System

The control diagram for phase-leg a is depicted in Figure 7.35, which consists of four different blocks. The output voltage control computes the value of the gain k. This gain and the estimated variables obtained from the KF block are used to obtain the sliding mode surfaces. The SMC block uses a hysteresis band generator and a switching decision algorithm to improve the switching frequency

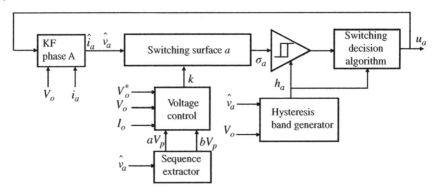

Figure 7.35 Control system for phase-leg a.

spectrum. In the next sections, these four blocks of the control system will be explained in detail.

7.4.4.1 Kalman Filter

Since the system is clearly nonlinear, an extended KF (EKF) should be used, with the drawback of the high computational burden needed by the algorithm. This computational burden can be drastically reduced if the model defined in (7.110)–(7.111) is used for each phase-leg. As explained before, this model can be considered linear in a switching period, and a KF can be used instead of an EKF, leading to an important reduction of the computational load. The main features provided by the KF can be summarized below:

1) The three controllers can be decoupled since the model defined in (7.110)–(7.111) has no dependence on the neutral point voltage. The KF will use this model to estimate the voltages and currents.
2) All the variables defined in the control algorithm are estimated and free of noise; the Kalman estimator produces a filtered version of the inductor currents providing an improved sliding motion.
3) The grid voltages can be estimated, so the number of measurement sensors can be reduced.

For the digital implementation of the KF, the state-space model must be discretized. The discrete model and measurement equations are given by the following expressions:

$$x_i(k+1) = \mathbf{A}_d x_i(k) + \mathbf{B}_d V_o(k) u_i(k) + \mathbf{\eta}_i(k) \tag{7.114}$$

$$y_i(k) = \mathbf{C} x_i(k) + \mathbf{w}_i(k) \tag{7.115}$$

where

$$\mathbf{A}_d = e^{\mathbf{A}T_s} \cong \mathbf{I} + \mathbf{A}T_s = \begin{pmatrix} 1 & T_s/L & 0 \\ 0 & 1 & \omega_o T_s \\ 0 & -\omega_o T & 1 \end{pmatrix} \tag{7.116}$$

$$\mathbf{B}_d = \int_0^{T_s} \mathbf{B}e^{\mathbf{A}\lambda} d\lambda \cong \mathbf{B}T_s = \begin{bmatrix} -\dfrac{T_s}{2L} & 0 & 0 \end{bmatrix}^T \tag{7.117}$$

being $\mathbf{I}$ the identity matrix and T_s the sampling time. Here, it is assumed that $V_o(k)$ is constant between samples of the same switching period, it means that $V_o(k+1) \cong V_o(k)$, and as a consequence, the discrete model can be considered linear over the same switching period.

The process and measurement of noise vectors are $\boldsymbol{\eta}_i(k)$ and $\mathbf{w}_i(k)$, respectively, with covariance matrices given by:

$$\mathbf{Q}_i(k) = E\{\boldsymbol{\eta}_i(k)\boldsymbol{\eta}_i^T(k)\} \tag{7.118}$$

$$\mathbf{R}_i(k) = E\{\mathbf{w}_i(k)\mathbf{w}_i^T(k)\} \tag{7.119}$$

The recursive Kalman algorithm computation is divided into two parts: (1) time updating and (2) measurement updating. The following equations show the recursive steps:

1) The time updating predicts the state ahead and the error covariance ahead using (7.120) and (7.121), respectively:

$$\hat{\mathbf{x}}_i^-(k) = \mathbf{A}_d\hat{\mathbf{x}}_i^-(k-1) + \mathbf{B}_d V_o(k-1)u_i(k-1) \tag{7.120}$$

$$\mathbf{P}_i^-(k) = \mathbf{A}_d\mathbf{P}_i^-(k-1)\mathbf{A}_d^T + \mathbf{Q}_i(k-1) \tag{7.121}$$

2) In the measurement update step, an update for the predicted values is made using the Kalman gain, which is computed using (7.122):

$$\mathbf{L}_i(k) = \mathbf{P}_i^-(k)\mathbf{C}^T\left(\mathbf{C}\mathbf{P}_i^-(k)\mathbf{C}^T + \mathbf{R}_i(k)\right)^{-1} \tag{7.122}$$

The gain is calculated recursively to minimize the mean square error between the measured values and the predicted ones for the system states. Once the gain is computed, the state estimation and the error covariance can be updated using (7.123) and (7.124), respectively:

$$\hat{\mathbf{x}}_i(k) = \hat{\mathbf{x}}_i^-(k) + \mathbf{L}_i(k)\left(i_i(k) - \hat{i}_i^-(k)\right) \tag{7.123}$$

$$\mathbf{P}_i(k) = (\mathbf{I} - \mathbf{L}_i(k)\mathbf{C})\mathbf{P}_i^-(k) \tag{7.124}$$

Then, from (7.123), the estimated vector $\hat{\mathbf{x}}_i(k) = \begin{bmatrix} \hat{i}_i & \hat{v}_i & \hat{v}_{iq} \end{bmatrix}$.

7.4.4.2 Practical Considerations: Election of Q and R Matrices

For the correct implementation of the Kalman filter algorithm, the matrices $\mathbf{R}_i$ and $\mathbf{Q}_i$ must be computed. The matrix $\mathbf{R}_i$, dimension of which is equal to the dimension of the product $\mathbf{CP}_i^-(k)\mathbf{C}^T$, can be estimated from measures. According to the value of $\mathbf{C}$, the size of matrix $\mathbf{A}_d$ and with the consideration that $\mathbf{P}_i^-(k)$ should be a square matrix, $\mathbf{R}_i$ is reduced to a scalar. Hence, $R_i(k)$ can be estimated using the unbiased mean power estimator:

$$R_i(k) = \frac{1}{NT_s} \sum_{n=0}^{N-1} w_i(n) \tag{7.125}$$

where $w_i(k)$ is an additive white Gaussian noise at phase-leg i and N is the number of samples used to perform the computation. To obtain the values of the noise, a reference DC-voltage $V_{ref}(k)$ used in the sensing system is employed. This voltage is connected to one analog input of the digital signal processor (DSP) and is contaminated with the system noise when the rectifier is switching. The noise samples can be obtained by subtracting from $V_{ref}(k)$, the exact value V_{ref}^* (i.e. the value of the reference voltage when the rectifier is not switching), yielding

$$R_i(k) = \frac{1}{NT_s} \sum_{n=0}^{N-1} \left| V_{ref}(n) - V_{ref}^* \right| \tag{7.126}$$

Matrix $\mathbf{Q}_i$ is more difficult to estimate and usually a tuning method is used. In this application, $\mathbf{Q}_i$ is a diagonal matrix of dimension 3 and its value has been obtained by means of simulations. In the simulation results, an approximate value of $\mathbf{Q}_i = 0.005\mathbf{I}_3$ has been obtained, where $\mathbf{I}_3$ is the three-dimensional identity matrix. All the conditions in the Kalman filter algorithm are initialized to zero except for the covariance matrix, which is initialized to the identity matrix.

7.4.4.3 Practical Considerations: Computational Burden Reduction

A high computational load is usually a drawback in an EKF algorithm implementation. The first important reduction was achieved when a KF with a reduced model have been used instead of the EKF with the nonlinear model. The possibility to reduce even more the time used in the computation will be an important improvement in the controller design. Especially, (7.122) contains a matrix inversion, which needs an important computational load. Using the simplified model (7.110) instead of (7.94), only one measurement per phase can be used, and the matrix inversion of (7.122) is reduced to a scalar inversion. Moreover, the noise $R_i(k)$ can be considered so similar in each phase (i.e. $\mathbf{L}_a(k) = \mathbf{L}_b(k) = \mathbf{L}_c(k)$), and the Kalman gain (7.122) is computed only for one phase. With these considerations the total time employed for the algorithm is around 4.9 μs, which makes the algorithm feasible.

7.4.5 Sliding Mode Control

The conventional UPFR control scheme consist of a fast inner input current loop, which ensures sinusoidal input currents in phase with the line voltages, $\mathbf{i} = k\mathbf{v}$, and a slow outer loop usually a PI controller, main task of which is to regulate the output voltage, modifying the input current amplitudes. In this section, a control design methodology is presented to guarantee unity power factor and provides a regulated output DC-voltage with fast dynamics response against sudden changes in the load. Besides, the amplitude control k is conceived in order to linearize the output voltage dynamics.

7.4.5.1 Inner Control Loop

The current controllers can be derived using the following switching surfaces:

$$\sigma_a = k\hat{v}_a - \hat{i}_a \tag{7.127}$$

$$\sigma_b = k\hat{v}_b - \hat{i}_b \tag{7.128}$$

$$\sigma_c = k\hat{v}_c - \hat{i}_c \tag{7.129}$$

where $\hat{\mathbf{v}} = [\hat{v}_a \quad \hat{v}_b \quad \hat{v}_c]^T$ and $\hat{\mathbf{i}} = [\hat{i}_a \quad \hat{i}_b \quad \hat{i}_c]$ are the estimated grid voltages and inductor currents by the KF algorithm shown in the previous section.

In the ideal case of sliding motion, that is infinite switching frequency, the average value of the control signal vector $\mathbf{u}$ is known as equivalent control $\mathbf{u}_{eq}$. The equivalent control is deduced by imposing the sliding regime condition $\dot{\mathbf{S}} = 0$, where $\dot{\mathbf{S}} = [S_a \quad S_b \quad S_c]$. Using (7.107) and (7.127)–(7.129) the expression of the equivalent control for each phase leg i is expressed as

$$u_{eqi} = \frac{2}{V_o}\left(\hat{v}_i - kL\frac{d\hat{v}_i}{dt} - L\hat{v}_i\frac{dk}{dt}\right) \tag{7.130}$$

The main requirement in the design of SMC is to satisfy the reaching conditions and guarantee the existence of a sliding regime in the switching surfaces $\boldsymbol{\sigma} = 0$. The reaching conditions for each phase-leg i are given by

$$\sigma_i\dot{\sigma}_i < 0 \tag{7.131}$$

Now, defining the control action

$$u_i = \begin{cases} u_i^+ & \text{if} \quad \sigma_i > 0 \\ u_i^- & \text{if} \quad \sigma_i < 0 \end{cases} \tag{7.132}$$

where $u_i \in \{-1, 1\}$ and considering (7.131) one has:
If $\sigma_i > 0 \rightarrow \dot{\sigma}_i < 0$

$$\frac{d(k\hat{v}_i)}{dt} - \frac{1}{L}\hat{v}_i + \frac{V_o}{2L}u_i^+ < 0 \tag{7.133}$$

If $\sigma_i < 0 \rightarrow \dot{\sigma}_i > 0$

$$\frac{d(k\hat{v}_i)}{dt} - \frac{1}{L}\hat{v}_i + \frac{V_o}{2L}u_i^- > 0 \tag{7.134}$$

From (7.133) and (7.134), the following expression can be deduced for each phase-leg i

$$\frac{v_0}{2L}\left(u_i^+ - u_i^-\right) < 0 \tag{7.135}$$

which allows to determine the switching action

$$u_i = \begin{cases} u_i^+ = -1 & \text{if} \quad \sigma_i > 0 \\ u_i^- = 1 & \text{if} \quad \sigma_i < 0 \end{cases} \tag{7.136}$$

7.4.5.2 Outer Control Loop

The outer control loop is designed to exhibit a slow dynamics behavior to avoid input currents distortion, but with a fast-dynamic response in case of a sudden load change. To achieve it, an appropriate control of the input current amplitude k is presented. This control has two parts, the first one is a slow proportional-integral (PI) controller, which is responsible to regulate the output voltage, and the second one is an output-current feedforward control that ensures a very fast response when a step change in the load is produced. The amplitude control k is also conceived in order to linearize in a large signal sense of the output voltage dynamics. Next, the control design procedure is described.

The zero dynamics concept [16] is used to design the outer loop. The zero dynamics analyzes the dynamics of v_o if the control objective $\mathbf{i} = k\mathbf{v}$ is achieved. The output-voltage differential equation in sliding regime can be found using (7.130) in (7.92):

$$C\frac{dV_o}{dt} = \frac{k}{V_o}(\hat{v}_a + \hat{v}_b + \hat{v}_c) - \frac{kL}{V_o}\frac{dk}{dt}(\hat{v}_a + \hat{v}_b + \hat{v}_c) - \frac{k^2L}{V_o}\left(v_a\frac{dv_a}{dt} + v_b\frac{dv_b}{dt} + v_c\frac{dv_c}{dt}\right) - I_o \tag{7.137}$$

where the grid voltages are sinusoidal waveforms defined by:

$$\hat{v}_a(t) = V_p \cos(\omega_o t) \tag{7.138}$$

$$\hat{v}_b(t) = V_p \cos\left(\omega_o t - \frac{2\pi}{3}\right) \tag{7.139}$$

$$\hat{v}_c(t) = V_p \cos\left(\omega_o t - \frac{4\pi}{3}\right) \tag{7.140}$$

with V_p the peak value. Using (7.138)–(7.140) in (7.137), the following equation is obtained:

$$C\frac{dV_o}{dt} = \frac{3V_p^2 k}{2V_o}\left(1 - L\frac{dk}{dt}\right) - I_o \tag{7.141}$$

As shown in [17, 18], the term dk/dt can be neglected since the energy stored in the inductor is smaller than the energy stored in the capacitor. A detailed demonstration can be found in [17]. Then, (7.141) can be approximated to

$$C\frac{dV_o}{dt} \cong \frac{3V_p^2 k}{2V_o} - I_o \tag{7.142}$$

The output-voltage differential Eq. (7.142) is clearly nonlinear and the output-voltage dynamics will depend on a proper design of k. The output voltage dynamics can be linearized if the following nonlinear control of the gain k is used:

$$k = \frac{2V_o}{3V_p^2}\left(k_i\int_{-\infty}^{t}\left(V_o^* - V_o\right)d\tau + k_p\left(V_o^* - V_o\right) + k_o I_o\right) \tag{7.143}$$

where k_p and k_i are the proportional and integral gains, respectively, and k_o can take the values 0 or 1. Now, by replacing (7.143) in (7.142) yields:

$$C\frac{dV_o}{dt} = \left(k_i\int_{-\infty}^{t}\left(V_o^* - V_o\right)d\tau + k_p\left(V_o^* - V_o\right) - (1 - k_o)I_o\right) \tag{7.144}$$

The dynamics of V_o can be obtained by taking the time derivative of (7.144), which in case of a resistive load takes the following form:

$$C\frac{d^2V_o}{dt^2} + \left(k_p + \frac{1 - k_o}{R_L}\right)\frac{dV_o}{dt} + k_i V_o = k_i V_o^* \tag{7.145}$$

The aforementioned equation is a linear differential equation, dependence of which on the load value can be removed by taking the value $k_o = 1$. In this case, the current feedforward term is introduced, and the system will be robust against sudden load variations. In other cases, if only a PI controller is used (i.e. $k_o = 0$), the output voltage dynamics will be affected by the output current variations producing an output voltage drop. Stability can be assured if k_p and k_i are positive values.

In case of asymmetrical faults in the grid voltages, the output voltage dynamics is also affected. The grid voltages defined in (7.138)–(7.140) are modified due to a negative sequence component. In this case, the grid voltages are redefined as follows:

$$\hat{v}_a = aV_p\cos(\omega_o t) + bV_p\cos(\omega_o t + \phi) \tag{7.146}$$

$$\hat{v}_b = aV_p \cos\left(\omega_o t - \frac{2\pi}{3}\right) + bV_p \cos\left(\omega_o t + \frac{2\pi}{3} + \phi\right) \tag{7.147}$$

$$\hat{v}_c = aV_p \cos\left(\omega_o t - \frac{4\pi}{3}\right) + bV_p \cos\left(\omega_o t + \frac{4\pi}{3} + \phi\right) \tag{7.148}$$

The grid voltages are represented by a positive and negative sequence, where ϕ is the angle between sequences and a, b specify the degree of imbalance. For instance, a balanced three-phase voltage is a special case of (7.146)–(7.148), where $a = 1$ and $b = 0$.

Using (7.146)–(7.148) and (7.137), the following expression for the output voltage dynamics in the presence of a voltage sag can be found:

$$C\frac{dv_o}{dt} = \frac{3k}{2V_o}\left((aV_p)^2 + (bV_p)^2 + 2(aV_p)(bV_p)\cos(2\omega_o t + \phi)\right)$$
$$+ \frac{3kL\omega_o}{v_o}(aV_p)(bV_p)\sin(2\omega_o t + \phi) - I_o \tag{7.149}$$

Equation (7.149) shows that ripple component of $2\omega_o$ appears in the output voltage when an asymmetrical fault occurs. If the aforementioned expression is averaged in a half grid period, the following expression can be deduced:

$$C\frac{d\langle V_o\rangle}{dt} = \frac{3k}{2\langle V_o\rangle}\left((aV_p)^2 + (bV_p)^2\right) - \langle I_o\rangle \tag{7.150}$$

Where the symbol $\langle\rangle$ means averaged value. It is worth mentioning that the above expression matches to (7.142) if $a = 1$ and $b = 0$ are selected. Like what was explained before, Eq. (7.150) can be linearized if the expression for k, defined in (7.143), is modified as follows:

$$k = \frac{2V_o}{3\left((aV_p)^2 + (bV_p)^2\right)}\left(k_i\int_{-\infty}^{t}(V_o^* - V_o)d\tau + k_p(V_o^* - V_o) + k_o I_o\right) \tag{7.151}$$

Note that the positive and negative sequences (values for a and b) can be obtained using a sequence detector [19, 20].

7.4.6 Hysteresis Band Generator with Switching Decision Algorithm

It is well known that the SMC is characterized by a variable switching frequency, which is usually not desired in most industry applications. The aim of this section is to design a fully digital hysteresis modulator that fixes the switching frequency by modifying the hysteresis band. In analog hysteretic control, the expression of the hysteresis band is presented in [21]. Then, for each phase-leg i, the expression for the hysteresis band can be defined as follows:

$$h_i = \frac{V_o}{8Lf_{sw}} \left(1 - \left(\frac{2\hat{v}_i}{V_o} \right)^2 \right) \qquad (7.152)$$

where f_{sw} is the switching frequency. While the sliding surface is moving inside the hysteresis bands $\pm h_i$, the switching frequency f_{sw} will remain constant.

If a digital control is used, an error in the switching frequency is produced due to the sampling process. Figure 7.36 shows this phenomenon, where some samples can be out of the hysteresis bands and the switching frequency obtained will be lower than the desired one. To eliminate this error and concentrate the switching frequency spectrum around the desired frequency, a switching decision algorithm (SDA) has been implemented.

Assuming that the hysteresis band has a slow variation in a sampling period, $h_i(t_k) \cong h_i(t_{k+1})$, the switching surfaces with the corresponding hysteresis bands can be represented as shown in Figure 7.36.

When a sample is acquired, the time between the sample and the hysteresis band is calculated according to the SDA algorithm. The algorithm improves the switching spectrum by selecting an appropriate instant of time for the switching. According to Figure 7.36, there are two different slopes, an increasing slope and a decreasing slope. According to the switching surfaces defined in (7.97)–(7.99), the slope m of the switching surface for each phase-leg i, σ_i, is given by:

$$m = -\frac{1}{L} \left(\hat{v}_i - \frac{V_o}{2} u_i \right). \qquad (7.153)$$

Two different cases must be analyzed:

1) The sample is on the increasing slope of the switching surface, m_1. In this case, according to (7.136), $\dot{\sigma} > 0 \rightarrow \sigma_i < 0 \rightarrow u_i = 1$, and m_1 is expressed as:

$$m_1 = \frac{1}{L} \left(\frac{V_o}{2} - \hat{v}_i \right) \qquad (7.154)$$

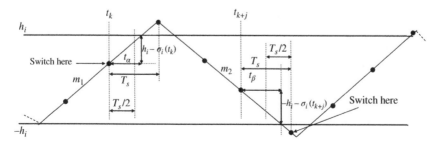

Figure 7.36 Hysteresis band and switching surface.

According to Figure 7.36, the time t_α can be computed as

$$t_\alpha = \frac{h_i - \sigma_i(t_k)}{m_1} = L\frac{h_i - \sigma_i(t_k)}{V_o/2 - \hat{v}_i} \tag{7.155}$$

2) The sample is on the decreasing slope of the switching surface, m_2. Once again, according to (7.86), $\dot{\sigma} < 0 \rightarrow \sigma_i > 0 \rightarrow u_i = -1$, and m_2 is expressed as

$$m_2 = -\frac{1}{L}\left(\hat{v}_i + \frac{V_o}{2}\right) \tag{7.156}$$

and according to Figure 7.36, the time t_α can be computed as

$$t_\beta = \frac{-h_i - \sigma_i(t_{k+j})}{m_2} = L\frac{h_i + \sigma_i(t_k)}{\hat{v}_i + V_o/2} \tag{7.157}$$

When the estimated times, t_α and t_β are known, the SDA operates as follows:

i) In the increasing slope, if $t_\alpha < \dfrac{T_s}{2}$, the switching is done, otherwise a new sample is taken and the time t_α is recalculated. This procedure is finished when the condition $t_\alpha < \dfrac{T_s}{2}$ is accomplished.

ii) In the decreasing slope, if $t_\beta < \dfrac{T_s}{2}$, the switching is done, otherwise a new sample is taken and the time t_β is recalculated. This procedure is finished when the condition $t_\beta < \dfrac{T_s}{2}$ is accomplished.

This procedure is done recursively for each sampling period.

A MATLAB code to implement the switching decision algorithm can be written as follows:

```
taa=L*(ha-Sa)/(0.5*Vo-va);
tba=L*(ha+Sa)/(va+0.5*Vo);

tab=L*(hb-Sb)/(0.5* Vo-vb);
tbb=L*(hb+Sb)/(vb+0.5*Vo);

tac=L*(hc-Sc)/(0.5* Vo-vc);
tbc=L*(hc+Sc)/(vc+0.5*Vo);

if ((ua==1) && (taa>0)&& (taa<0.5*Ts))
            ua=-1;uan=1;
    elseif ((ua==-1) && (tba>0) && (tba<0.5*Ts))
            ua=1; uan=-1;
        end
```

```
if ((ub==1) && (tab>0)&& (tab<0.5*Ts))
            ub=-1;ubn=1;
   elseif ((ub==-1) && (tbb>0)&& (tbb<0.5*Ts))
            ub=1; ubn=-1;
      end
   if ((uc==1) && (tac>0)&& (tac<0.5*Ts))
            uc=-1;ucn=1;
   elseif ((uc==-1) && (tbc>0)&& (tbc<0.5*Ts))
            uc=1; ucn=-1;
      end
```

The above MATLAB code computes the times to reach the switching surface on both increasing and decreasing slopes. Then, the computed time is compared to $T_s/2$, and in case the computed time is less than $T_s/2$ the switching is done, otherwise the control signal is maintained to the last value.

7.4.7 Experimental Study

An experimental three-phase rectifier prototype was built using a 4.5-kVA SEMI-KRON full-bridge as the power converter and a TMS320F28M36 floating-point DSP as the control platform with a sampling frequency of 40 kHz. The grid voltages have been generated using a PACIFIC 360-AMX source. The system parameters are listed in Table 7.5. The control scheme was shown in Figure 7.35.

Figure 7.37 shows the control signals u_a, u_b, u_c, and their corresponding switching surfaces, σ_a, σ_b, and σ_c in the case of decoupled controllers. As can be seen the

Table 7.5 System parameters.

Symbol	Description	Value
L	Filter input inductance	5 mH
C	Output capacitor	340 μF
V_O	DC-link voltage	250 V
f_s	Sampling frequency	40 kHz
f_{sw}	Switching frequency	4 kHz
f_{grid}	Grid frequency	60 Hz
V_p	Grid peak voltage	50 V_{rms}
k_p	Proportional gain	0.03
k_i	Integral gain	2
R_L	Load	135 Ω

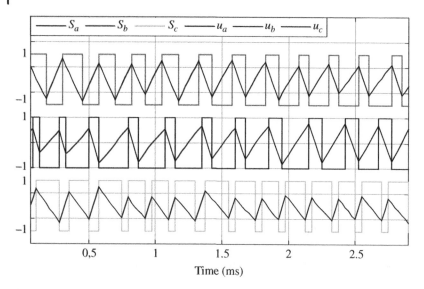

Figure 7.37 Experimental control signals and their corresponding sliding surfaces.

result obtained is like the simulated one in Figure 7.34. Therefore, we can conclude that the presented controller provides perfect axis-decoupling with a fixed switching frequency, as expected.

Figure 7.38 shows the inductor current from phase-leg a and the grid voltage v_a. Both signals are in phase, providing a unity power factor.

With the aim of analyzing the effect of the feedforward term in the output voltage control, two cases are presented:

1) In Figure 7.39, the output voltage, the three-phase grid currents, and the value of the gain k are depicted. In this case, the feedforward term is not used ($k_o = 0$). Figure 7.39a shows the dynamics of the output voltage (7.145) and the three-phase currents. An important voltage drop appears in the output voltage when a sudden load change from 460 to 920 W is produced. Figure 7.39b shows the dynamics of the gain k, where the settling time is around 0.008 seconds due to the slow dynamics of the PI controller.

2) The same waveforms are shown in Figure 7.40, but in this case the feedforward term is used ($k_o = 1$). The dynamics of v_o is now independent of the load change as it is shown in Figure 7.40a. In this case, the settling time of the gain k is drastically reduced, making the voltage drop inappreciable.

Figure 7.41a shows the output voltage and the three-phase grid currents during an asymmetrical grid fault. The control of the gain k is done by means of the

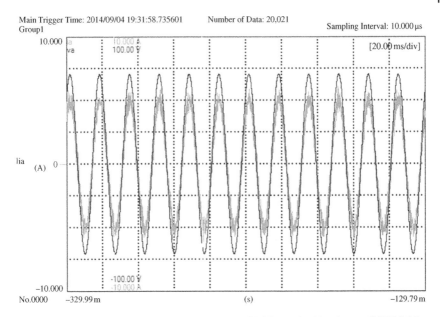

Figure 7.38 Experimental three-phase currents (5 A/div) and grid voltages (20 V/div) for phase-leg *a*.

expression (7.143), and in this case a voltage drop appears in the output voltage, similar as in Figure 7.39a. Note that a ripple frequency component of $2\omega_o$ appears in the output voltage, as expected.

In Figure 7.41b, the output voltage and the three-phase currents are shown. As can be seen, the current tracks the asymmetrical grid voltages during the sag.

The same waveforms are shown in Figure 7.42 but in this case using the control expressed in (7.151) for the gain k. Figures 7.42a and b show that the voltage drop is insignificant. As expected the ripple of $2\omega_o$ is maintained. The hysteresis band with its switching surface of phase-leg *a* is shown in Figure 7.43. If a hysteresis band without the SDA algorithm is used, an error on the desired switching frequency appears. In Figure 7.43a, this problem can be clearly seen, where several samples are out of the hysteresis limits provoking an error in the desired switching frequency. The problem is solved using the SDA algorithm. Figure 7.43b shows how the number of samples that are out of bounds of the hysteresis limits have been reduced using this algorithm, and the error is solved as it will be seen in the next results. Finally, the spectrum of the switching frequency is shown in the following figures. In Figure 7.44a, the spectrum of the control signal is depicted without the use of estimated currents. In this case, the switching frequency is

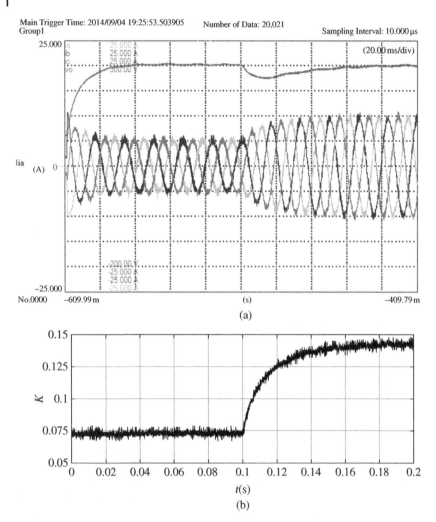

Main Trigger Time: 2014/09/04 19:25:53.503905   Number of Data: 20,021
Group1   Sampling Interval: 10.000 μs

Figure 7.39 Experimental results: (a) output voltage (50 V/div) and three-phase inductor currents (5 A/div) without feedforward term when a load step change is produced from 460 to 960 W, (b) value of the gain.

concentrated between 8 and 10 kHz. Figure 7.44b shows the spectrum using estimated currents but without the hysteresis modulator. As can be seen the switching frequency is increased, starting at 11–20 kHz. The use of the KF increases the switching speed but with a spread spectrum. Figure 7.44c shows the switching spectrum with hysteresis bands but without the SDA algorithm. The spectrum

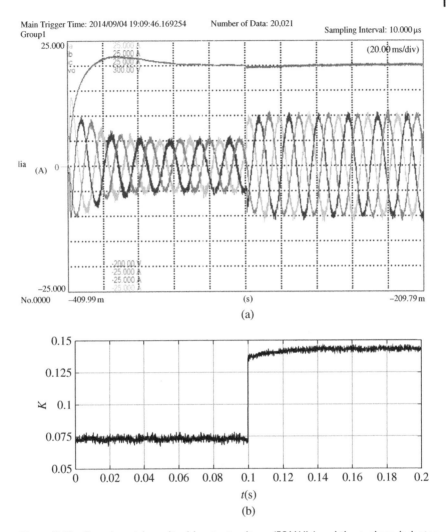

Figure 7.40 Experimental results: (a) output voltage (50 V/div) and three-phase inductor currents (5 A/div) with feedforward term when a load step change is produced from 460 to 960 W, (b) value of the gain k.

is concentrated around a fixed switching frequency of 3 kHz but with an error, since the desired switching frequency is 4 kHz. This problem is overcome using the SDA algorithm, and the result is depicted on Figure 7.44d, where the switching frequency is now around 4.

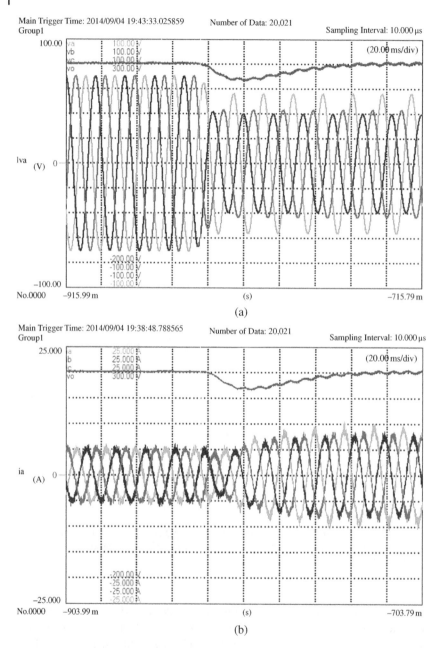

Figure 7.41 (a) Output voltage (50 V/div) and three-phase grid voltages (20 V/div) with asymmetrical fault (a = 0.65, b = 0.5) using (7.143) (b) Output voltage and three-phase currents (5A/div).

Main Trigger Time: 2014/09/05 23:39:54.171292   Number of Data: 20,021   Sampling Interval: 10.000 μs
Group1

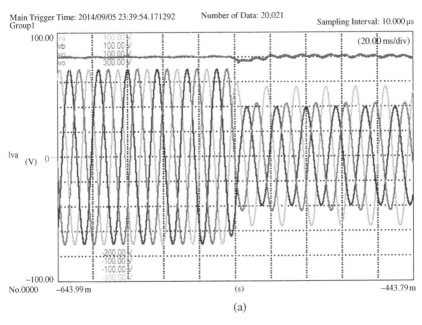

(a)

Main Trigger Time: 2014/09/05 23:33:08.934141   Number of Data: 20,021   Sampling Interval: 10.000 μs
Group1

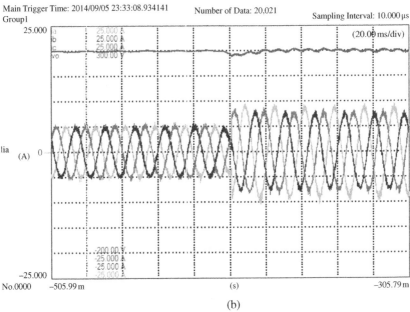

(b)

Figure 7.42 (a) Output voltage (50 V/div) and three-phase grid voltages (20 V/div) with asymmetrical fault ($a = 0.65$, $b = 0.5$) using (7.151) (b) Output voltage and three-phase currents (5 A/div).

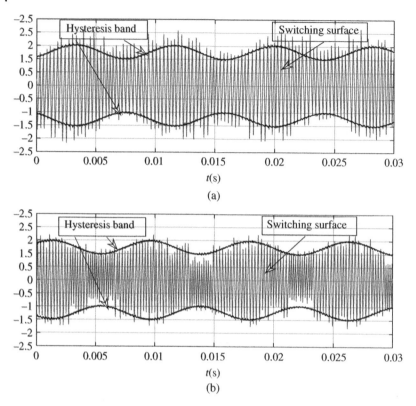

Figure 7.43 Experimental switching surface with its hysteresis band for phase-leg a (a) without SDA, (b) with SDA.

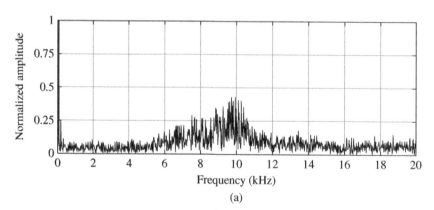

Figure 7.44 Experimental switching frequency spectrum with amplitude relative to fundamental (a) without estimated signals (b) with estimated signals but without hysteresis modulator, (c) with estimated signals and hysteresis modulator and (d) with estimated signals, hysteresis modulator and SDA algorithm.

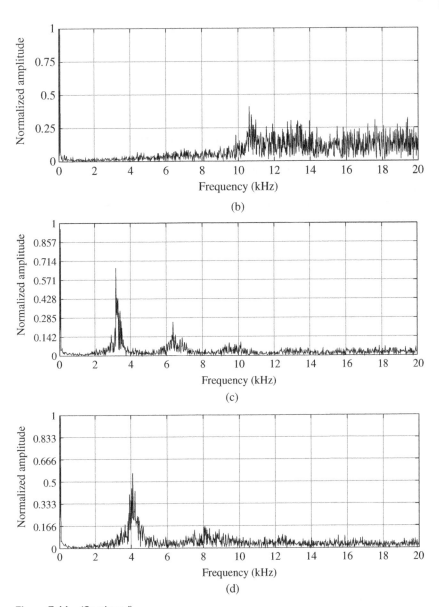

Figure 7.44 (Continued)

7.5 Three-Phase Transformerless Dynamic Voltage Restorer

7.5.1 Mathematical Modeling of Transformerless Dynamic Voltage Restorer

Figure 7.45 shows the block diagram of a three-phase transformerless dynamic voltage restorer (TDVR) introduced in [22]. Each phase of TDVR consists of a half-bridge voltage source inverter (VSI), a coupling inductor L_f, and a series-connected capacitor C_{se}. The output terminals of the half-bridge VSI are connected across C_{se} so as to inject a compensation voltage ($v_{se, k}$) when needed. The operation of the system can be described by the following equations:

$$L_f \frac{di_{fk}}{dt} = u_k V_{dc} + v_{se,k} \tag{7.158}$$

$$C_{se} \frac{dv_{se,k}}{dt} = -i_{sk} - i_{fk} \tag{7.159}$$

$$u_k = \begin{cases} 1 & S_k \quad \text{on} \\ -1 & \overline{S}_k \quad \text{on} \end{cases}, \quad k = a, b, c \tag{7.160}$$

where u_k is the control input. The switching of S_k and $\overline{S}_k$ should be done in such a way that a compensation voltage is produced across C_{se} with the aim of keeping the

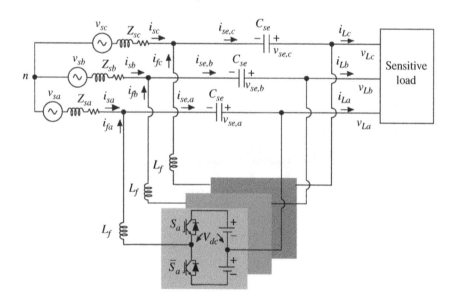

Figure 7.45 Three-phase TDVR.

load voltage sinusoidal and at a desired level for all times during the voltage distortions and sags occurring in the grid voltage.

7.5.2 Design of Sliding Mode Control for TDVR

Let us define the state variables in terms of the compensation voltage error and its time derivative defined in [23] as follows:

$$x_{1k} = v_{se,k} - v_{se,k}^* \tag{7.161}$$

$$x_{2k} = \dot{x}_{1k} = \dot{v}_{se,k} - \dot{v}_{se,k}^* \tag{7.162}$$

where $v_{se,k}^* = v_{Lk} - v_{sk}$ denotes the reference of $v_{se,k}$. The sliding surface function and its derivative with respect to time can be defined as

$$\sigma_k = \lambda x_{1k} + x_{2k} \tag{7.163}$$

$$\dot{\sigma}_k = \lambda \dot{x}_{1k} + \dot{x}_{2k} \tag{7.164}$$

where λ is the sliding coefficient, which should be positive due to stability. Now, taking derivative of (7.162) and making use of (7.159), one can obtain the following equation:

$$\dot{x}_{2k} = -\left(\frac{1}{C_{se}} \frac{di_{sk}}{dt} + \frac{1}{C_{se}} \frac{di_{fk}}{dt} + \frac{d^2 v_{se,k}^*}{dt^2} \right) \tag{7.165}$$

From (7.158) solving for di_{fk}/dt and substituting it into (7.165) yields

$$\dot{x}_{2k} = -\left(\frac{1}{C_{se}} \frac{di_{sk}}{dt} + \frac{1}{L_f C_{se}} (v_{se,k} + u_k V_{dc}) + \frac{d^2 v_{se,k}^*}{dt^2} \right) \tag{7.166}$$

Since $v_{se,k} = x_{1k} + v_{se,k}^*$, then we can obtain

$$\dot{x}_{2k} = -\omega_o^2 (x_{1k} + u_k V_{dc}) - \omega_o^2 D_k \tag{7.167}$$

where $\omega_o^2 = 1/L_f C_{se}$ and D_k represents disturbance that is given by

$$D_k = -L_f \frac{di_{sk}}{dt} - L_f C_{se} \frac{d^2 v_{se,k}^*}{dt^2} + v_{se,k}^* \tag{7.168}$$

Substituting (7.167) into (7.164), we obtain

$$\dot{\sigma}_k = \lambda x_{2k} - \omega_o^2 (x_{1k} + u_k V_{dc}) - \omega_o^2 D_k \tag{7.169}$$

Now, let the control input be defined as

$$u_k = sign(\sigma_k) = \begin{cases} 1 & \text{if} \quad \sigma_k > 0 \\ -1 & \text{if} \quad \sigma_k < 0 \end{cases} \tag{7.170}$$

When the system enters the sliding mode, the state trajectory moves along $\sigma_k = 0$ by making a zigzag motion independent of the system parameters. The SMC possesses two modes that are known as the reaching mode and sliding mode. In the reaching mode, the error variables are forced toward the sliding surface. In the sliding mode, the error variables slide along the sliding surface until they reach the equilibrium point at which the error variables are zero. In order to assure the stability of the reaching mode, the reaching condition (also known as the existence condition) $\sigma_k \dot{\sigma}_k < 0$ must be satisfied. Hence, the following relations can be obtained for each σ_k value.

i) When $\sigma_k < 0$, $u_k = -1$, which means that $\overline{S}_k$ is turned on. In this case, we obtain

$$\dot{\sigma}_k > 0 \quad \Rightarrow \quad l_{1k} = \dot{\sigma}_k = -\omega_o^2 x_{1k} + \lambda x_{2k} + d_{1k} > 0 \tag{7.171}$$

ii) When $\sigma_k > 0$, $u_k = 1$, which means that S_k is turned on. In this case, we obtain

$$\dot{\sigma}_k < 0 \quad \Rightarrow \quad l_{2k} = \dot{\sigma}_k = -\omega_o^2 x_{1k} + \lambda x_{2k} + d_{2k} < 0 \tag{7.172}$$

where d_{1k} and d_{2k} are given by

$$d_{1k} = \omega_o^2 (V_{dc} - D_k) \tag{7.173}$$
$$d_{2k} = -\omega_o^2 (V_{dc} + D_k) \tag{7.174}$$

Region of stability of the reaching mode bounded by lines l_{1k} and l_{2k} is shown in Figure 7.46. When a voltage sag occurs in the grid, x_{1k} becomes nonzero. Thereafter, the reaching phase starts where the controller forces the trajectory to move toward $\sigma_k = 0$ line. When the trajectory hits this line, the sliding mode starts and

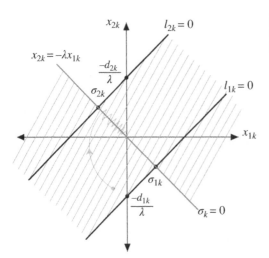

Figure 7.46 Region of stability of the reaching mode. *Source:* Komurcugil and Biricik [22]/IEEE.

the trajectory moves along the line until the equilibrium point ($x_{1k} = 0$ and $x_{2k} = 0$) is reached. Since the sliding surface function is zero ($\sigma_k = 0$) during the sliding mode, the following equation can be obtained:

$$\dot{x}_{1k} + \lambda x_{1k} = 0 \tag{7.175}$$

It is evident that (7.175) is a first-order differential equation whose solution is $x_{1k}(t) = x_{1k}(0)e^{-\lambda t}$. It is easy to see that $x_{1k}(t) \to 0$ for $\lambda > 0$. With $x_{1k}(t) = 0$, it follows from (7.162) that $x_{2k}(t) \to 0$ as well. Hence, x_{1k} and x_{2k} can converge to zero independent from the system parameters, which confirms the inherent robustness feature of the SMC. On the other hand, the main function of TDVR is to inject a compensating voltage ($v_{se,\,k}$) across capacitor C_{se} when anomalies occur in the grid voltage. In the absence of these voltage anomalies (i.e. $v_{Lk} = v_{sk}$), both state variables are zero ($x_{1k} = 0$ and $x_{2k} = 0$) since the actual and reference capacitor voltages are zero ($v_{se,k}^* = v_{Lk}^* - v_{sk} = 0$).

7.5.3 Time-Varying Switching Frequency with Single-Band Hysteresis

It should be noted that the implementation of the control input described in (7.170) results in a very high switching frequency, which in turn causes undesired chattering. Therefore, in order to control the switching frequency and suppress the chattering, the sign function is usually replaced by a hysteresis function with a boundary layer as follows:

$$u_k = \begin{cases} 1 & \text{if} \quad \sigma_k > +h \\ -1 & \text{if} \quad \sigma_k < -h \end{cases} \tag{7.176}$$

Figure 7.47 shows the evolution of the sliding surface function within the boundaries and the switching logic. The sliding surface function is forced to slide along $\sigma_k = 0$ within lower ($-h$) and upper ($+h$) boundaries. When σ_k goes below $-h$, $\overline{S}_k$ is

Figure 7.47 Evolution of the sliding surface function and the switching logic. *Source:* Adapted from Komurcugil and Biricik [22]/IEEE.

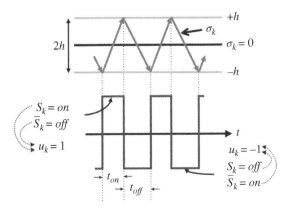

turned on and $-V_{dc}$ is applied to the $L_f C_{se}$ filter. On the other hand, when σ_k goes above $+h$, S_k is turned on and $+V_{dc}$ is applied to the filter.

The switching frequency expression in terms of hysteresis band and other parameters plays an important role in the realization of the control method in real time. Here, the switching frequency derivation is done for S_a and $\overline{S}_a$. Assuming that the state variables are negligibly small $(x_{1a} = x_{2a} = 0)$ in the steady state, (7.169) can be written as

$$\dot{\sigma}_a = -\omega_0^2 (u_a V_{dc} + D_a) \tag{7.177}$$

The reference voltage for the series connected capacitor can be written as

$$v_{se,a}^* = v_{La} \sin(\omega t) - V_{sa} \sin(\omega t) = V_{se} \sin(\omega t) \tag{7.178}$$

where V_{se} is the amplitude of $v_{se,a}^*$. Taking second derivative of $v_{se,a}^*$ with respect to time yields

$$\frac{d^2 v_{se,a}^*}{dt^2} = -\omega^2 V_{se} \sin(\omega t) \tag{7.179}$$

If the load draws the following sinusoidal current from the grid,

$$i_{sa} = i_{La} = I_s \sin(\omega t - \theta) \tag{7.180}$$

The derivative of (7.180) is given by

$$\frac{di_{sa}}{dt} = \omega I_s \cos(\omega t - \theta) \tag{7.181}$$

Substituting (7.178), (7.177), and (7.181) into (7.168), one obtains

$$D_a = K \cos(\omega t - \phi) \tag{7.182}$$

where

$$K = \sqrt{K_1^2 + K_2^2 + 2K_1 K_2 \sin(\theta)} \tag{7.183}$$

$$K_1 = -V_{se} \left(\frac{(\omega_0^2 + \omega^2)}{\omega_0^2} \right) \tag{7.184}$$

$$K_2 = \omega L_f I_s \tag{7.185}$$

$$\phi = \tan^{-1} \left(\frac{(K_1 + K_2 \sin(\theta))}{K_2} \right) \tag{7.186}$$

Now, substituting (7.182) into (7.177), we obtain

$$\dot{\sigma}_a = -\omega_0^2 (u_a V_{dc} + K \cos(\omega t - \phi)) \tag{7.187}$$

It is worth noting that for the values used in the simulation and experimental studies, the phase shift can be approximated as $\phi \cong -90°$. This implies that $K \cos(\omega t + 90°) = -K\sin(\omega t)$ and

$$\dot{\sigma}_a \cong -\omega_o^2(u_a V_{dc} - K\sin(\omega t)) \tag{7.188}$$

The on and off durations of switches S_a and $\overline{S}_a$ can be written as

$$t_{on} = \frac{-2h}{\dot{\sigma}_a|_{u_a = 1}} = \frac{2h}{\omega_o^2(V_{dc} - K\sin(\omega t))} \tag{7.189}$$

$$t_{off} = \frac{2h}{\dot{\sigma}_a|_{u_a = -1}} = \frac{2h}{\omega_o^2(V_{dc} + K\sin(\omega t))} \tag{7.190}$$

The expression of the switching frequency can be written as

$$f_{sw} = \frac{1}{t_{on} + t_{off}} = \frac{\omega_o^2 V_{dc}}{4h} - \frac{\omega_o^2 K^2}{4hV_{dc}}\sin^2(\omega t) \tag{7.191}$$

It is evident from (7.191) that the switching frequency is time-varying and dependent on the filter parameters, hysteresis band, and DC input voltage. Although the switching frequency can be controlled by the hysteresis band, the effect of chattering is not completely eliminated, and the time-varying switching frequency still exists. The average switching frequency can be obtained by integrating (7.191) as follows:

$$f_{sw,av} = \frac{1}{\pi}\int_0^\pi f_{sw}d\omega t = \frac{\omega_o^2}{8hV_{dc}}\left(2V_{dc}^2 - K^2\right) \tag{7.192}$$

The desired $f_{sw,av}$ can be obtained by selecting the appropriate values of V_{dc}, h, L_f, and C_{se}. When V_{dc} is fixed, increasing h would result in smaller $f_{sw,av}$ and vice versa.

7.5.4 Constant Switching Frequency with Boundary Layer

An alternative way to eliminate the chattering effect and achieve a constant switching frequency is to employ a pulse width modulation (PWM) instead of the hysteresis modulation. The PWM is based on comparing the desired control signal with a triangular carrier signal. The comparison process produces an output signal whose frequency is equal to the frequency of the carrier signal. The desired control signal is obtained from a smoothing operation applied to the control discontinuity within a narrow boundary layer introduced in the vicinity of the sliding surface function [22, 24]. It is worth noting that the smoothing operation is performed to eliminate the chattering. All trajectories starting inside the boundary layer should be maintained inside the boundary layer. On the other hand, the trajectories starting outside the boundary layer are directed toward the boundary layer.

In order to achieve these requirements, the sliding surface function is interpolated inside the boundary layer with a thickness Φ by replacing the discontinuous control in (7.170) by σ_k/Φ. However, as mentioned in [24], the result of smoothing does not guarantee a perfect tracking and, therefore, Φ is to be selected to make a compromise between tracking error and smoothing operation. Replacing σ_k by σ_k/Φ results in the following equation:

$$\frac{\sigma_k}{\Phi} = \frac{\lambda x_{1k} + x_{2k}}{\Phi} = \frac{\lambda x_{1k}}{\Phi} + \frac{1}{\Phi C_{se}}\left(i_{se,k} - i_{se,k}^*\right) \tag{7.193}$$

The capacitor current error can be written as

$$i_{se,k} - i_{se,k}^* = -\left(i_{fk} + i_{Lk}\right) - C_{se}\frac{dv_{se,k}^*}{dt} \tag{7.194}$$

Assuming that the load is resistive (R_L) and substituting the load current $i_{Lk} = \left(\Delta v_{Lk} + v_{Lk}^*\right)/R_L$ into (7.194), we obtain

$$i_{se,k} - i_{se,k}^* = -\left(i_{fk} - i_{fk}^*\right) - \frac{\Delta v_{Lk}}{R_L} \tag{7.195}$$

where $\Delta v_{Lk} = v_{Lk} - v_{Lk}^*$. Since the main objective of TDVR is to maintain the load voltage at desired level, then the load voltage error can be considered zero for all times ($\Delta v_{Lk} = 0$). Now, considering that $\Delta v_{Lk} = 0$ and substituting (7.195) into (7.193), one can obtain

$$\frac{\sigma_k}{\Phi} = \frac{\lambda x_{1k}}{\Phi} - \frac{1}{\Phi C_{se}}\left(i_{fk} - i_{fk}^*\right) \tag{7.196}$$

The λx_{1k} term in (7.196) represents the multiplication of capacitor voltage error with the slope of the sliding line. Since the inductor current ripple is much greater than the capacitor voltage ripple, then (7.196) can be approximated as

$$\frac{\sigma_k}{\Phi} \simeq -\frac{1}{\Phi C_{se}}\left(i_{fk} - i_{fk}^*\right) \tag{7.197}$$

The slope of (7.197) can be written as [24]

$$Slope_{\sigma_k/\Phi} = \frac{V_{dc}}{4L_f C_{se}\Phi} \tag{7.198}$$

On the other hand, considering the triangular carrier and pulse width modulated (PWM) signals in Figure 7.48, the slope of a triangular carrier signal can be easily written as [25]

$$Slope_{carrier} = 4V_p f_{sw} \tag{7.199}$$

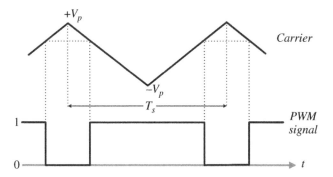

Figure 7.48 Triangular carrier and pulse width modulated signals.

where V_p and f_{sw} are the amplitude and frequency of the carrier signal, respectively. In order to generate the PWM signals properly, the following condition should be satisfied:

$$\frac{V_{dc}}{4L_f C_{se} \Phi} < 4V_p f_{sw} \tag{7.200}$$

Thus, the lower bound of Φ can easily be determined from (7.200) as follows:

$$\frac{V_{dc}}{16L_f C_{se} V_p f_{sw}} < \Phi \tag{7.201}$$

It is worthy noting that the value of Φ should be determined accurately in order to achieve the desired chattering suppression.

7.5.5 Simulation Study

The effectiveness of SMC method under time-varying and constant switching frequency operation is investigated through simulation studies performed by using MATLAB/Simulink model designed according to the block diagram shown in Figure 7.49. The simulation studies are carried out by using the system and control parameters listed in Table 7.6.

Figure 7.50 shows the simulation results obtained by the time-varying switching frequency-based SMC method for the switching frequency, injected and load voltages when a voltage sag occurs in the grid voltage from 230 V (rms) to 160 V (rms) and back to 230 V (rms). The system injects zero compensating voltage (v_{se}) across C_{se} when there is no voltage variation in the grid voltage (v_s). In this case, the simulated switching frequency is around 12.5 kHz, which agrees well with the computed value (17.74 kHz) obtained by using the appropriate parameters in Table 7.6 in Eq. (7.191). However, when the grid voltage is reduced from 230 V (rms) to 160 V (rms), the TDVR immediately generates and injects compensating voltage for

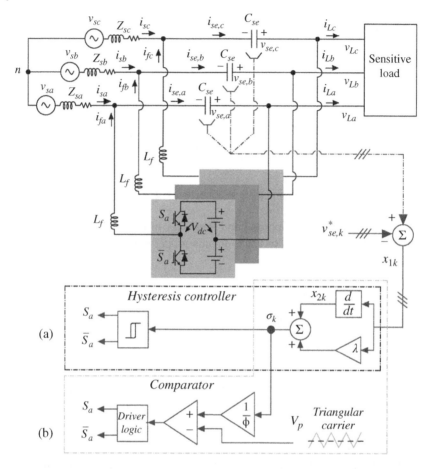

Figure 7.49 Block diagram of three-phase TDVR with: (a) Time-varying switching frequency-based SMC, (b) Constant switching frequency-based SMC. *Source:* Komurcugil and Biricik [22]/IEEE.

each phase as shown in Figure 7.50a. As a consequence of this voltage injection, the load voltages are almost not affected from the voltage sag. Also, it can be noticed that the switching frequency of the inverter is time-varying during the voltage sag period. According to the switching frequency equation in (7.191), the time-varying component of the switching frequency is a sinusoidal signal with 2ω angular frequency. It is obvious from Figure 7.50b that the time-varying component of the switching frequency is in good agreement with Eq. (7.191). When the voltage sag period is over, while the injected voltage becomes zero again, the switching frequency becomes constant.

Figure 7.51 shows the simulation results obtained by the constant switching frequency-based SMC method for the switching frequency, injected and load voltages

Table 7.6 System and control parameters.

Symbol	Value
DC-link voltage, V_{in}	350 V
Inductance, L_f	0.6867 mH
Capacitance, C_{se}	50 μF
Grid voltage amplitude, V_g	$230\sqrt{2}$V
Grid frequency, f_g	50 Hz
Sliding coefficient, λ	1900
Hysteresis band, h	200 000
Boundary layer thickness, Φ	282 000
Carrier amplitude, V_p	1 V
Sensitive load, series RL	54.16 Ω, 50 mH

when a voltage sag occurs in the grid voltage from 230 V (rms) to 160 V (rms) and back to 230 V (rms). At first glance, the load voltages are not affected by the voltage sag. However, after a close inspection to the load voltages, one can easily notice that there is a small steady-state error, which means that the amplitude of load voltages is less than 230 V (rms). It is clear from Figure 7.51b that the switching frequency during the voltage sag is constant at the expense of steady-state error in the load voltages. It is worthy pointing out that this steady-state error can be reduced by reducing the boundary layer thickness (Φ) value.

7.5.6 Experimental Study

The efficacy and feasibility of the SMC method under time-varying and constant switching frequency operation is verified through experimental results by using OPAL-RT real-time platform and its associated tools. The system and control parameters listed in Table 7.6 were used in experimental studies.

Figure 7.52 shows the performance of time-varying switching frequency-based SMC when voltage sag occurs from 230 V (rms) to 160 V (rms). It is evident that the TDVR does not generate any compensating voltage when there is no voltage sag. In this case, the load voltages are sinusoidal and equal to the grid voltages. However, when a sudden voltage sag occurs, the TDVR reacts to this voltage sag by generating the required compensating voltages as shown in Figure 7.52b. As a result of these copensating voltages, the load voltages are always maintained at 230 V (rms) as shown in Figure 7.52c. The THD of load voltage is measured as 1.92%.

Figure 7.53 shows the performance of constant switching frequency-based SMC when voltage sag occurs from 230 V (rms) to 160 V (rms). It is obvious that the load voltages are sinusoidal and equal to the grid voltages when there is no voltage sag.

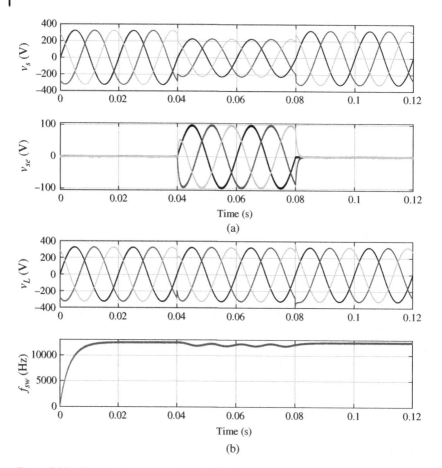

Figure 7.50 Simulated responses obtained by the time-varying switching frequency-based SMC for the switching frequency, injected and load voltages for a voltage sag in the grid voltage from 230 V (rms) to 160 V (rms) and back to 230 V (rms). (a) Grid voltages (v_s), injected voltages (v_{se}), (b) load voltages (v_L), switching frequency (f_{sw}).

However, when voltage sag exists in the grid voltage, the TDVR generates the required compensating voltages as shown in Figure 7.52b to keep the load voltage sinusoidal and 230 V (rms) as shown in Figure 7.52c. The THD of load voltage is measured as 1.34%. Comparing the THD values in the voltage sag case, it can be seen that the constant switching frequency-based SMC method yields smaller THD for the load voltages.

Figure 7.54 shows the performance of time-varying switching frequency-based SMC under distorted grid voltages. The THD of distorted grid voltages is measured to be 12.38%. It is obvious from Figure 7.54b that the TDVR does not inject any

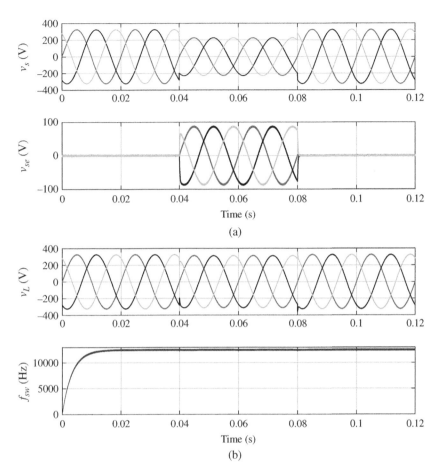

Figure 7.51 Simulated responses obtained by the constant switching frequency-based SMC for the switching frequency, injected and load voltages for a voltage sag in the grid voltage from 230 V (rms) to 160 V (rms) and back to 230 V (rms). (a) Grid voltages (v_s), injected voltages (v_{se}), (b) load voltages (v_L), switching frequency (f_{sw}).

voltage when the grid voltages are not distorted. In this case, the load voltages are equal to the grid voltages. However, when the grid voltages become distorted, the TDVR reacts to this distortion and injects the required compensating voltages across C_{se} as shown in Figure 7.54b. As a consequence of these compensating voltages, the load voltages are almost not affected from the distorted grid voltages as shown in Figure 7.54c. The THD of load voltages is measured as 1.83%.

Figure 7.55 shows the performance of constant switching frequency-based SMC under distorted grid voltages in Figure 7.54a. It is obvious that the load voltages in Figure 7.55c are maintained sinusoidal at desired level (230 V rms). Thanks to

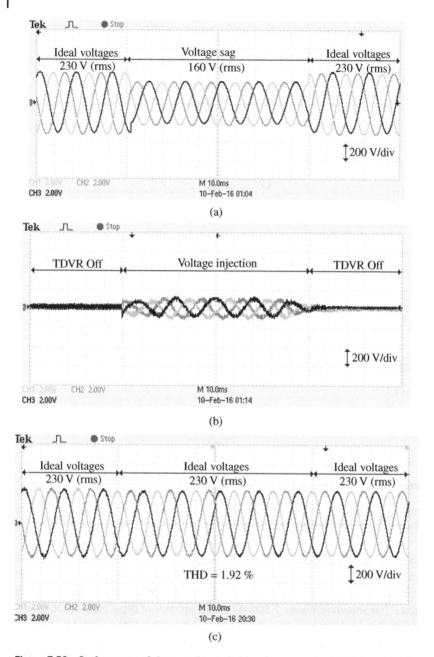

(a)

(b)

(c)

Figure 7.52 Performance of time-varying switching frequency-based SMC under voltage sag. (a) Grid voltages, (b) Capacitor voltages, and (c) Load voltages. *Source:* Komurcugil and Biricik [22]/IEEE.

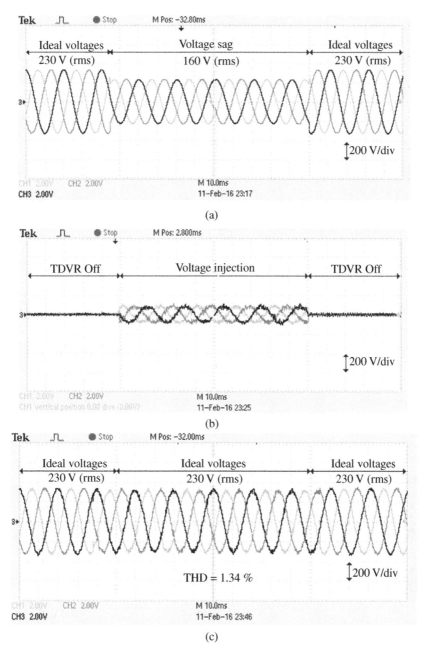

Figure 7.53 Performance of constant switching frequency-based SMC under voltage sag. (a) Grid voltages, (b) Capacitor voltages, and (c) Load voltages. *Source:* Komurcugil and Biricik [22]/IEEE.

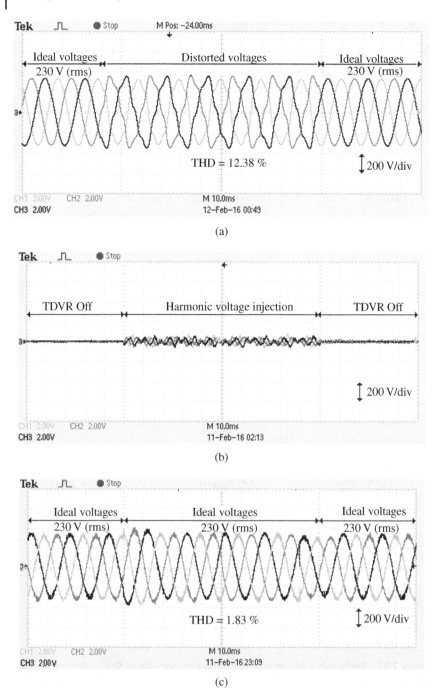

Figure 7.54 Performance of time-varying switching frequency-based SMC under distorted grid voltages. (a) Grid voltages, (b) Capacitor voltages, and (c) Load voltages. *Source:* Komurcugil and Biricik [22]/IEEE.

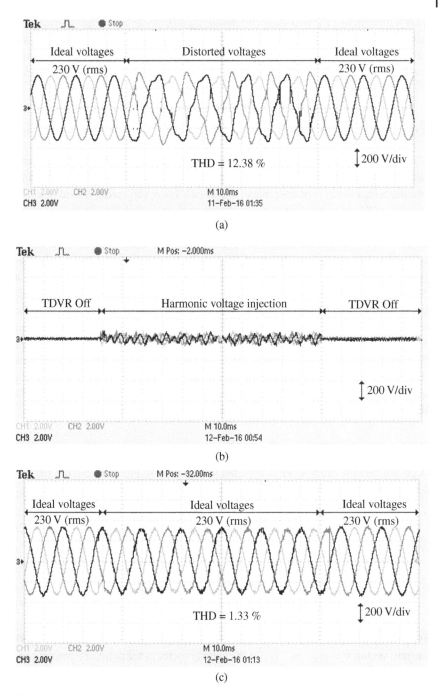

Figure 7.55 Performance of constant switching frequency-based SMC under distorted grid voltages. (a) Grid voltages, (b) Capacitor voltages, and (c) Load voltages. *Source:* Komurcugil and Biricik [22]/IEEE.

TDVR for injecting the necessary compensating voltages so as to achieve sinusoidal load voltages. The THD of load voltages is measured as 1.33%. Comparing the measured THDs for the time-varying and constant switching frequency-based SMC methods, one can see that the latter provides smaller THD for the load voltages.

7.6 Three-Phase Shunt Active Power Filter

This section presents a robust model-based control in natural frame for a three-phase shunt active power filter (SAPF), which was presented previously in [6]. For the control method, a linear converter model is deduced. Then, this model is used in a Kalman filter (KF) in order to estimate the system state-space variables. Even though the states estimation do not match the variables of the real system, it has allowed to design three sliding mode controllers providing the following features to the closed loop system: (i) robustness due to the fact that control specifications are met independently of any variation in the system parameters; (ii) noise immunity, since a KF is applied; (iii) a lower total harmonic distortion (THD) of the current delivered by the grid compared with the standard solution using measured variables; (iv) the fundamental component of the voltage at point of common coupling is estimated even in the case of a distorted grid; and (v) a reduction in the number of sensors. Thanks to this solution the sliding surfaces for each controller are independent. This decoupling property of the three controllers allows using a fixed switching frequency algorithm that ensures a perfect current control. Finally, experimental results validate the control strategy and illustrate all its interesting features.

7.6.1 Nonlinear Model of the SAPF

A circuit scheme of an SAPF is depicted in Figure 7.56. From the circuit, the following differential equations can be written:

$$L_F \frac{d\mathbf{i}_F}{dt} = v - \frac{V_{DC}}{2}\mathbf{u} + v_n \mathbf{1}^T \tag{7.202}$$

$$C\frac{dV_{DC}}{dt} = i_{Fa}\frac{u_a}{2} + i_{Fb}\frac{u_b}{2} + i_{Fc}\frac{u_c}{2} \tag{7.203}$$

$$v_n = \frac{V_{DC}}{6}(u_a + u_b + u_c) \tag{7.204}$$

where v_n is the neutral point voltage, $\mathbf{i}_F = \begin{bmatrix} i_{Fa} & i_{Fb} & i_{Fc} \end{bmatrix}^T$ is the inductor filter current vector, $\mathbf{v} = \begin{bmatrix} v_a & v_b & v_c \end{bmatrix}^T$ is the voltage vector at the point of common coupling (PCC), $\mathbf{u} = \begin{bmatrix} u_a & u_b & u_c \end{bmatrix}^T$ is the control variable vector with $u_{a,b,c} \in \{\pm 1\}$,

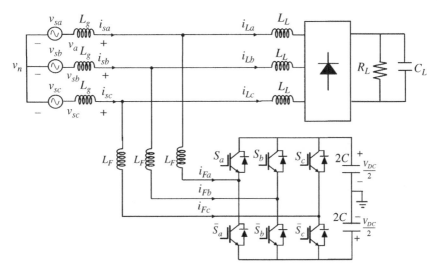

Figure 7.56 Circuit diagram of a three-phase SAPF.

V_{DC} is the *dc* voltage at the SAPF capacitor, and $\mathbf{1}^T = [1 \quad 1 \quad 1]^T$. The previous equations can be written as a nonlinear state-space model:

$$\frac{dx}{dt} = \mathbf{f}(x, u) + \mathbf{E}(v + v_n) \tag{7.205}$$

$$y = \mathbf{h}(x, u) \tag{7.206}$$

where

$$\mathbf{x} = \begin{bmatrix} i_{Fa} & i_{Fb} & i_{Fc} & v_{dc} \end{bmatrix}^T \tag{7.207}$$

$$\mathbf{f} = \begin{pmatrix} 0 & 0 & 0 & -\dfrac{u_a}{2L_F} \\[2mm] 0 & 0 & 0 & -\dfrac{u_b}{2L_F} \\[2mm] 0 & 0 & 0 & -\dfrac{u_c}{2L_F} \\[2mm] \dfrac{u_a}{2C} & \dfrac{u_b}{2C} & \dfrac{u_c}{2C} & 0 \end{pmatrix} \tag{7.208}$$

$$\mathbf{E} = \frac{1}{L_F}[1 \quad 1 \quad 1 \quad 0]^T \tag{7.209}$$

and the output matrix is given by **h**, where

$$
\mathbf{H} = \frac{\partial \mathbf{h}}{\partial \mathbf{x}} = \begin{pmatrix} 1 & 0 & 0 & 0 \\ 0 & 1 & 0 & 0 \\ 0 & 0 & 1 & 0 \\ 0 & 0 & 0 & 1 \end{pmatrix}
\tag{7.210}
$$

7.6.2 Problem Formulation

The main objective of an SAPF is to guarantee sinusoidal grid currents in phase with the grid voltages in presence of nonlinear loads. As a first approach, the sliding surfaces can be generated using the grid currents vector in order to obtain a grid current in phase with the voltage source:

$$
\boldsymbol{\sigma} = \mathbf{i}_s^* - \mathbf{i}_s
\tag{7.211}
$$

where $\mathbf{i}_s^* = k\mathbf{v} = \begin{bmatrix} i_{sa}^* & i_{sb}^* & i_{sc}^* \end{bmatrix}^T$ is the reference current vector, and $\mathbf{i}_s = \mathbf{i}_F + \mathbf{i}_L$ being $\mathbf{i}_L = \begin{bmatrix} i_{La} & i_{Lb} & i_{Lc} \end{bmatrix}^T$ the load currents vector.

Using (7.202), (7.204), and (7.211), the dynamics for each sliding surface can be obtained as follows:

$$
\frac{d\sigma_a}{dt} = \frac{di_{sa}^*}{dt} - \frac{1}{L_F}\left(v_a + \frac{V_{DC}}{2}\frac{-2u_a + u_b + u_c}{3}\right) - \frac{di_{La}}{dt}
\tag{7.212}
$$

$$
\frac{d\sigma_b}{dt} = \frac{di_{sb}^*}{dt} - \frac{1}{L_F}\left(v_b + \frac{V_{DC}}{2}\frac{u_a - 2u_b + u_c}{3}\right) - \frac{di_{Lb}}{dt}
\tag{7.213}
$$

$$
\frac{d\sigma_c}{dt} = \frac{di_{sc}^*}{dt} - \frac{1}{L_F}\left(v_c + \frac{V_{DC}}{2}\frac{u_a + u_b - 2u_c}{3}\right) - \frac{di_{Lc}}{dt}
\tag{7.214}
$$

The aforementioned expressions show two different issues to be considered: (i) a cross coupling term introduced by the neutral point voltage (7.160), which is expressed as a linear combination of the control signals $u_{a,b,c}$ in each sliding surface dynamics. This term may produce severe irregularities in the ordinary hysteresis operation since each phase-leg cannot be controlled independently [5], and (ii) the reference currents are generated using the measured PCC voltages, and for this reason the distortion in the grid voltages will involve grid currents with a higher harmonic content. As a solution to the aforementioned problems, a model-based control methodology will be used with the aim of obtaining three independent sliding mode controllers with filtered references. To do so, a KF can be used using a converter model where the neutral point voltage is obviated. This fact has a negligible effect in the averaged dynamics of the closed loop system. In addition, the fundamental component of the voltages at the PCC can be estimated to generate sinusoidal reference currents even in the case of a distorted grid. An interesting side effect of this control scheme is the fact that the sliding surfaces for each controller will be designed independently. This decoupling property between the three current controllers allows the use of a fixed switching frequency algorithm. The model to achieve all such aims will be introduced in the next section.

7.6.3 Control System

The control diagram for phase-leg a is depicted in Figure 7.57. Phase-legs b and c have the same control scheme. A KF is used together with the converter model, as explained in the next subsection, in order to estimate the states. A PI controller is used to regulate the filter output voltage and to obtain the value of the gain k. This gain and the state estimation are the inputs of a sliding surface, which computes the error between estimated grid current and its reference. Finally, this error is considered in a variable hysteresis comparator with a switching decision algorithm. This algorithm was presented in Section 7.4.6. This combination allows to obtain a fixed switching frequency, with an improved switching spectrum.

7.6.3.1 State Model of the Converter

A linear simplified model of the SAPF can be obtained from the equations presented before (7.202)–(7.204). This linear model will be obtained with the aim of solving the problems exposed in Section 7.6.2. The first problem is regarding the coupling among controllers due to the neutral point voltage. It can be solved by eliminating this effect from the controller dynamics (7.212)–(7.214). By obtaining this new space-state model, it is also an opportunity to reduce the number of sensors by adding two additional states, the voltage at the PCC and its quadrature, to estimate the fundamental component of this voltage. Due to the grid inductance and the grid voltage harmonics, the measured voltage at the PCC can be distorted. Thus, this estimation will improve the quality in the generation of the reference currents, leading to a reduction of the THD of the grid currents. With this approach, the second problem can be solved.

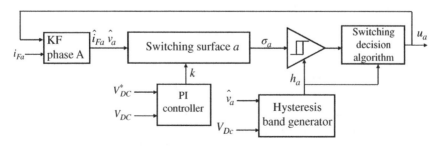

Figure 7.57 Control system for phase-leg a.

Hence, to derive the adequate SAPF model, some considerations must be considered. First, it is necessary to ensure that the model is decoupled (i.e. each phase-leg only depends on its own control variable) and second, the model must be linear, accurate, and simple to reduce the computational time. For these reasons the following assumptions are made:

1) The neutral point voltage, v_n in the filter current Eq. (7.202) is not considered in the model, since it is a high-frequency signal, and it only affects the inductors current ripple. Thus, this approximation has a negligible effect on the averaged dynamics of the closed-loop system.

2) The SAPF capacitor is usually large, and its voltage has a slow dynamic. Hence, the sate variable V_{DC} can be assumed constant between different sampling instants of the same switching period. Besides, since this variable is measured, it can be considered as a parameter and its dynamics can be neglected. With this consideration, the new state-space model for each phase-leg could be considered linear in a switching period [14].

3) The voltage at the PCC v_i and its quadrature v_{iq} are added to the model in order to estimate the fundamental component of the PCC voltages. Then, the reference currents generation does not rely on the noise or harmonic distortion of the grid.

The aforementioned considerations lead to redefine new differential equations for each phase-leg i:

$$\frac{di_{Fi}}{dt} = \frac{1}{L_F}v_i - \frac{V_{DC}}{2L_F}u_i \tag{7.215}$$

$$\frac{dv_i}{dt} = \omega_o v_{iq} \tag{7.216}$$

$$\frac{dv_{iq}}{dt} = -\omega_o v_i \tag{7.217}$$

where ω_o is the grid frequency. From the last equations, the space-state model of the converter can be obtained as

$$\frac{dx_i}{dt} = \mathbf{A}x_i + \mathbf{B}V_{DC}\mathbf{u}_i \tag{7.218}$$

$$y_i = \mathbf{C}x_i \tag{7.219}$$

where $x_i = [\,i_{Fi} \quad v_i \quad v_{iq}\,]^T$, $A = \begin{pmatrix} 0 & 1/L_F & 0 \\ 0 & 0 & \omega_o \\ 0 & -\omega_o & 0 \end{pmatrix}$ and $B = [\,-1/2L_F \quad 0 \quad 0\,]^T$.

As mentioned, using this model, a KF can be used to estimate the filter currents without the effect of the neutral point voltage and to extract the fundamental component of the PCC voltages. This fact allows to generate sinusoidal reference currents even in case of a distorted grid [8]. The control algorithm will not use grid voltage sensors. Hence, the value of vector $\mathbf{C}$ can defined as $\mathbf{C} = \begin{bmatrix} 1 & 0 & 0 \end{bmatrix}$. According to matrix $\mathbf{C}$ only the filter current $\mathbf{i}_{Fi}$ is measured. Then, it is necessary to prove if the system is observable and controllable using only the measured variable $\mathbf{i}_{Fi}$. The observability matrix is defined by

$$\mathbf{O} = \begin{bmatrix} \mathbf{C} & \mathbf{CA} & \mathbf{CA}^2 \end{bmatrix}^T \tag{7.220}$$

and the controllability matrix is

$$\mathbf{\Gamma} = \begin{bmatrix} \mathbf{B} & \mathbf{AB} & \mathbf{A}^2\mathbf{B} \end{bmatrix} \tag{7.221}$$

Matrices $\mathbf{O}$ and $\mathbf{\Gamma}$ are both of full rank (i.e. rank$\{\mathbf{O}\}$ = rank $\{\mathbf{\Gamma}\}$ = 3), so that the system is controllable and can be observed using only the measured current $\mathbf{i}_{Fi}$. In the next sections, these four blocks of the control system will be explained in detail.

7.6.3.2 Kalman Filter

It is well known that in nonlinear systems an extended KF (EKF) should be used to estimate the states, with the drawback of the high computational burden needed by the algorithm. This is the case of the SAPF, which is a clearly nonlinear system. However, the model defined in (7.174)–(7.175). As explained before, this model can be considered linear in a switching period, and a KF can be used instead of an EKF, leading to an important reduction of the computational load.

For the digital implementation of the KF, the state-space model must be discrete. The discrete model and measurement equations are given by the following expressions:

$$\mathbf{x}_i(k + 1) = \mathbf{A}_d\mathbf{x}_i(k) + \mathbf{B}_d V_{DC}(k)u_i(k) + \mathbf{\eta}_i(k) \tag{7.222}$$

$$y_i(k) = \mathbf{C}\mathbf{x}_i(k) + \mathbf{w}_i(k) \tag{7.223}$$

where

$$\mathbf{A}_d = e^{\mathbf{A}T_s} \cong \mathbf{I} + \mathbf{A}T_s = \begin{pmatrix} 1 & T_s/L_F & 0 \\ 0 & 1 & \omega_o T_s \\ 0 & -\omega_o T & 1 \end{pmatrix} \tag{7.224}$$

$$\mathbf{B}_d = \int_0^{T_s} \mathbf{B}e^{\mathbf{A}\lambda}d\lambda \cong \mathbf{B}T_s = \begin{bmatrix} -\dfrac{T_s}{2L_F} & 0 & 0 \end{bmatrix}^T \tag{7.225}$$

being $\mathbf{I}$ the identity matrix and T_s the sampling time. Here, it is assumed that $v_{dc}(k)$ is constant between samples of the same switching period, it means that $v_{dc}(k + 1) \cong v_{dc}(k)$, and as a consequence, the discrete model can be considered linear over the same switching period.

The process and measurement of noise vectors are $\boldsymbol{\eta}_i(k)$ and $\mathbf{w}_i(k)$, respectively, with covariance matrices are given by

$$\mathbf{Q}_i(k) = E\{\boldsymbol{\eta}_i(k)\boldsymbol{\eta}_i^T(k)\} \tag{7.226}$$

$$\mathbf{R}_i(k) = E\{\mathbf{w}_i(k)\mathbf{w}_i^T(k)\} \tag{7.227}$$

The recursive Kalman algorithm computation was explained in Section 7.4.4.1 and can be applied in the same way in this application. Then, according to the KF estimation, the estimated vector $\hat{\mathbf{x}}_i(k) = \begin{bmatrix} \hat{i}_{Fi} & \hat{v}_i & \hat{v}_{iq} \end{bmatrix}$ is obtained.

7.6.3.3 Sliding Mode Control

The conventional SAPF control scheme consists of a fast inner input-current loop, which ensures sinusoidal input currents in phase with the line voltages, $\mathbf{i}_s = k\mathbf{v}$, and a slow-outer loop usually a PI controller, whose main task is to regulate the output voltage, modifying the input-current amplitudes. In this application, the inner loop is designed using estimated variables as explained in the previous sections. With this aim, the following switching surfaces are introduced:

$$\sigma_a = \hat{i}^*_{sa} - \hat{i}_{sa} \tag{7.228}$$

$$\sigma_b = \hat{i}^*_{sb} - \hat{i}_{sb} \tag{7.229}$$

$$\sigma_c = \hat{i}^*_{sc} - \hat{i}_{sc} \tag{7.230}$$

where $\hat{\mathbf{i}}^*_s = \begin{bmatrix} \hat{i}^*_{sa} & \hat{i}^*_{sb} & \hat{i}^*_{sc} \end{bmatrix}^T$ and $\hat{\mathbf{i}}_s = \begin{bmatrix} \hat{i}_{sa} & \hat{i}_{sb} & \hat{i}_{sc} \end{bmatrix}$ are the estimated grid current references and the estimated filter currents by the KF algorithm. Note that the estimated current references are obtained from the estimated PCC voltages as $\hat{\mathbf{i}}^*_s = k\hat{\mathbf{v}}$, where $\mathbf{v} = \begin{bmatrix} \hat{v}_a & \hat{v}_b & \hat{v}_c \end{bmatrix}$. Note that the gain k can be obtained by means of a PI controller [26]:

$$k = k_p\left(V^*_{DC} - V_{DC}\right) + k_i \int_0^t \left(V^*_{DC} - V_{DC}\right) d\tau \tag{7.231}$$

As mentioned before, it should be noted that using this model, the dynamics of the controllers will only depend on their control variable u_a, u_b, and u_c:

$$\frac{d\sigma_a}{dt} = \frac{di^*_{sa}}{dt} - \frac{1}{L_F}\left(\hat{v}_a - \frac{V_{DC}}{2}u_a\right) - \frac{di_{La}}{dt} \tag{7.232}$$

$$\frac{d\sigma_b}{dt} = \frac{di^*_{sb}}{dt} - \frac{1}{L_F}\left(\hat{v}_b - \frac{V_{DC}}{2}u_b\right) - \frac{di_{Lb}}{dt} \tag{7.233}$$

$$\frac{d\sigma_c}{dt} = \frac{di^*_{sc}}{dt} - \frac{1}{L_F}\left(\hat{v}_c - \frac{V_{DC}}{2}u_c\right) - \frac{di_{Lc}}{dt} \tag{7.234}$$

The main requirement in the design of SMC is to satisfy the reaching conditions and also guarantee the existence of a sliding regime in the switching surfaces $\mathbf{S} = 0$. The reaching conditions for each phase-leg i are given by

$$\sigma_i \dot{\sigma}_i < 0 \tag{7.235}$$

Now, defining the control action

$$u_i = \begin{cases} u_i^+ & \text{if} & \sigma_i > 0 \\ u_i^- & \text{if} & \sigma_i < 0 \end{cases} \tag{7.236}$$

where $u_i \in \{-1, 1\}$ and using a similar way as described in Section 7.4.6, one has

$$\frac{v_0}{2L} \left(u_i^- - u_i^+ \right) < 0 \tag{7.237}$$

which allows to determine the switching action

$$u_i = \begin{cases} u_i^+ = 1 & \text{if} & \sigma_i > 0 \\ u_i^- = -1 & \text{if} & \sigma_i < 0 \end{cases} \tag{7.238}$$

7.6.3.4 Hysteresis Band Generator with SDA

This section deals with a fully digital hysteretic modulator to fix the switching frequency of the SAPF. Conventionally, the expression for the hysteresis band is given by the following equation [21]:

$$h_i = \frac{V_{DC}}{8L_F f_{sw}} \left(1 - \left(\frac{2\hat{v}_i}{V_{DC}} \right)^2 \right) \tag{7.239}$$

where f_{sw}, is the switching frequency.

It is well known that a finite sampling frequency can cause deviations in the desired switching frequency due to the samples that appear out of the hysteresis limits. Then, while the sliding surface is moving inside the hysteresis bands $\pm h_i$, the switching frequency f_{sw} will remain constant. A switching decision algorithm has been shown in Section 7.3. Then, following the same steps of the algorithm, two different cases must be analyzed:

1) The sample is on the increasing slope of the switching surface. In this case, $\sigma_i > 0 \rightarrow u_i = 1$, and the time between the sample and the upper hysteresis limit is computed as

$$t_\alpha = L_F \frac{h_i - \sigma_i(t_k)}{V_{DC}/2 - \hat{v}_i} \tag{7.240}$$

2) The sample is on the decreasing slope of the switching surface. In this case $\sigma_i < 0 \rightarrow u_i = -1$, and the time between the sample and the lower hysteresis limit is computed as

$$t_\beta = L_F \frac{h_i + \sigma_i(t_k)}{\hat{v}_i + V_{DC}/2} \qquad (7.241)$$

where $\sigma_i(t_k)$ is the value of the switching surface for phase-leg i at the instant t_k. When the estimated times, t_α and t_β, are known, the SDA operates as follows:

i) In the increasing slope, if $t_\alpha < \dfrac{T_s}{2}$, the control signal u_i is changed from 1 to −1; otherwise, the control signal u_i remains to the actual value.

ii) In the decreasing slope, if $t_\beta < \dfrac{T_s}{2}$, the control signal u_i is changed from −1 to 1; otherwise, the control signal u_i remains to be the actual value.

This procedure is done recursively for each sampling period.

7.6.4 Experimental Study

The prototype of the SAPF was built using a 4.5-kVA SEMIKRON full-bridge as the power converter and a TMS320F28M36 floating-point DSP as the control platform. The grid voltages have been generated using a PACIFIC 360-AMX source. Some experimental results have been exported to MATLAB for the final representation. The system parameters are listed in Table 7.7. The computational load of

Table 7.7 System parameters.

Symbol	Description	Value
L_F	Filter input inductance	5 mH
L_g	Grid inductance	0.5 mH
C	Output capacitor	1500 μF
V_{DC}	DC-link voltage	400 V
f_s	Sampling frequency	40 kHz
f_{sw}	Switching frequency	4 kHz
f_{grid}	Grid frequency	60 Hz
V_p	Grid peak voltage	110 V_{rms}
k_p	Proportional gain	0.03
k_i	Integral gain	0.5
R_L	Load	48 Ω − 24 Ω
L_L	Load inductance	5 mH
C_L	Load capacitor	100 μF

the KF algorithm is around 3 μs. Note that in each phase-leg, different KF is used, but the Kalman gain is the same for each phase-leg as explained in Section 7.4.4.3. This is an important reduction in the total algorithm time, which is now about 23 μs. This fact allows a sampling frequency $f_s = 40$ kHz, and it makes possible the implementation of this method in the DSP.

The switching frequency when an SMC is used can be chosen as $f_{sw} = f_s/10$ to ensure enough samples for a proper performance of the variable hysteresis comparator. This relation provides a good tradeoff between the use of the computational load in a switching period and the switching-frequency accuracy. The following figures show the good performance of the control system.

In Figure 7.58a, the sliding surfaces with their corresponding control signals in the case of coupled controllers (7.211) are shown. In this case, the three controllers are implemented with measured variables instead of the estimated ones. This figure shows the effect of the neutral-point voltage (7.204) in the dynamics of the controllers. As can be seen, the slope of each sliding surface changes with any of the control signals, losing synchronization and making it impossible to design a current control with variable hysteresis band operating at fixed switching frequency.

In Figure 7.58b, the sliding surfaces with their corresponding control signals in the case of decoupled controllers (7.228)–(7.230) are shown. In this case, the sliding mode controllers are designed using the estimated variables by means of the KF. As it can be seen, the axis-decoupling is effective, and the slope of each sliding surface only changes with its control variable. Besides, using this model, the three controllers are independent, and the variable hysteresis control can be performed ensuring a fixed switching frequency.

7.6.4.1 Response of the SAPF to Load Variations

Figure 7.59 represents the main waveforms of the SAPF when a sudden load step change occurs, from no load to full load, and from full load to half load. This figure shows, from top to bottom, the grid currents, the nonlinear load currents, the filter currents, and the output voltage. Figure 7.59a shows the waveforms when the conventional SMC is used, while the waveforms in Figure 7.59b are obtained using the presented control scheme. Although in both figures the transient response is similar, it can be observed that when this controller is used, the THD of the grid currents is improved.

Figure 7.60 shows the harmonic spectrum of the grid current for phase-leg a. The spectrum before the compensation is shown in Figure 7.60a with a THD around 28.45%. When the conventional SMC is used, the THD is reduced to 5.36%, as shown in Figure 7.60b. Conversely, if the improved controller is applied, the THD is reduced to 2.51%, as shown in Figure 7.60c. A reduction of 46.8% in the THD is achieved with respect to the conventional SMC, which proves the

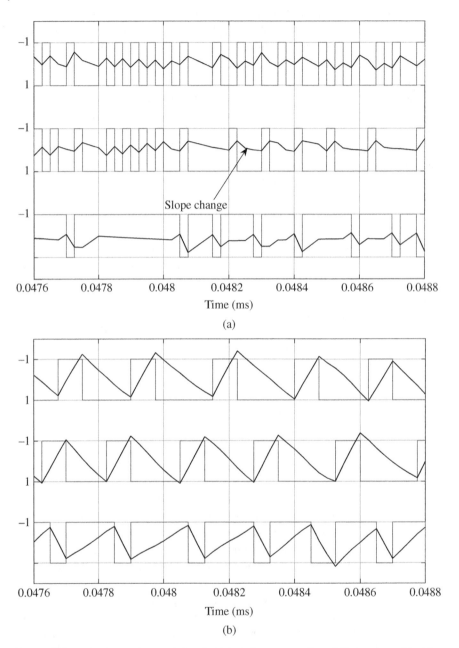

Figure 7.58 Experimental control signals with their corresponding sliding surfaces (a) with coupled controllers and (b) with decoupled controllers operating at fixed switching frequency.

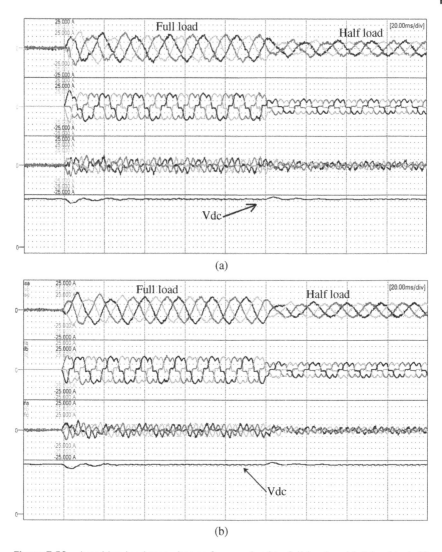

(a)

(b)

Figure 7.59 A sudden load step change from no load to full load and full load to half load. From top to bottom, grid currents (5 A/div), load currents (5 A/div), filter currents (5 A/div), output voltage (50 V/div): (a) using conventional SMC, and (b) using the improved SMC.

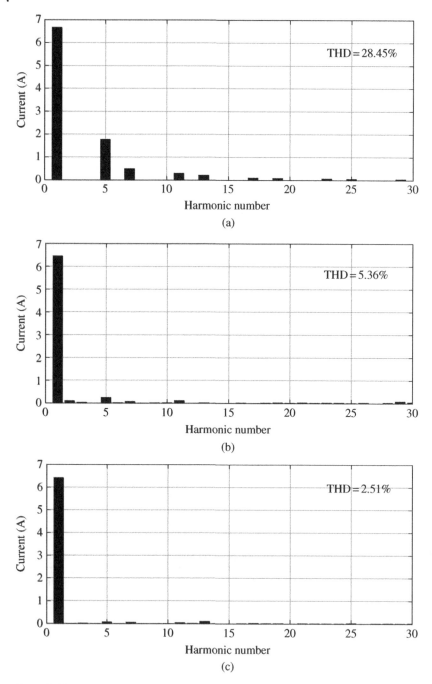

Figure 7.60 Grid current harmonics for phase-leg *a*: (a) Before compensation, (b) after compensation but using conventional SMC and (c) after compensation but using the presented control.

effectiveness of the presented control method. Note that in both cases the fundamental component is not represented in order to represent with a higher resolution the remaining harmonics.

7.6.4.2 SAPF Performances Under a Distorted Grid

Figure 7.61 compares the three-phase grid currents using a conventional SMC and the controller presented before in the case of a distorted grid. This figure illustrates from top to bottom: the distorted PCC voltages, being the THD about 14%, the load currents, and the grid currents. As it can be seen, when the conventional SMC is used, from 0 to 50 ms, the grid currents are distorted since the reference current generation method uses distorted PCC voltages. In contrast, when the presented controller is used, from 50 to 100 ms, the grid currents are practically sinusoidal since the reference current is generated by using only the fundamental component of the PCC voltage obtained from the KF.

A detail of the distorted PCC voltage for phase-leg a together with the estimated fundamental component is shown in Figure 7.62. Using the estimated voltage at the PCC, the reference current is sinusoidal. In addition, the estimated voltage is perfectly in phase with the measured one showing a good synchronization, at least as good as when using PLL.

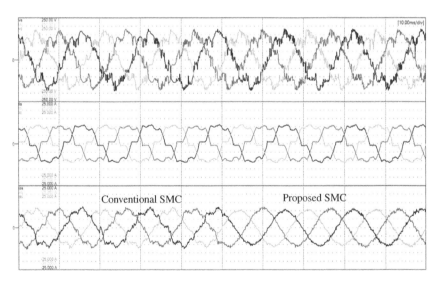

Figure 7.61 From top to bottom: PCC voltages (50 V/div) with THD = 14%, load current (5 A/div) and grid currents (5 A/div).

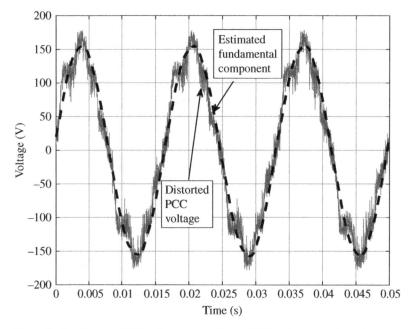

Figure 7.62 Measured PCC voltage and estimated PCC voltage for phase-leg *a*.

7.6.4.3 SAPF Performances Under Grid Voltage Sags

Figure 7.63 shows the SAPF performances under grid voltage sags. Two different sag types are analyzed in this test: a one-phase fault and a two-phase fault represented in Figure 7.63a,b, respectively. In both cases, the reference currents are obtained using the positive sequence of the PCC voltages using $\mathbf{i}_s^* = k\mathbf{v}^+$. The positive sequence is computed from the estimated PCC voltages and their quadrature obtained from the KF. With this solution, the use of a specific PLL algorithm for extracting the positive and negative sequence grid voltage components is not necessary. Even without using this algorithm, the quality of the grid currents waveforms is good in both normal and abnormal (voltage sag) conditions. Since the grid current tracks only the positive sequence of the PCC voltage, a ripple frequency component of $2\omega_o$ appears in the output voltage, as expected.

7.6.4.4 Spectrum of the Control Signal

The hysteresis band with its switching surface of phase-leg *a* is shown in the following figures. If a hysteresis band without the switching decision algorithm is used, an error on the desired switching frequency appears. In Figure 7.64, this problem can be clearly seen, where several samples are out of the hysteresis limits.

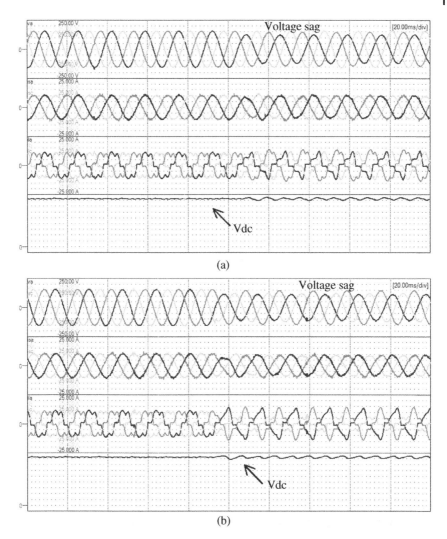

Figure 7.63 From top to bottom: grid voltages (50 V/div), grid currents (5 A/div), load currents (5 A/div), and v_{dc} (50 V/div) under unbalanced grid fault: (a) one-phase fault and (b) two-phase fault.

The problem is solved using the switching decision algorithm presented in Section 7.4.6. Figure 7.64b shows how the number of samples that are out of the bounds of the hysteresis limits have been reduced using this algorithm, and the error has disappeared. Finally, the spectrum of the switching frequency is

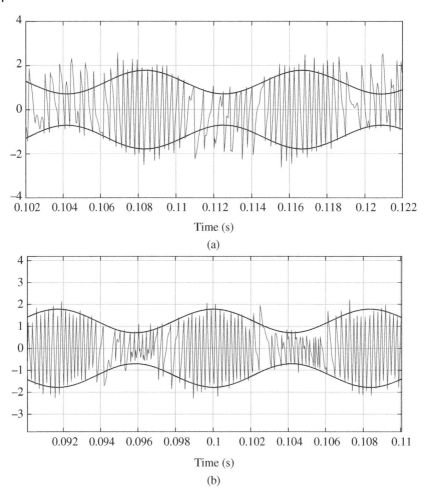

Figure 7.64 Experimental hysteresis band and its switching surface for phase-leg *a* : (a) without switching decision algorithm and (b) with switching decision algorithm.

shown in the following figures. Figure 7.65a shows the switching spectrum with hysteresis bands but without the switching decision algorithm. The spectrum is concentrated around a fixed-switching frequency of 3 kHz with an error around 1 kHz, since the desired switching frequency is 4 kHz. This problem is solved using the switching decision algorithm. Its results are depicted on Figure 7.65b. Now the switching frequency is the desired one, 4 kHz.

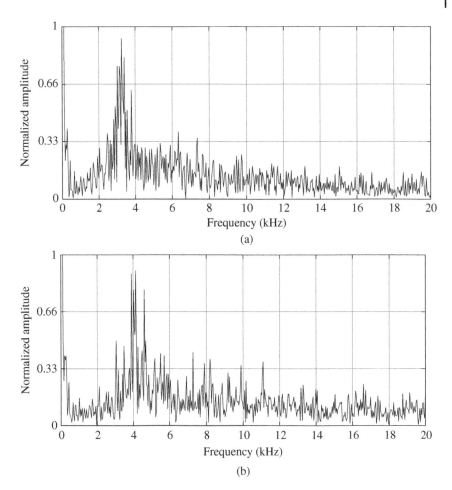

Figure 7.65 Switching spectrum: (a) without switching decision algorithm and (b) with switching decision algorithm.

References

1 H. Komurcugil, S. Ozdemir, I. Sefa, N. Altin, and O. Kukrer, "Sliding-mode control for single-phase grid-connected LCL-filtered VSI with double-band hysteresis scheme," *IEEE Trans. Ind. Electron.*, vol. 63, no. 2, pp. 864–873, Feb. 2016.

2 H. Komurcugil, S. Biricik, S. Bayhan, and Z. Zhang, "Sliding mode control: overview of its applications in power converters," *IEEE Ind. Electron. Mag.*, vol. 15, no. 1, pp. 40–49, Mar. 2021.

3 O. Kukrer, H. Komurcugil, and A. Doganalp, "A three-level hysteresis function approach to the sliding mode control of single phase UPS inverters," *IEEE Trans. Ind. Electron.*, vol. 56, no. 9, pp. 3477–3486, Sept. 2009.

4 R. Guzman, L. G. de Vicuña, M. Castilla, J. Miret, and J. de la Hoz, "Variable structure control for three-phase LCL-filtered inverters using a reduced converter model," *IEEE Trans. Ind. Electron.*, vol. 65, no. 1, pp. 5–15, Jan. 2018.

5 R. Guzman, L. G. de Vicuna, J. Morales, M. Castilla, and J. Matas, "Sliding-mode control for a three-phase unity power factor rectifier operating at fixed switching frequency," *IEEE Trans. Power Electron.*, vol. 31, no. 1, pp. 758–769, Jan. 2016.

6 R. Guzman, L. G. de Vicuna, J. Morales, M. Castilla, and J. Miret, "Model-based control for a three-phase shunt active power filter," *IEEE Trans. Ind. Electron.*, vol. 63, no. 7, pp. 3998–4007, Jul. 2016.

7 R. Guzman, L. G. de Vicuna, J. Morales, M. Castilla, and J. Miret, "Model-based active damping control for three-phase voltage source inverters with LCL filter," *IEEE Trans. Power Electron.*, vol. 32, no. 7, pp. 5637–5650, Jul. 2017.

8 J. Kanieski, R. Cardoso, H. Pinheiro, and H. Grundling, "Kalman filter-based control system for power quality conditioning devices," *IEEE Trans. Ind. Electron.*, vol. 60, no. 11, pp. 5214–5227, Nov. 2013.

9 G. F. Franklin, J. D. Powell, and M. L. Workman, *Digital Control of Dynamic Systems*. Reading, MA, USA: Addison-Wesley, 1997.

10 Q. Xu and Y. Li, "Model predictive discrete-time sliding mode control of a nanopositioning piezostage without modeling hysteresis," *IEEE Trans. Control Systems Technol.*, vol. 20, no. 4, pp. 983–994, Jul. 2012.

11 R. Peña Alzola, M. Liserre, F. Blaabjerg, M. Ordonez, and T. Kerekes, "A self-commissioning notch filter for active damping in a three-phase LCL-filter-based grid-tie converter," *IEEE Trans. Power Electron.*, vol. 29, no. 12, pp. 6754–6761, Dec. 2014.

12 R. Peña Alzola, M. Liserre, F. Blaabjerg, M. Ordonez, and Y. Yang, "LCL-filter design for robust active damping in grid-connected converters," *IEEE Trans. Ind. Inform.*, vol. 10, no. 4, pp. 2192–2203, Nov. 2014.

13 H. P. Park and J. H. Jung, "PWM and PFM hybrid control method for LLC resonant converters in high switching frequency operation," *IEEE Trans. Ind. Electron.*, vol. 64, no. 1, pp. 253–263, Jan. 2017.

14 S. Mariethoz and M. Morari, "Explicit model-predictive control of a PWM inverter with an LCL filter," *IEEE Trans. Ind. Electron.*, vol. 56, no. 2, pp. 389–399, Feb. 2008.

15 N. B. Youssef, K. Al-Haddad, and H. -Y. Kanaan, "Sensorless nonlinear control of a three-phase/ switch/ level Vienna rectifier based on a numerical reconstruction of DC and AC voltages," *IEEE Ind. Appl. Soc. Annu. Meet.*, vol. 2008, pp. 1–7, May 2008.

16 A. Isidori, *Nonlinear Control Systems*. New York, NY, USA: Springer Verlag, 1995.

17 M. Eissa, S. Leeb, G. C. Verghese, and A. Stankovic, "Fast controller for a unity-power-factor PWM rectifier," *IEEE Trans. Power Electron.*, vol. 11, no. 1, pp. 1–6, Jan. 1996.

18 O. Lopez, L. Garcia De Vicuna, M. Castilla, J. Matas, and M. Lopez, "Sliding mode control design of a high-power factor buck-boost rectifier," *IEEE Trans. Ind. Electron.*, vol. 46, no. 3, pp. 604–612, Jun. 1999.

19 P. Rodriguez, A. Luna, I. Candela, R. Mujal, R. Teodorescu, and F. Blaabjerg, "Multiresonant frequency-locked loop for grid synchronization of power converters under distorted grid conditions," *IEEE Trans. Power Electron.*, vol. 58, no. 1, pp. 127–138, Jan. 2011.

20 J. Matas, M. Castilla, J. Miret, L. Garcia de Vicuna, and R. Guzman, "An adaptive prefiltering method to improve the speed/accuracy tradeoff of voltage sequence detection methods under adverse grid conditions," *IEEE Trans. Power Electron.*, vol. 61, no. 5, pp. 2139–2151, May 2014.

21 D. Holmes, R. Davoodnezhad, and B. McGrath, "An improved three-phase variable-band hysteresis current regulator," *IEEE Trans. Power Electron.*, vol. 28, no. 1, pp. 441–450, Jan. 2013.

22 H. Komurcugil and S. Biricik, "Time-varying and constant switching frequency-based sliding-mode control methods for transformerless DVR employing half-bridge VSI," *IEEE Trans. Ind. Electron.*, vol. 64, no. 4, pp. 2570–2579, Apr. 2017.

23 H. Komurcugil, "Rotating-sliding-line based sliding-mode control for single-phase UPS inverters," *IEEE Trans. Ind. Electron.*, vol. 59, no. 10, pp. 3719–3726, Oct. 2012.

24 A. Abrishamifar, A. Ahmad, and M. Mohamadian, "Fixed switching frequency sliding mode control for single-phase unipolar inverters," *IEEE Trans. Power Electron.*, vol. 27, no. 5, pp. 2507–2514, May 2012.

25 N. R. Zargari, P. D. Ziogas, and G. Joos, "A two-switch high-performance current regulated DC/AC converter module," *IEEE Trans. Ind. Appl.*, vol. 31, no. 3, pp. 583–589, May/Jun. 1995.

26 S. Rahmani, A. Hamadi, and K. Al-Haddad, "A Lyapunov-function-based control for a three-phase shunt hybrid active filter," *IEEE Trans. Ind. Electron.*, vol. 59, no. 3, pp. 1418–1429, Mar. 2012.

8

Design of Lyapunov Function-Based Control of Various Power Converters

8.1 Introduction

This chapter presents the Lyapunov function-based control of various power converters which include single-phase grid-connected inverter with LCL filter, single-phase quasi-Z-source grid-connected inverter with LCL filter, single-phase uninterruptible power supply inverter, and three-phase AC–DC rectifier. The design of Lyapunov-function-based control for each converter is presented in detail. The MATLAB/Simulink model of each converter together with the Lyapunov function-based control is made available to help the practicing engineers, researchers as well as graduate students who are interested in Lyapunov function based control design.

8.2 Single-Phase Grid-Connected Inverter with LCL Filter

8.2.1 Mathematical Modeling and Controller Design

The mathematical modeling of grid-connected inverter with LCL filter and controller design explained in this chapter are based on the method presented in [1]. Figure 8.1 shows a single-phase grid-connected inverter with LCL filter. The LCL filter is connected between inverter and grid and it consists of an inverter side inductor (L_1), a capacitor (C), and a grid side inductor (L_2).

The differential equations describing the operation of the system can be written as [1]

$$L_1 \frac{di_1}{dt} + r_1 i_1 = v_i - v_C \tag{8.1}$$

Advanced Control of Power Converters: Techniques and MATLAB/Simulink Implementation, First Edition. Hasan Komurcugil, Sertac Bayhan, Ramon Guzman, Mariusz Malinowski, and Haitham Abu-Rub.
© 2023 The Institute of Electrical and Electronics Engineers, Inc.
Published 2023 by John Wiley & Sons, Inc.
Companion website: www.wiley.com/go/komurcugil/advancedcontrolofpowerconverters

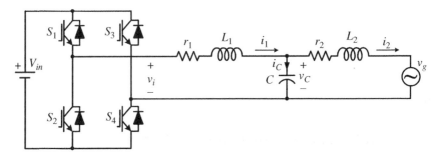

Figure 8.1 Single-phase grid-connected inverter with LCL filter. *Source:* Adapted from Komurcugil et al. [1].

$$L_2 \frac{di_2}{dt} + r_2 i_2 = v_C - v_g \qquad (8.2)$$

$$C \frac{dv_C}{dt} = i_1 - i_2 \qquad (8.3)$$

where $v_i = uV_{in}$, u is the control input, $v_g = V_m \sin(\omega t)$ is the grid voltage, r_1 and r_2 are the parasitic resistances of inductors L_1 and L_2, respectively. It is obvious that (8.1)–(8.3) are linear.

The control input can be written as

$$u = U_o + \Delta u \qquad (8.4)$$

where U_o and Δu denote the steady-state and perturbed values, respectively. The expression of control input in the steady-state can be obtained from (8.1) as

$$U_o = \frac{1}{V_{in}} \left(L_{1e} \frac{di_1^*}{dt} + r_{1e} i_1^* + v_C^* \right) \qquad (8.5)$$

where v_C^* and i_1^* are the references for v_C and i_1 given by

$$v_C^* = L_{2e} \frac{di_2^*}{dt} + r_{2e} i_2^* + v_g \qquad (8.6)$$

$$i_1^* = i_C^* + i_2^* = L_{2e} C_e \frac{d^2 i_2^*}{dt^2} + r_{2e} C_e \frac{di_2^*}{dt} + C_e \frac{dv_g}{dt} + i_2^* \qquad (8.7)$$

where L_{1e}, r_{1e}, L_{2e}, r_{2e} and C_e are the estimated values of L_1, r_1, L_2, r_2 and C, respectively. Since the current injected into the grid is required to be sinusoidal and in phase with the grid voltage, then reference grid current should be selected as follows

$$i_2^* = I_2^* \sin(\omega t) \qquad (8.8)$$

The state variables and control inputs can be defined as follows

$$x_1 = i_1 - i_1^*, x_2 = i_2 - i_2^*, x_3 = v_C - v_C^* \qquad (8.9)$$

Substituting $i_1 = x_1 + i_1^*$, $i_2 = x_2 + i_2^*$, $v_C = x_3 + v_C^*$, and $u = U_o + \Delta u$ into (8.1)–(8.3) and extracting the terms that involve state variables, the following equations can be obtained

$$L_1 \frac{dx_1}{dt} = \varepsilon_{L_1} \frac{di_1^*}{dt} + \varepsilon_{r_1} i_1^* - r_1 x_1 - x_3 + \Delta u V_{in} \tag{8.10}$$

$$L_2 \frac{dx_2}{dt} = \varepsilon_{L_2} \frac{di_2^*}{dt} + \varepsilon_{r_2} i_2^* - r_2 x_2 + x_3 \tag{8.11}$$

$$C \frac{dx_3}{dt} = \varepsilon_C \frac{dv_C^*}{dt} + x_1 - x_2 \tag{8.12}$$

where $\varepsilon_{L_1} = L_{1e} - L_1$, $\varepsilon_{r_1} = r_{1e} - r_1$, $\varepsilon_{L_2} = L_{2e} - L_2$, $\varepsilon_{r_2} = r_{2e} - r_2$, and $\varepsilon_C = C_e - C$ are the estimation errors.

Now, let us consider the Lyapunov function and its rate of change as follows

$$V(x) = \frac{1}{2}L_1 x_1^2 + \frac{1}{2}L_2 x_2^2 + \frac{1}{2}Cx_3^2 \tag{8.13}$$

$$\frac{dV(x)}{dt} = x_1 L_1 \frac{dx_1}{dt} + x_2 L_2 \frac{dx_2}{dt} + x_3 C \frac{dx_3}{dt} \tag{8.14}$$

Substituting (8.10)–(8.12) into (8.14) yields

$$\frac{dV(x)}{dt} = x_1 \varepsilon_{L_1} \frac{di_1^*}{dt} + x_1 \varepsilon_{r_1} i_1^* + x_2 \varepsilon_{L_2} \frac{di_2^*}{dt} + x_2 \varepsilon_{r_2} i_2^* - r_1 x_1^2 - r_2 x_2^2 + x_1 \Delta u V_{in} \tag{8.15}$$

It is worthy to remark that the estimation errors become zero when the estimated parameters equal to the actual parameters. In this case, the negative definiteness of $\dfrac{dV(x)}{dt}$ can be guaranteed if the perturbed control input is selected as

$$\Delta u = K_{i1} V_{in} x_1, \quad K_{i1} < 0 \tag{8.16}$$

where K_{i1} is a constant which determines the performance of controller. It is worthy to remark that $|K_{i1}|$ should be selected as large as possible so that the term $x_1 \Delta u V_{in}$ in (8.15) dominates the right hand side of (8.15) and assures $\dfrac{dV(x)}{dt} < 0$. Combining the steady-state and perturbed control inputs (U_o and Δu), one can write the following expression

$$u = \frac{1}{V_{in}}\left(L_{1e}\frac{di_1^*}{dt} + r_{1e}i_1^* + v_C^*\right) + K_{i1}V_{in}x_1 \tag{8.17}$$

When the inverter is controlled with (8.17), the closed-loop system is globally asymptotically stable under perturbations away from the operating point. However, the grid current is oscillatory due to the closed-loop poles located close to the imaginary axis. These oscillations can be removed by including the capacitor voltage feedback into (8.16). On the other hand, the computation of (8.17) requires i_1^* and v_C^*. While v_C^* can be generated by using (8.6), the generation of i_1^* using (8.7) is challenging since it involves first and second derivative operations. Moreover, since i_1^* is sensitive to L_{2e}, r_{2e} and C_e, the generation of i_1^* becomes inaccurate which may adversely affect the performance of the closed-loop system and cause steady-state error in the grid current. As a remedy to this, i_1^* can be generated by using a proportional-resonant (PR) controller that has the ability to track ac signals with zero steady-state error. The modification of the Lyapunov-based control with the capacitor voltage feedback and the description of the PR controller are presented in the next sub-section.

8.2.2 Controller Modification with Capacitor Voltage Feedback

In order to achieve the desired damping performance, the control approach in (8.16) is modified by adding the following capacitor voltage feedback

$$\Delta u = K_{i1} V_{in} x_1 - K_v x_3, \quad K_{i1} < 0 \quad \text{and} \quad K_v > 0 \tag{8.18}$$

Substitution of (8.18) into (8.15) results in

$$\begin{aligned}
\frac{dV(x)}{dt} &= x_1 \varepsilon_{L_1} \frac{di_1^*}{dt} + x_1 \varepsilon_{r_1} i_1^* + x_2 \varepsilon_{L_2} \frac{di_2^*}{dt} + x_2 \varepsilon_{r_2} i_2^* - r_1 x_1^2 - r_2 x_2^2 \\
&\quad + V_{in}^2 K_{i1} x_1^2 - x_1 x_3 K_v V_{in}
\end{aligned} \tag{8.19}$$

Under perfect parameter match (i.e.: $\varepsilon_{L_1} = 0$, $\varepsilon_{r_1} = 0$, $\varepsilon_{L_2} = 0$, $\varepsilon_{r_2} = 0$, and $\varepsilon_C = 0$), the negative definiteness of $\dfrac{dV(x)}{dt}$ can be assured if K_v is chosen as

$$K_v > \left(K_{i1} V_{in} - \frac{r_1}{V_{in}} - \frac{r_2 x_2^2}{V_{in} x_1^2} \right) \frac{x_1}{x_3} \tag{8.20}$$

However, under parameter mismatch, K_{i1} and K_v should be selected as large as possible such that the terms $V_{in}^2 K_{i1} x_1^2$ and $x_1 x_3 K_v V_{in}$ in (8.19) guarantee the negative definiteness of $\dfrac{dV(x)}{dt}$. Now, combining (8.5) and (8.18), the overall expression of the control input involving capacitor voltage feedback can be written as

$$u = \frac{1}{V_{in}} \left(L_{1e} \frac{di_1^*}{dt} + r_{1e} i_1^* + v_C^* \right) + K_{i1} V_{in} x_1 - K_v x_3 \tag{8.21}$$

8.2.3 Inverter-Side Current Reference Generation Using Proportional-Resonant Controller

In this sub-section, the generation of i_1^* by using a PR controller will be explained. The transfer function of a PR controller in the s-domain is given by [2]

$$G_{PR}(s) = K_p + \frac{2K_r \omega_c s}{s^2 + 2\omega_c s + \omega^2} \tag{8.22}$$

where ω is the resonant frequency, ω_c is the cut-off frequency, K_p and K_r are the proportional and resonant gains, respectively. When the input to the PR controller is grid current error $(i_2^* - i_2)$, the PR controller produces i_1^*. The expression of i_1^* in s-domain can be written as

$$I_1^*(s) = \left(I_2^*(s) - I_2(s)\right)\left(\frac{K_p s^2 + 2\omega_c\left(K_p + K_r\right)s + K_p\omega^2}{s^2 + 2\omega_c s + \omega^2}\right) \qquad (8.23)$$

If ω is set to the frequency of reference grid current (i.e.: 100π rad/s), then the PR controller defined in (8.22) introduces a very high gain at the resonant frequency ω so that the grid current tracks its reference with zero error in the steady-state. It is worth noting that the desired dynamic and steady-state responses can be determined by tuning K_p and K_r. The block diagram of Lyapunov function-based control with capacitor voltage feedback with the grid-connected inverter with LCL filter is shown in Figure 8.2. It can be seen that there are two inner loops that combine v_C and i_1 feedbacks and an outer loop. While the inner loops achieve the global stability of the closed-loop system and improve the damping performance of the system, the outer loop employs PR controller that guarantees zero steady-state error in the grid current. Hence, the Lyapunov function-based control strategy relies on cascading the inverter current loop (inner loop) with the outer

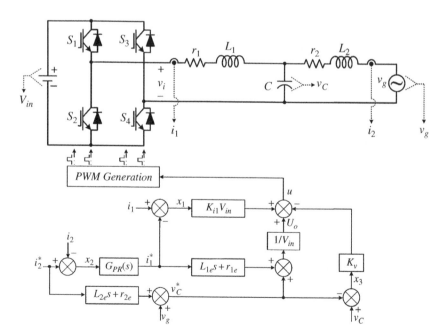

Figure 8.2 Block diagram of Lyapunov function-based control strategy with capacitor voltage feedback with the grid-connected inverter with LCL filter.

loop. The main advantage of such control technique is the improvement in the dynamic response of the control loop. It is clear from the block diagram that the effect of any disturbance within the inner loop is corrected by the outer loop before it reaches the output. As a consequence, the dynamic response of the overall system is improved.

8.2.4 Grid Current Transfer Function

The closed-loop transfer function relating the reference grid current (i_2^*) to the actual grid current (i_2) is useful in investigating system's behavior in the frequency domain. For the sake of simplifying the derivation of transfer function, the inductor resistances (r_1 and r_2) are neglected. Nevertheless, since r_1 and r_2 provide additional passive damping, the transfer function without r_1 and r_2 would represent the worst case. Now, substituting (8.21) into (8.1) yields

$$L_1 \frac{di_1}{dt} = L_{1e} \frac{di_1^*}{dt} + v_C^* + K_{i1} V_{in}^2 \left(i_1 - i_1^*\right) - K_v V_{in} \left(v_C - v_C^*\right) - v_C \tag{8.24}$$

The expression of v_C can be solved from (8.2) as

$$v_C = L_2 \frac{di_2}{dt} + v_g \tag{8.25}$$

It should be noted that (8.25) is derived by considering $r_2 = 0$ due to the reasons explained above. Now, substituting (8.25) and (8.6) with $r_{2e} = 0$ into (8.24), taking Laplace transform of resulting equation and replacing $I_1^*(s)$ by (8.23), one can obtain the closed-loop transfer function from i_2^* to i_2 as follows

$$H(s) = \frac{I_2(s)}{I_2^*(s)} = \frac{N_3 s^3 + N_2 s^2 + N_1 s + N_0}{D_5 s^5 + D_4 s^4 + D_3 s^3 + D_2 s^2 + D_1 s + D_0} \tag{8.26}$$

where the coefficients in the nominator and denominator are given by

$$\begin{aligned}
N_3 &= L_{1e} K_p + L_{2e}(1 + K_v V_{in}) \\
N_2 &= 2\omega_c L_{1e} \left(K_p + K_r\right) - K_p K_{i1} V_{in}^2 + 2\omega_c L_{2e}(1 + K_v V_{in}) \\
N_1 &= \omega^2 L_{1e} K_p - 2\omega_c K_{i1} V_{in}^2 \left(K_p + K_r\right) + \omega^2 L_{2e}(1 + K_v V_{in}) \\
N_0 &= -\omega^2 K_p K_{i1} V_{in}^2
\end{aligned} \tag{8.27}$$

$$\begin{aligned}
D_5 &= L_1 L_2 C \\
D_4 &= 2\omega_c L_1 L_2 C - L_2 C K_{i1} V_{in}^2 \\
D_3 &= L_1 + L_2 + \omega^2 L_1 L_2 C - 2\omega_c L_2 C K_{i1} V_{in}^2 + L_{1e} K_p + L_2 K_v V_{in} \\
D_2 &= 2\omega_c \left(L_{1e} K_p + L_{1e} K_r + L_1 + L_2 + L_2 K_v V_{in}\right) - K_{i1} V_{in}^2 \left(1 + K_p + \omega^2 L_2 C\right) \\
D_1 &= \omega^2 \left(L_1 + L_2 + L_{1e} K_p + L_2 K_v V_{in}\right) - 2\omega_c K_{i1} V_{in}^2 \left(K_p + K_r + 1\right) \\
D_0 &= -\omega^2 K_{i1} V_{in}^2 \left(1 + K_p\right)
\end{aligned} \tag{8.28}$$

It can be noticed that $H(s)$ is not sensitive to the variations in C. The stability of Lyapunov function-based control can also be investigated by examining the characteristic equation which appears in the denominator of $H(s)$. Since $K_{i1} < 0$, $K_v > 0$, $K_p > 0$ and $K_r > 0$, all coefficients (D_0, D_1, D_2, D_3, D_4, D_5) in the characteristic equation are positive. According to Routh–Hurwitz stability criterion (see Chapter 1), the closed-loop system is stable if the following conditions hold

$$D_1 D_4 - D_0 D_5 > 0$$

$$D_3 D_4 - D_2 D_5 > 0$$

$$(D_3 D_4 - D_2 D_5)[D_1 D_2 D_4 - D_0 D_5(D_2 + D_4)] + (D_1 D_4 - D_0 D_5)(D_0 D_4 D_5 - D_1 D_4^2) > 0$$

$$(8.29)$$

However, it is worth noting that these conditions cannot provide the desired values for the controller gains. Since the characteristic equation is fifth order, it is extremely difficult to obtain the optimum controller gains analytically.

The effectiveness of the Lyapunov function-based control with and without capacitor voltage feedback under parameter variations in the LCL filter is investigated. Using $V_{in} = 400$ V, $V_g = 230\sqrt{2}$ V , $L_1 = 1.436$ mH, $L_2 = 0.6867$ mH, $C = 49.404$ μF, $\omega = 100\pi$ rad/s, $\omega_c = 1$ rad/s, $K_p = 10$, $K_r = 400$, and considering 15% variation in all LCL parameters at the same time, the magnitude and phase responses of $H(s)$ are plotted as shown in Figure 8.3. It can be seen from the magnitude response (see Figure 8.3a) that a peak occurs when the control does not involve capacitor voltage feedback (i.e.: $K_v = 0$). This peak cannot be damped even if different K_{i1} value is used as shown in Figure 8.3b. However, when the capacitor voltage loop is employed (i.e.: $K_v \neq 0$), the peak is damped as shown in Figure 8.3a. The desired damping can be achieved by varying the value of K_v. On the other hand, in spite of 15% variation in the LCL filter parameters, no steady-state error (0 dB magnitude at 50 Hz) and phase shift (0° phase at 50 Hz) occur in the grid current. This result clearly demonstrates the effect of using PR controller in achieving zero steady-state error and phase shift in the grid current.

8.2.5 Harmonic Attenuation and Harmonic Impedance

The harmonic attenuation performance of LCL filter can be investigated if a transfer function relating i_2 and i_1 is obtained. The open-loop transfer function of the undamped LCL filter (i.e.: $r_1 = 0$ and $r_2 = 0$) can be written as follows

$$\frac{I_2(s)}{I_1(s)} = \frac{1}{L_2 C s^2 + 1} \tag{8.30}$$

It is obvious from (8.30) that an s term is missing in the denominator. This missing term causes a sharp peak at the resonant frequency of $L_2 C$ which endangers the stability according to the Routh–Hurwitz stability criterion. In order to obtain a stable system, the sharp peak should be attenuated.

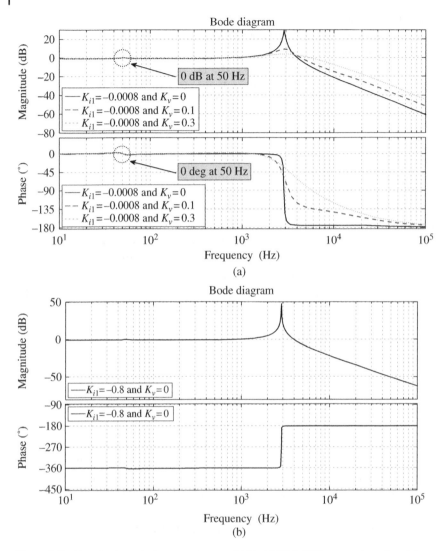

Figure 8.3 Magnitude and phase responses of $H(s)$: (a) With 15% parameter mismatch. (b) With 15% parameter mismatch using $K_{i1} = -0.8$ and $K_v = 0$. *Source:* Komurcugil et al. [1]/IEEE.

Now, let us investigate attenuation performance of the LCL filter with the Lyapunov function-based control. The closed-loop transfer function relating i_2 and i_1^* can be written as follows

$$\frac{I_2(s)}{I_1^*(s)} = \frac{L_1 s - K_{i1} V_{in}^2}{D_5 s^3 - \underbrace{K_{i1} V_{in}^2 L_2 C s^2}_{damping} + (L_1 + L_2 + \underbrace{K_v V_{in} L_2)s}_{damping} - \underbrace{K_{i1} V_{in}^2}_{damping}} \tag{8.31}$$

In the derivation of (8.31), it is assumed that i_1^* is already generated by the PR controller and there is perfect match between the estimated and actual parameters. It is obvious that the Lyapunov function-based control introduces additional terms D_5, $K_{i1} V_{in}^2$, $K_v V_{in} L_2$ and $K_{i1} V_{in}^2$ in the denominator of (8.31). This means that the Lyapunov function-based control method with capacitor voltage feedback not only guarantees the global stability, but also provides damping whose extent can be adjusted by K_{i1} and K_v. The magnitude responses of (8.31) obtained with different K_{i1} and K_v values are shown in Figure 8.4. It is clear that the Lyapunov function-based control suppresses the high frequency harmonics in i_2. Although different K_{i1} and K_v values result in different attenuation levels in the 100–10000 Hz range, the attenuation level remains the same for higher frequencies.

On the other hand, the distorted grid voltage containing various harmonics causes a serious distortion in the grid current. Therefore, the ability of mitigating

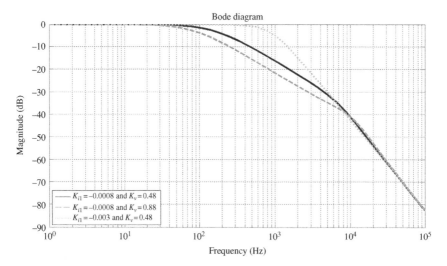

Figure 8.4 Magnitude responses of $I_2(s)/I_1^*(s)$ obtained with different K_{i1} and K_v values. *Source:* Komurcugil et al. [1]/IEEE.

the current distortion can be investigated by using the harmonic impedance which is defined as

$$Z_h(s) = \frac{V_g(s)}{I_2(s)} = \frac{D_5 s^5 + D_4 s^4 + D'_3 s^3 + D'_2 s^2 + D'_1 s + D_0}{-L_1 C s^2 + K_{i1} V_{in}^2 C s} \tag{8.32}$$

where D'_1, D'_2, and D'_3 are the coefficients defined in (8.28) with L_{1e} replaced by L_1. The magnitude of $Z_h(s)$ can be used to determine the harmonics in the current. For instance, the grid voltage harmonics result in a relatively small harmonic current when the magnitude of $Z_h(s)$ is large. The magnitude of $Z_h(s)$ obtained with different K_p values, while the values of other control parameters are same as given before, is depicted in Figure 8.5. It is clear that $Z_h(s)$ exhibits its minimum (which is above 140 dB) around 50 Hz and increases for higher frequencies as K_p increases. This clearly shows that the Lyapunov function-based control strategy is capable of rejecting the grid current distortion caused by distorted grid voltage for low-frequency as well as high-frequency.

8.2.6 Results

The effectiveness of the Lyapunov function-based control method has been verified by simulations and experiments on a 3.3 kW LCL-filter based single-phase grid-connected VSI prototype. Simulations were carried out by MATLAB/

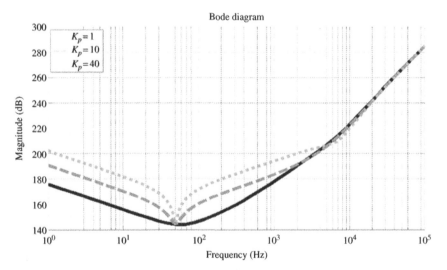

Figure 8.5 Magnitude responses of $Z_h(s)$ with different K_p values. *Source:* Komurcugil et al. [1]/IEEE.

Simulink. In the experimental study, the Lyapunov function-based control method was implemented by using TMS320F28335 floating point digital signal processor (DSP). The sensed signals are then applied to the DSP through its analog-to-digital converter (ADC). A phase locked loop (PLL) is used to generate the current reference synchronized with the grid voltage and frequency. After generating the required reference signals in the DSP (i_1^*, i_2^*, and v_C^*), the control input is computed. Then, the PWM signals are produced by comparing the computed switching function with a carrier wave. The system and control parameters are given in Table 8.1.

Figure 8.6 shows the simulated and experimental responses of the grid voltage and current in the steady-state obtained by the Lyapunov function-based control method with PR control and without capacitor voltage feedback. Here, it is assumed that the estimated LCL parameters match with the actual parameters. It is clear that the injected grid current is in phase with the grid voltage. The oscillations on the grid current clearly show that the controller is not able to remove the oscillations in the grid current.

Figure 8.7 shows the simulated and experimental responses of the grid voltage and current in the steady-state obtained by the Lyapunov function-based control method with the capacitor voltage feedback under perfect parameter match.

Table 8.1 System and control parameters.

Symbol	Value
DC link voltage, V_{in}	400 V
Inverter-side inductance, L_1	1.436 mH
Filter capacitance, C	49.404 μF
Grid-side inductance, L_2	0.6867 mH
Inductor resistors, r_1 and r_2	0.17 and 0.076 Ω
Grid voltage amplitude, V_g	$230\sqrt{2}$V
Switching, sampling and grid frequencies, f_{sw}, f_s, and f_g	12.5 kHz, 12.5 kHz, and 50 Hz
Current gain, K_{i1}	-3×10^{-4}
Voltage gain, K_v	0.48
PR gains, K_p and K_r	10 and 500
Cut-off frequency, ω_c	1 rad/s
Resonant frequency, ω	314.16 rad/s

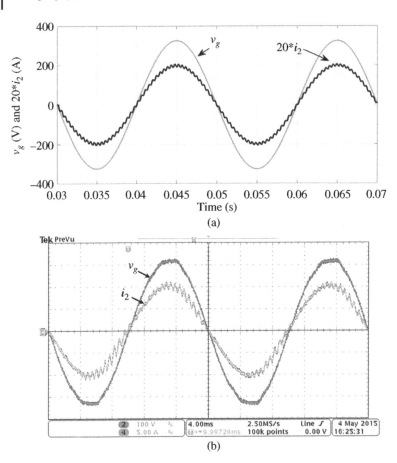

Figure 8.6 Simulated and experimental response of v_g and i_2 obtained by the Lyapunov function-based control method without using capacitor voltage feedback. (a) Simulation, (b) Experiment. *Source:* Komurcugil et al. [1]/IEEE.

Comparing the grid currents in Figures 8.6 and 8.7, one can see the oscillations in the grid current are removed when the capacitor voltage feedback is employed.

Figure 8.8 shows the simulated and experimental responses of the grid voltage and current in the steady-state obtained by the Lyapunov function-based control method with the capacitor voltage loop when the estimated parameters differ from the actual parameters by 15% ($L_{1e} = 1.6514$ mH, $r_{1e} = 0.1955\ \Omega$, $L_{2e} = 0.789705$ mH, and $r_{2e} = 0.0874\ \Omega$). The worst case parameter variation impact to the control system occurs when all the parameter variations exist at the same time. Despite these parameter variations, the grid current has almost the same amplitude and is in

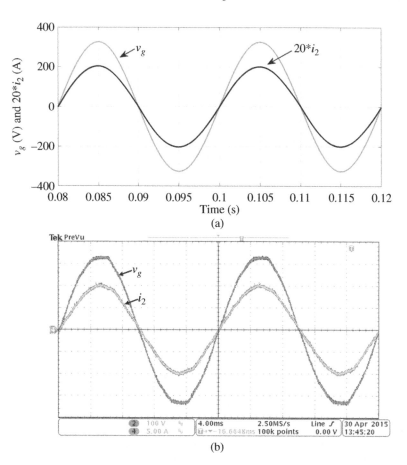

Figure 8.7 Simulated and experimental response of v_g and i_2 obtained by the Lyapunov function-based control method with using capacitor voltage feedback. (a) Simulation, (b) Experiment. *Source:* Komurcugil et al. [1]/IEEE.

phase with the grid voltage compared to that of presented in Figure 8.7. This clearly shows the effectiveness of the PR controller in tracking the grid current reference in the case of parameter variations.

Figure 8.9 shows the simulated response of $\frac{dV(x)}{dt}$ in one cycle that corresponds to the case presented in Figure 8.8. Clearly, despite of 15% parameter variation in LCL filter parameters, $\frac{dV(x)}{dt}$ stays negative which means that the Lyapunov function-based control method with capacitor voltage feedback is globally asymptotically stable.

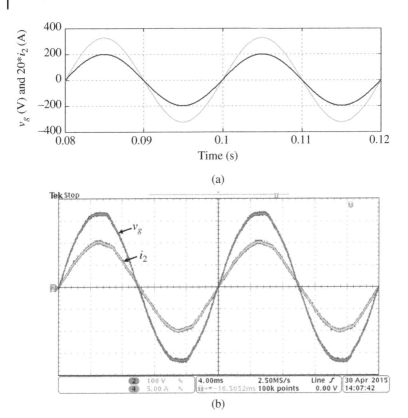

(a)

(b)

Figure 8.8 Simulated and experimental response of v_g and i_2 obtained by the Lyapunov function-based control method with using capacitor voltage feedback when there is 15% variation in LCL filter parameters. (a) Simulation, (b) Experiment. *Source:* Komurcugil et al. [1]/IEEE.

Figure 8.10 shows the simulated and experimental dynamic responses of the grid current for a step change in the grid current reference amplitude (I_2^*) from 10 to 20 A obtained by the Lyapunov function-based control method with capacitor voltage feedback under perfect parameter match. The measured spectrum of grid current is also depicted for $I_2^* = 20$ A. It can be seen that the Lyapunov function-based control method exhibits a fast dynamic response to this step change without having a stability problem. The THD of the grid current is 2.5% which is reasonable.

Figure 8.11 shows experimental steady-state responses of the grid voltage, grid current and grid voltage spectrum obtained by the Lyapunov function-based control method with capacitor voltage feedback under distorted grid voltage. Despite

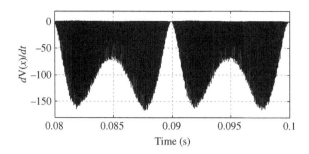

Figure 8.9 Simulated response of $\dfrac{dV(x)}{dt}$ in one cycle. *Source:* Komurcugil et al. [1]/IEEE.

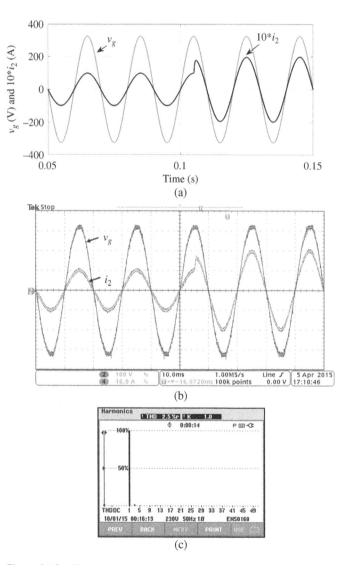

Figure 8.10 Simulated and experimental dynamic responses of i_2 for a step change in I_2^* from 10 to 20 A obtained by the Lyapunov function-based control method with using capacitor voltage feedback under perfect parameter match. (a) Simulation, (b) Experiment, (c) Measured spectrum of grid current at $I_2^* = 20$ A. *Source:* Komurcugil et al. [1]/IEEE.

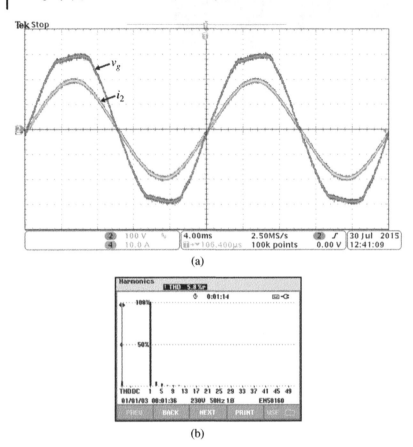

(a)

(b)

Figure 8.11 (a) Experimental response of v_g, i_2 and spectrum of v_g obtained by the Lyapunov function-based control method with capacitor voltage feedback under distorted grid voltage, (b) Measured spectrum of v_g. *Source:* Komurcugil et al. [1]/IEEE.

of THD = 5.8% with 3rd and 5th harmonic components in the grid voltage, the controller rejects these harmonic components and injects a sinusoidal grid current into the grid.

The Lyapunov function-based control method with capacitor voltage feedback is compared with the existing linear and nonlinear control methods such as the active damping method presented in [3], the SMC based control method presented in [4], the composite nonlinear feedback control method presented in [5], and feedback based PR current control method presented in [6]. A summary of comparison between four control methods and the Lyapunov function-based control method is shown in Table 8.2. Clearly, the Lyapunov function-based control method which is combination of linear and nonlinear control strategies offers many advantages

Table 8.2 Comparisons of four control strategies with the Lyapunov-function based control method.

Comparison category	[3]	[4]	[5]	[6]	Lyapunov function based control method
Number of sensors	2	4	4	3	4
Robustness	Not reported	Very good	Not reported	Not reported	Very good
Steady-state error	Not reported	Does not exist	Not reported	Does not exist	Does not exist
Global stability	Does not exist	Not reported	Does not exist	Does not exist	Guaranteed
Damping resistor	Not needed	Not needed	Not needed	Needed	Not needed
Dynamic response	Fast	Very fast	Very fast	Fast	Very fast
Harmonic rejection	Good	Not reported	Not reported	Not reported	Very good

compared with the existing linear and nonlinear control strategies separately. Especially, comparison of the Lyapunov function-based control with the feedback based PR current control strategy [6] shows that the Lyapunov function-based control strategy offers a novelty since it guarantees the global stability, improves the damping by making use of capacitor voltage loop rather than using a damping resistor which causes a power loss, and offers a very good harmonic rejection capability (sensitivity of the control strategy to grid voltage harmonics) for low and high frequencies rather than for a few preset low frequencies which are determined by the harmonic compensators in [6]. Recent control strategies that have been devised for grid connected VSIs are based on active damping rather than passive damping. Since the feedback based PR current control strategy employs a damping resistor, it cannot be considered as an active damping method which makes it out of date.

8.3 Single-Phase Quasi-Z-Source Grid-Connected Inverter with LCL Filter

8.3.1 Quasi-Z-Source Network Modeling

Figure 8.12 shows a single-phase quasi-Z-source grid-connected inverter (qZSI) with an LCL filter. It is clear that the quasi-Z-source (qZS) network is connected between DC the supply and dc input of the inverter [7]. The qSZ contains two

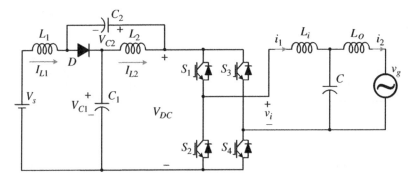

Figure 8.12 Single-phase grid-tied qZSI with LCL filter. *Source:* Komurcugil et al. [7]/IEEE.

inductors (L_1 and L_2), two capacitors (C_1 and C_2), and a diode (D). The operation of qZS network can be analyzed in two states known as the active state and shoot-through state.

Figure 8.13 depicts these states in which the inverter is represented by a constant current source during the active state and short-circuit in the shoot-through state. The active state equations can be obtained from Figure 8.13a as follows [7]

$$V_{L_1} = V_s - V_{C_1}, \quad V_{L_2} = -V_{C_2} \tag{8.33}$$

$$V_{DC} = V_{C_1} - V_{L_2} = V_{C_1} + V_{C_2}, \quad V_{diode} = 0 \tag{8.34}$$

The shoot-through state equations can be written as

$$V_{L_1} = V_s + V_{C_2}, \quad V_{L_2} = V_{C_1} \tag{8.35}$$

$$V_{DC} = 0, \quad V_{diode} = V_{C_1} + V_{C_2} \tag{8.36}$$

The average voltage of L_1 and L_2 in the steady-state over one switching period is zero. Using (8.33) and (8.35), one can obtain

$$V_{C_1} = \frac{T_a}{T_a - T_{ST}} V_s, \quad V_{C_2} = \frac{T_{ST}}{T_a - T_{ST}} V_s \tag{8.37}$$

where T_a and T_{ST} are the durations in the active and shoot-through states, respectively. The peak value of DC-link voltage (V_{DC}) at the input terminals of the inverter can be obtained from (8.34), (8.36) and (8.37) as follows

$$V_{DC} = V_{C_1} + V_{C_2} = \frac{T}{T_a - T_{ST}} V_s = BV_s \tag{8.38}$$

where B and T are the boost factor and switching period given by

$$T = T_a + T_{ST} \tag{8.39}$$

$$B = \frac{1}{1 - 2D_{ST}}, \quad D_{ST} = \frac{T_{ST}}{T} \tag{8.40}$$

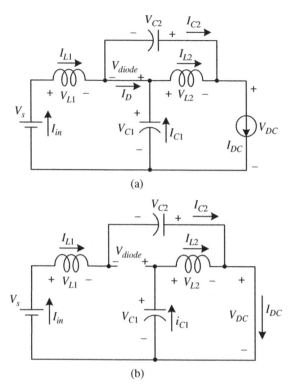

Figure 8.13 Equivalent circuit of the qZSI in [7]: (a) Active state, (b) Shoot through state. *Source:* Komurcugil et al. [7]/IEEE.

where D_{ST} is the duty ratio of shoot-through state which should satisfy the following inequality

$$0 < D_{ST} < 1, \quad D_{ST} \neq \frac{1}{2} \tag{8.41}$$

Therefore, for a given B value, the output voltage of qZS network (V_{DC}) can be greater than the input voltage (V_s). On the other hand, the average currents flowing through L_1 and L_2 can be written as follows

$$I_{L_1} = I_{L_2} = I_{in} = \frac{P_{in}}{V_s} \tag{8.42}$$

where P_{in} denotes the input power. Applying Kirchhoff's current law to Figure 8.13a and making use of (8.42), one can easily obtain

$$I_{C_1} = I_{C_2} = I_{DC} - I_{L_1}, \quad I_D = 2I_{L_1} - I_{DC} \tag{8.43}$$

8.3.2 Grid-Connected Inverter Modeling

It is evident from Figure 8.12 that the output voltage of qZS network is connected as the input dc-link voltage of the H-bridge inverter which is connected to the grid via an LCL filter. Applying Kirchhoff's voltage and current laws to the output of inverter, one can obtain the following differential equations

$$L_i \frac{di_1}{dt} + r_i i_1 = v_i - v_C \tag{8.44}$$

$$L_o \frac{di_2}{dt} + r_o i_2 = v_C - v_g \tag{8.45}$$

$$C \frac{dv_C}{dt} = i_1 - i_2 \tag{8.46}$$

where $v_g = V_g \sin(\omega t)$ is the grid voltage, $v_i = \mathbf{d} V_{DC}$ is the output voltage of the inverter, $\mathbf{d}$ is the switching function, r_i and r_o are the resistances of inductors L_i and L_o, respectively. The switching function can be written in terms of steady-state (d_{ss}) and perturbed (d_p) components as follows

$$\mathbf{d} = d_{ss} + d_p \tag{8.47}$$

Assuming that the system operates at the steady-state (i.e.: actual variables converged to the their references), the steady-state component of the switching function can be solved from (8.44) as

$$d_{ss} = \frac{1}{V_{DC}} \left(L_i \frac{di_1^*}{dt} + r_i i_1^* + v_C^* \right) \tag{8.48}$$

where v_C^* and i_1^* denote the references of v_C and i_1 given by

$$v_C^* = L_o \frac{di_2^*}{dt} + r_o i_2^* + v_g \tag{8.49}$$

$$i_1^* = i_C^* + i_2^* = L_o C \frac{d^2 i_2^*}{dt^2} + r_o C \frac{di_2^*}{dt} + C \frac{dv_g}{dt} + i_2^* \tag{8.50}$$

where i_2^* is the references of i_2. In grid-connected inverter systems, the injected current is required to be sinusoidal and in phase with the grid voltage. Therefore, the reference grid current can be written as

$$i_2^* = I_2^* \sin(\omega t) \tag{8.51}$$

It is worth noting that v_C^* and i_2^* can be generated accurately if the values of L_o, C, and r_o are known accurately. However, it is not possible to generate v_C^* and i_2^* accurately since the values of L_o, C, and r_o vary by heat, operating condition, aging and other factors.

8.3.3 Control of Quasi-Z-Source Network

The control of qZS network can be accomplished by controlling the shoot-through duty ratio (D_{ST}). In order to achieve this, the dc capacitor voltages and inductor currents in the qZS network should be controlled. In this study, the inductor current I_{L_1} and capacitor voltage V_{C_1} are controlled to produce D_{ST}. In literature, such control is referred to as the simple boost control [8]. It involves two proportional-integral (PI) controllers which operate with the capacitor voltage and inductor current errors as follows [7]

$$I_{L_1}^* = K_{p1}^{PI}\left(V_{C_1}^* - V_{C_1}\right) + K_{i1}^{PI}\int\left(V_{C_1}^* - V_{C_1}\right)dt \tag{8.52}$$

$$D_{ST} = K_{p2}^{PI}\left(I_{L_1}^* - I_{L_1}\right) + K_{i2}^{PI}\int\left(I_{L_1}^* - I_{L_1}\right)dt \tag{8.53}$$

The shoot-through duty ratio will be combined with the switching function in the next sub-section.

8.3.4 Control of Grid-Connected Inverter

The control of inverter will be achieved by using the Lyapunov-function-based control which aims to develop a control law such that the time derivative of Lyapunov-function is always negative. In [1], the conventional Lyapunov function-based control approach for a single-phase grid-connected inverter with LCL filter has been studied extensively. It is pointed out that, although the global stability is guaranteed, the conventional Lyapunov function-based control should be extended by involving the capacitor voltage feedback into the control loop such that the LCL resonance is damped as well as the closed-loop stability is secured.

The state variables used in this study are defined as

$$x_1 = i_1 - i_1^*, \quad x_2 = i_2 - i_2^*, \quad x_3 = v_C - v_C^* \tag{8.54}$$

Substituting (8.54) into (8.44)–(8.46) yields

$$L_i\frac{dx_1}{dt} = -r_i x_1 - x_3 + d_p V_{DC} \tag{8.55}$$

$$L_o\frac{dx_2}{dt} = -r_o x_2 + x_3 \tag{8.56}$$

$$C\frac{dx_3}{dt} = x_1 - x_2 \tag{8.57}$$

The formation of Lyapunov function is usually based on the energy dissipation of the converter under consideration and its details are available in Chapter 4. For the single-phase inverter system shown in Figure 8.12, the following Lyapunov function can be considered

$$V(x) = \frac{1}{2}L_i x_1^2 + \frac{1}{2}L_o x_2^2 + \frac{1}{2}C x_3^2 \tag{8.58}$$

The derivative of $V(x)$ can be written as

$$\frac{dV(x)}{dt} = x_1 L_i \frac{dx_1}{dt} + x_2 L_o \frac{dx_2}{dt} + x_3 C \frac{dx_3}{dt} \tag{8.59}$$

Substituting (8.55)–(8.57) into (8.59) gives

$$\frac{dV(x)}{dt} = -r_i x_1^2 - r_o x_2^2 + x_1 d_p V_{DC} \tag{8.60}$$

Apparently, $\dfrac{dV(x)}{dt}$ is always negative if the perturbed switching function is chosen as follows

$$d_p = K_i V_{DC} x_1, \quad K_i < 0 \tag{8.61}$$

However, as explained above, the damping performance of closed-loop system using (8.61) is insufficient. For this reason, the perturbed switching function is modified by including the capacitor voltage feedback (i.e.: x_3) as follows

$$d_p = K_i V_{DC} x_1 - K_v x_3, \quad K_v > 0 \tag{8.62}$$

Combining (8.48) and (8.62) yields the total switching function as follows

$$\mathbf{d} = \frac{1}{V_{DC}} \left(L_i \frac{di_1^*}{dt} + r_i i_1^* + v_C^* \right) + K_i V_{DC} x_1 - K_v x_3 \tag{8.63}$$

When (8.62) is substituted in (8.60), the derivative of Lyapunov function can be obtained as follows

$$\frac{dV(x)}{dt} = -r_i x_1^2 - r_o x_2^2 + V_{DC}^2 K_i x_1^2 - x_1 x_3 K_v V_{DC} \tag{8.64}$$

Clearly, the terms $-r_i x_1^2$, $-r_o x_2^2$ and $V_{DC}^2 K_i x_1^2$ are always negative. Hence, the term $-x_1 x_3 K_v V_{DC}$ can ensure $\dfrac{dV(x)}{dt} < 0$ by using appropriate K_v value. On the other hand, the implementation of control law in (8.63) requires references i_1^* and v_C^*.

8.3.5 Reference Generation Using Cascaded PR Control

As mentioned in sub-section 8.2.3, the PR controllers are widely used in tracking the reference sinusoidal signals. In [1], the inverter current reference (i_1^*) is generated by a PR current controller whose transfer function is defined as

$$G_{PR_1}(s) = K_{p1}^{PR} + \frac{2K_{r1}^{PR}\omega_c s}{s^2 + 2\omega_c s + \omega^2} \qquad (8.65)$$

where ω_c and ω denote the cut-off and resonant frequencies as defined before, and K_{p1}^{PR} and K_{r1}^{PR} are the proportional and resonant gains, respectively. The input applied to the PR controller is the grid current error $(i_2^* - i_2)$. In [1], v_C^* is generated by using (8.49) which requires derivative operation as well as the knowledge of L_o and r_o. However, the generation of v_C^* may be inaccurate due to the lack of exact knowledge about the parameters L_o and r_o. Also, the use of derivative is not preferred.

As a remedy to this problem, v_C^* is generated by a second PR controller whose input is connected to the output of first one. Such connection leads to cascade connected PRs. The transfer function of the second PR controller is given by

$$G_{PR_2}(s) = K_{p2}^{PR} + \frac{2K_{r2}^{PR}\omega_c s}{s^2 + 2\omega_c s + \omega^2} \qquad (8.66)$$

where K_{p2}^{PR} and K_{r2}^{PR} are the proportional and resonant gains, respectively. The input of the second PR controller is the inverter current error $\left(i_1^* - i_1\right)$ as shown in Figure 8.14 below. The block diagram of the complete controller is depicted in Figure 8.15.

8.3.6 Results

The effectiveness and feasibility of the Lyapunov function-based control approach has been verified by simulations using the system and control parameters listed in Table 8.3.

Figure 8.16 shows the responses of the DC-side variables and AC-side currents in the steady-state. The reference voltage of V_{C_1} was selected as 400 V. The amplitude of the reference grid current was selected as $I_2^* = 15$ A. Despite $V_s = 250$ V, the control of V_{C_1} is achieved successfully at 400 V as shown in Figure 8.16a. This means that the qZS network operates in the boost mode ($B > 1$). It is clear that a double-line frequency (2ω) ripple exists on V_{C_1}, V_{C_2}, and V_{DC}. It is worth noting that 2ω is discernible in the single-phase systems. The AC-side currents together

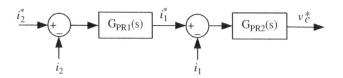

Figure 8.14 Reference function generation with cascaded PR controllers. *Source:* Komurcugil et al. [7]/IEEE.

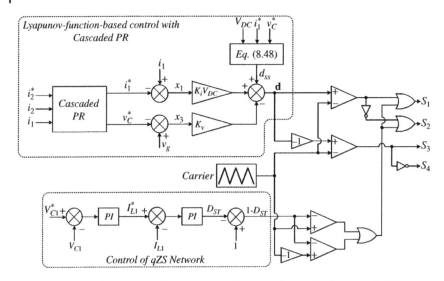

Figure 8.15 Block diagram of Lyapunov-function-based control with cascaded PR controllers.

Table 8.3 System and control parameters.

Description and symbol	Value
Dc-link voltage, V_s	250 V
qZS network inductances (L_1 and L_2)	3 mH
qZS network capacitors (C_1 and C_2)	1000 μF
Inverter-side inductance, L_i	1.4 mH
Filter capacitance, C	50 μF
Grid-side inductance, L_o	0.5 mH
Inductor resistors, r_i and r_o	0.08 and 0.05Ω
Grid voltage amplitude, V_g	$230\sqrt{2}$V
Switching and grid frequencies, f_{sw} and f_g	12.5 kHz and 50 Hz
Cut-off frequency, ω_c	1 rad/s
PI gains, K_{p1}^{PI}, K_{p2}^{PI}, K_{i1}^{PI} and K_{i2}^{PI}	0.051, 2, 0.5, and 20
PR gains, K_{p1}^{PR}, K_{p2}^{PR}, K_{r1}^{PR} and K_{r2}^{PR}	5, 1, 1000, and 1200
Control gains, K_i and K_v	−0.0008 and 0.875

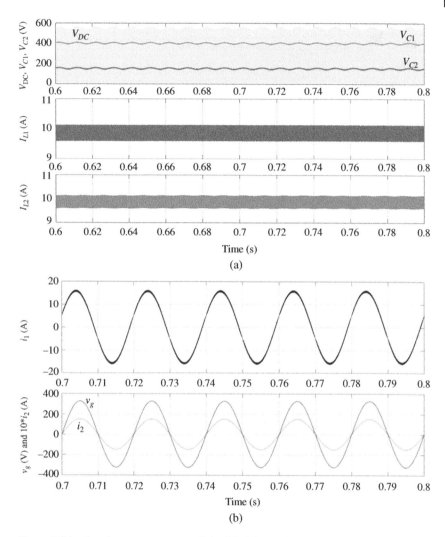

Figure 8.16 Steady-state responses of the DC-side variables and AC-side currents. (a) DC-side variables, (b) AC-side variables.

with the grid voltage are shown in Figure 8.16b. Clearly, the grid current is in phase with the grid voltage. The amplitude of grid current is 15 A which means that the controller has good tracking performance.

Figure 8.17 shows the dynamic responses of the DC-side variables and grid current for a step change in I_2^* from 15 to 20 A. It can be seen from Figure 8.17a that the DC-side variables can be controlled to their desired levels after the step change

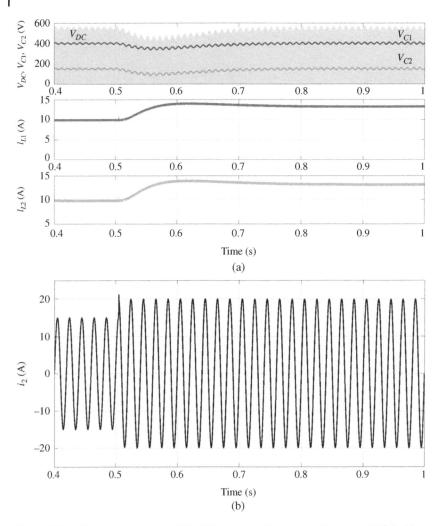

Figure 8.17 Dynamic responses of the DC-side variables and grid current. (a) DC-side variables, (b) Grid current.

occurs. Although V_{DC}, V_{C_1}, and V_{C_2} are affected from this step change, they settle down to their previous values before the step change. However, I_{L_1}, and I_{L_2} are changed to their new values. The settling time of these variables can be decreased further by fine tuning the PI gains. Nevertheless, the effect of slow convergence of the currents and voltages in the qZS network is not reflected to the AC-side currents as shown in Figure 8.17b. The dynamic response of the grid current is quite fast.

8.4 Single-Phase Uninterruptible Power Supply Inverter

8.4.1 Mathematical Modeling of Uninterruptible Power Supply Inverter

Figure 8.18 shows a single-phase voltage-source uninterruptible power supply (UPS) inverter. The equations describing the operation of the inverter can be written as [9]

$$L\frac{di_L}{dt} + ri_L = uV_{in} - v_o \tag{8.67}$$

$$C\frac{dv_o}{dt} = i_L - i_o \tag{8.68}$$

where r is the resistance of inductor L and u is the continuous signal in a closed interval $[-1, 1]$ which acts as the control input to the inverter system.

The control input can be defined in terms of its steady-state and perturbed terms as follows

$$u = U_o + \Delta u \tag{8.69}$$

where U_o and Δu are the steady-state and perturbed control inputs, respectively. Assuming that the inductor current and capacitor voltage track their references in the steady-state and control input is equal to the steady-state control input, (8.67) and (8.68) can be written as

$$L\frac{di_L^*}{dt} + ri_L^* = U_o V_{in} - v_o^* \tag{8.70}$$

$$C\frac{dv_o^*}{dt} = i_L^* - i_o \tag{8.71}$$

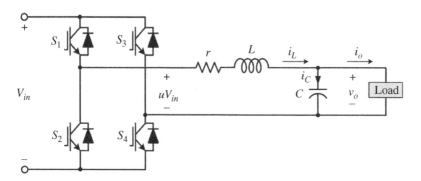

Figure 8.18 Single-phase voltage-source UPS inverter. *Source:* Adapted from Komurcugil et al. [9].

Defining the state variables as $x_1 = i_L - i_L^*$ and $x_2 = v_o - v_o^*$ and making use of (8.70) and (8.71), (8.67) and (8.68) can be written as

$$L \frac{dx_1}{dt} = \Delta u V_{in} - rx_1 - x_2 \qquad (8.72)$$

$$C \frac{dx_2}{dt} = x_1 \qquad (8.73)$$

8.4.2 Controller Design

The behavior of a system about its equilibrium point of interest can be investigated by using Lyapunov's stability theorem [10]. In the case of single-phase inverter, the equilibrium point of interest is ($x_1 = 0, x_2 = 0$). The main objective here is to achieve inverter's global asymptotic stability around this equilibrium point rather than its local stability. The former can be accomplished by using Lyapunov's direct method with the fact that the inverter system states converge to the equilibrium point if the total energy is continuously dissipated. When the state trajectory reaches to the equilibrium point, then the energy dissipation converges to zero. The typical energy distribution of the single-phase inverter is depicted in Figure 8.19. The input energy delivered by the DC power supply is denoted by E_{in}. The energy dissipated by the resistance of inductor and switching devices are denoted by E_r and E_{sw}, respectively. Hence, while part of E_{in} is dissipated by the resistance of inductor and switching devices, the rest of E_{in} is transferred to the load denoted by E_{out}. It is important to remark that the energy stored in the inverter is distributed in the inductor and capacitor due to the fact that these passive components do not dissipate energy. Therefore, part of E_{in} is

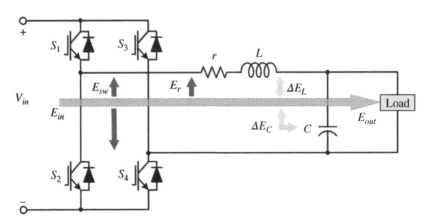

Figure 8.19 Distribution of energy in single-phase inverter. *Source:* Adapted from Komurcugil et al. [9].

exchanged by the energy stored (ΔE_L and ΔE_C) in these components in a bidirectional manner until the whole energy dissipation settles down to the equilibrium point of the inverter.

Now, let us consider the following Lyapunov function and its derivative

$$V(x) = \frac{1}{2}Lx_1^2 + \frac{1}{2}Cx_2^2 \tag{8.74}$$

$$\frac{dV(x)}{dt} = x_1 L \frac{dx_1}{dt} + x_2 C \frac{dx_2}{dt} \tag{8.75}$$

Substituting (8.72) and (8.73) into (8.75) yields

$$\frac{dV(x)}{dt} = V_{in}x_1\Delta u - rx_1^2 \tag{8.76}$$

It is evident that the derivative of Lyapunov function $\left(\dfrac{dV(x)}{dt}\right)$ is always negative if the perturbed control input is selected as

$$\Delta u = K_i V_{in}x_1, \quad K_i < 0 \tag{8.77}$$

where K_i is a real constant. The overall control input expression is given by

$$u = U_o + \Delta u = \frac{1}{V_{in}}\left(L\frac{di_L^*}{dt} + ri_L^* + v_o^*\right) + K_i V_{in}x_1 \tag{8.78}$$

where $v_o^* = V_m \sin(\omega t)$ is the reference of the output voltage (v_o), $i_L^* = C\dfrac{dv_o^*}{dt} + i_o$ is the reference of inductor current (i_L). Unlike the steady-state control input derivation in Section 8.2 where parameter mismatch was considered, the steady-state control input is derived assuming a perfect parameter match. Also, it can be noticed that the control input does not involve output voltage error. However, when the inverter is controlled with the control input in (8.78), the stability is guaranteed at the expense of steady-state error in the output voltage. For this reason, the control input should be modified by adding the output voltage feedback in to the control loop. Hence, the perturbed control input can be written as

$$\Delta u = K_i V_{in}x_1 - K_v x_2, \quad K_v > 0 \tag{8.79}$$

Substituting (8.79) into (8.76) gives

$$\frac{dV(x)}{dt} = K_i V_{in}^2 x_1^2 - rx_1^2 - K_v V_{in}x_1 x_2 \tag{8.80}$$

Since $K_i < 0$ and $r > 0$, the negative definiteness of (8.80) is determined by the value of K_v whose lower bound is given by

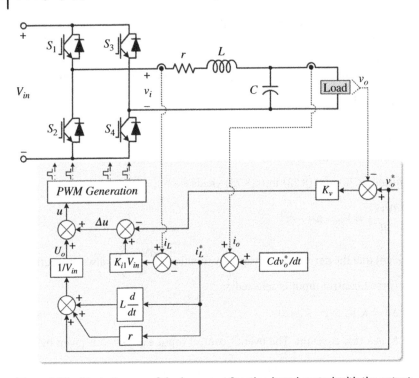

Figure 8.20 Block diagram of the Lyapunov-function based control with the output voltage feedback. *Source:* Adapted from Komurcugil et al. [9].

$$K_v > \left(\frac{K_i V_{in}^2 - r}{V_{in}} \right) \frac{x_1}{x_2} \tag{8.81}$$

Since (8.81) determines the lower bound of K_v, the value of K_v can be selected as large as possible to achieve the negative definiteness of (8.80). The overall control input with the output voltage feedback can be written as

$$u = \frac{1}{V_{in}} \left(L \frac{di_L^*}{dt} + ri_L^* + v_o^* \right) + K_i V_{in} x_1 - K_v x_2 \tag{8.82}$$

The block diagram of the Lyapunov-function based control with the output voltage feedback is depicted in Figure 8.20.

8.4.3 Criteria for Selecting Control Parameters

It is well known that the performance of a control method depends on many factors including the values of control parameters. Therefore, before attempting to select the appropriate controller parameter values, the effect of control parameters on the behavior of the closed-loop system should be investigated. Now, substituting (8.79) into (8.72) yields

$$L\frac{dx_1}{dt} = \left(K_iV_{in}^2 - r\right)x_1 - \left(K_vV_{in} + 1\right)x_2 \tag{8.83}$$

Equations (8.73) and (8.83) can be written in the following form

$$\frac{d}{dt}\begin{bmatrix} x_1 \\ x_2 \end{bmatrix} = \begin{bmatrix} \frac{1}{L}\left(K_iV_{in}^2 - r\right) & -\frac{1}{L}\left(K_vV_{in} + 1\right) \\ \frac{1}{C} & 0 \end{bmatrix}\begin{bmatrix} x_1 \\ x_2 \end{bmatrix} \tag{8.84}$$

The characteristic equation of (8.84) is the determinant given by

$$\det(sI - A) = s^2 + \frac{\left(r - K_iV_{in}^2\right)}{L}s + \frac{\left(K_vV_{in} + 1\right)}{L} \tag{8.85}$$

where A is the 2×2 matrix in (8.84) and s denotes the Laplace operator. Applying Routh-Hurwitz stability criterion, one can obtain the following conditions

$$\frac{\left(r - K_iV_{in}^2\right)}{L} > 0, \quad \frac{\left(K_vV_{in} + 1\right)}{LC} > 0 \tag{8.86}$$

Clearly, the first condition always holds since $K_i < 0$. Using second condition, the lower limit of K_v can obtained as

$$K_v > \frac{-1}{V_{in}} \tag{8.87}$$

The gains K_i and K_v can also be selected by taking into account the dynamics of the closed-loop system. From (8.85), the closed-loop poles can be obtained as

$$s_{1,2} = \frac{\dfrac{\left(K_iV_{in}^2 - r\right)}{L} \pm \sqrt{\left(\dfrac{r - K_iV_{in}^2}{L}\right)^2 - 4\left(\dfrac{K_vV_{in} + 1}{LC}\right)}}{2} \tag{8.88}$$

The satisfactory dynamic response can be obtained if the gains K_i and K_v are properly chosen in such a way that the poles are located at desired locations. Figure 8.21 shows the loci of the closed-loop poles obtained with the system parameters used the simulation and experimental study (see sub-section 8.4.4) as K_i and K_v change. First, the effect of K_i on the closed-loop system is investigated while K_v is kept constant. The initial values of poles obtained with $K_i = -2 \times 10^{-4}$ and $K_v = 2.65/400$ are located at $s_1 = -2.4297 \times 10^3$ and $s_2 = -4.4281 \times 10^4$ as shown in Figure 8.15 (top). It is clear that when $|K_i|$ is decreased gradually (from -2×10^{-4} to -2×10^{-5}), the poles move toward each other and become complex in compliance with (8.88). This clearly shows that K_i plays an important role in

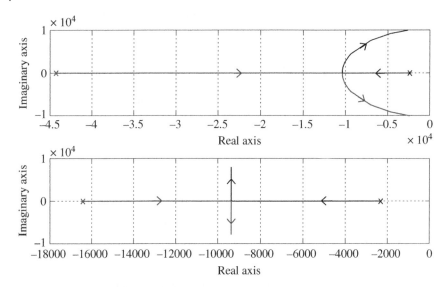

Figure 8.21 Root loci of closed-loop system when K_i and K_v are changed. *Source:* Komurcugil et al. [9].

determining the convergence time of closed-loop system. Secondly, the effect of K_v on the closed-loop system is investigated while K_i is kept constant. The initial values of poles obtained with $K_i = -8 \times 10^{-5}$ and $K_v = 0.3/400$ are located at $s_1 = -2.3342 \times 10^3$ and $s_2 = -1.6416 \times 10^4$ as shown in Figure 8.15 (bottom). When K_v is increased gradually (from 0.3/400 to $K_v = 4.1/400$), the poles move toward each other and become complex in compliance with (8.88). In the case of complex poles, any increment in K_v results in a larger magnitude of the imaginary part with no change in magnitude of the real part. This clearly shows that K_v does not have direct influence on the dynamic response when the closed-loop poles are complex.

8.4.4 Results

The effectiveness of the Lyapunov function-based control method with output voltage feedback is verified by simulation and experimental results. Simulations were carried out by MATLAB/Simulink. Experimental results were obtained from a 1 kW UPS inverter prototype in which the control method is implemented via TMS320F28335 floating point DSP. After generating the required reference signals (v_o^* and i_L^*) in the DSP, the control input presented in (8.82) is computed. The PWM signals are produced by comparing the control input with a carrier wave.

Table 8.4 System and control parameters.

Symbol	Value
Output voltage amplitude, V_m	$230\sqrt{2}$ V
DC input, V_{in}	400 V
Filter capacitance, C	49.404 μF
Inductance, L	0.6867 mH
Inductor resistance, r	0.076 Ω
Switching and sampling frequencies, f_{sw} and f_s	12.5 kHz
Angular frequency, ω	314.16 rad/s
Current gain, K_i	-6×10^{-5}
Voltage gain, K_v	0.006625

The system and control parameters are given in Table 8.4. Note that with these gains, the closed-loop poles are located at $s_{1,\,2} = -7.0453 \times 10^3 \pm j7.6126 \times 10^3$ which are reasonable for a satisfactory transient response.

Figure 8.22 shows the simulated and experimental waveforms of output voltage and load current for a resistive load (54.16 Ω). It is clear that the output voltage is sinusoidal in shape with a very low distortion. The total harmonic distortions (THD) of the output voltage obtained by simulation and experiment were 0.13% and 1.1%, respectively. Note that there is a discrepancy in the THDs between the simulation and the experimental results. The main reason of this difference comes from non-ideal effects such as additional distortion caused by noise disturbances in the practical system. This distortion is clearly seen on the experimental waveforms.

Figure 8.23 shows the simulated and experimental responses of the output voltage and the load current for a step change in load resistance from no load to 54.16 Ω. It is obvious that the output voltage is almost unaffected by the sudden change in the load current. In addition, the control method exhibits a very fast dynamic response and is quite successful in preserving the stability against this sudden disturbance in the load current.

Figure 8.24 shows the simulated response of $dV(x)/dt$ in one cycle that corresponds to the step change in load resistance presented in Figure 8.23. It can be observed that $dV(x)/dt$ stays always negative before, during and after the step change demonstrating that the closed-loop system is globally asymptotically stable.

Figure 8.25 shows the simulated and experimental waveforms of the output voltage and the load current for a diode bridge rectifier load case. It is obvious that the

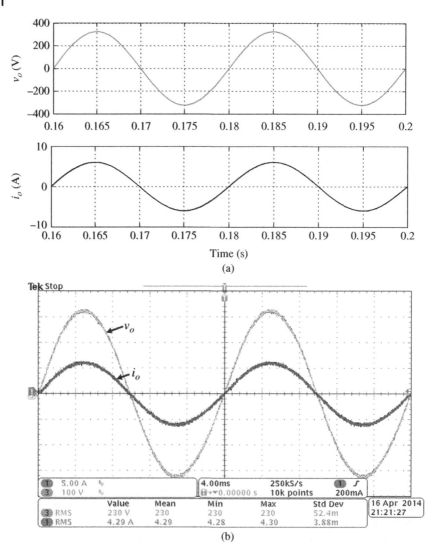

Figure 8.22 Simulated and experimental waveforms of v_o and i_o. (a) Simulation, (b) Experiment (CH1 (i_o): 5 A/div, CH3 (v_o): 100 V/div). *Source:* Komurcugil et al. [9]/IEEE.

output voltage is almost not affected by the fast changing of load current. The THD of the output voltage obtained by experiment is measured to be 1.3%.

Figure 8.26 shows the simulated and experimental waveforms of the capacitor current, the output voltage, the load current and the inductor current for the non-linear load case presented in Figure 8.25. It can be seen that the simulation and

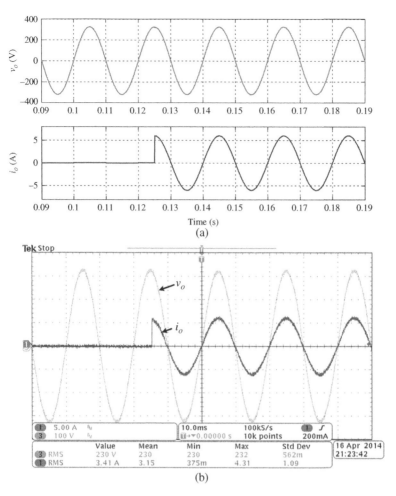

Figure 8.23 Simulated and experimental responses of v_o and i_o for a step change in load resistance from no load to 54.16 Ω with $K_i = -6 \times 10^{-5}$. (a) Simulation, (b) Experiment (CH1 (i_o): 5 A/div, CH3 (v_o): 100 V/div). *Source:* Komurcugil et al. [9]/IEEE.

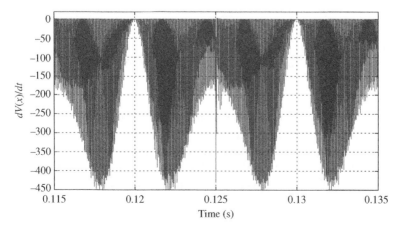

Figure 8.24 Simulated response of $dV(x)/dt$ for the step change in the load resistance. *Source:* Komurcugil et al. [9]/IEEE.

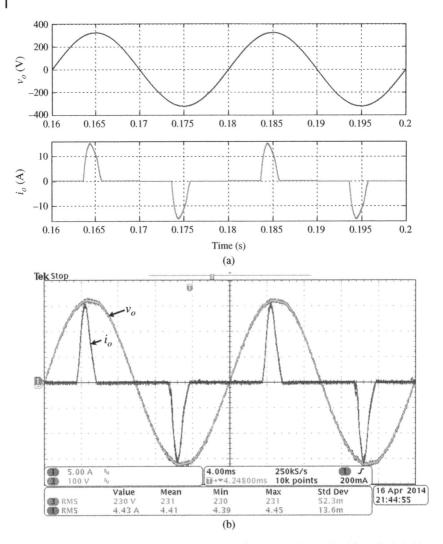

Figure 8.25 Simulated and experimental responses of v_o and i_o obtained for a diode bridge rectifier load. (a) Simulation, (b) Experiment (CH1 (i_o): 5 A/div, CH3 (v_o): 100 V/div). *Source:* Komurcugil et al. [9]/IEEE.

experimental results are in good agreement indicating the correct operation of the Lyapunov function-based control method.

Figure 8.27 shows the simulated response of $dV(x)/dt$ in one output cycle that corresponds to the case in Figure 8.25. Note that $dV(x)/dt$ is always negative which shows that the equilibrium point is globally asymptotically stable.

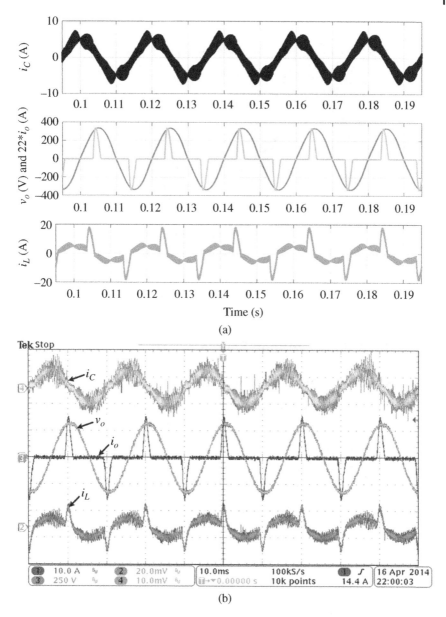

Figure 8.26 Simulated and experimental responses of i_C, v_o, i_o, and i_L obtained for a diode bridge rectifier load. (a) Simulation, (b) Experiment (CH1 (i_o): 10 A/div, CH2 (i_L):20 A/div, CH3 (v_o): 250 V/div, CH4 (i_C):10 A/div). *Source:* Komurcugil et al. [9]/IEEE.

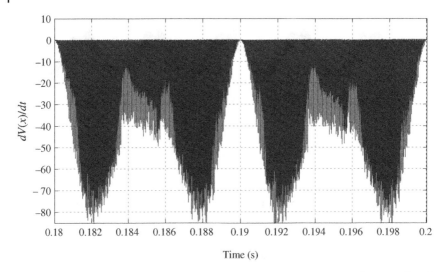

Figure 8.27 Simulated response of *dV(x)/dt* for the nonlinear load. *Source:* Komurcugil et al. [9]/IEEE.

8.5 Three-Phase Voltage-Source AC–DC Rectifier

8.5.1 Mathematical Modeling of Rectifier

Figure 8.28 shows the circuit diagram of a three-phase voltage-source AC–DC rectifier [11]. The three-phase utility grid voltages are given by

$$
\begin{aligned}
v_{ga} &= V_g \cos(\omega t) \\
v_{gb} &= V_g \cos(\omega t - 120°) \\
v_{gc} &= V_g \cos(\omega t - 240°)
\end{aligned}
\tag{8.89}
$$

The differential equations of the rectifier in the ac and dc sides can be written as

$$
L\frac{d\mathbf{i_g}}{dt} + r\mathbf{i_g} = \mathbf{v_g} - \mathbf{v_{in}}
\tag{8.90}
$$

$$
C\frac{dV_o}{dt} = I_{dc} - I_o
\tag{8.91}
$$

where the matrices are given by

$$
\begin{aligned}
\mathbf{v_g} &= \begin{bmatrix} v_{ga} & v_{gb} & v_{gc} \end{bmatrix}^T \\
\mathbf{i_g} &= \begin{bmatrix} i_{ga} & i_{gb} & i_{gc} \end{bmatrix}^T \\
\mathbf{v_{in}} &= \begin{bmatrix} v_{an} & v_{bn} & v_{cn} \end{bmatrix}^T
\end{aligned}
\tag{8.92}
$$

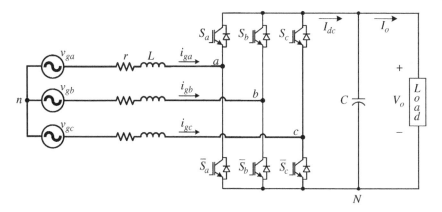

Figure 8.28 Three-phase voltage-source AC–DC rectifier. *Source:* Adapted from Komurcugil and Kukrer [11].

The pole voltages can be obtained as

$$v_{aN} = \frac{1}{2}(1 + u_a)V_o$$
$$v_{bN} = \frac{1}{2}(1 + u_b)V_o \qquad (8.93)$$
$$v_{cN} = \frac{1}{2}(1 + u_c)V_o$$

where u_a, u_b, and u_c denote bipolar control inputs which take the following values depending on the position of switches connected on each leg of the rectifier

$$u_k = \begin{cases} +1 & \text{if} \quad S_k \quad \text{is} \quad \text{closed} \\ -1 & \text{if} \quad \overline{S}_k \quad \text{is} \quad \text{closed} \end{cases}, \quad k = a, \ b, \ c \qquad (8.94)$$

The phase-to-neutral voltage expressions in terms of output voltage and bipolar control inputs can be written as

$$\mathbf{v_{in}} = \frac{1}{6}V_o \begin{bmatrix} 2 & -1 & -1 \\ -1 & 2 & -1 \\ -1 & -1 & 1 \end{bmatrix} \begin{bmatrix} u_a \\ u_b \\ u_c \end{bmatrix} \qquad (8.95)$$

Since $i_{ga} + i_{gb} + i_{gc} = 0$, the output current of the rectifier can be expressed in terms of grid currents as

$$I_{dc} = \frac{1}{2}\begin{bmatrix} u_a & u_b & u_c \end{bmatrix} \begin{bmatrix} i_{ga} \\ i_{gb} \\ i_{gc} \end{bmatrix} \qquad (8.96)$$

Usually, the controller design of three-phase systems is done in the synchronously rotating dq frame in which all AC quantities become DC. The three-phase grid voltages, grid currents and control inputs can be transformed into rotating dq frame as follows

$$\begin{bmatrix} v_{gd} \\ v_{gq} \end{bmatrix} = \frac{2}{3} \mathbf{T} \begin{bmatrix} v_{ga} \\ v_{gb} \\ v_{gc} \end{bmatrix} \tag{8.97}$$

$$\begin{bmatrix} i_{gd} \\ i_{gq} \end{bmatrix} = \frac{2}{3} \mathbf{T} \begin{bmatrix} i_{ga} \\ i_{gb} \\ i_{gc} \end{bmatrix} \tag{8.98}$$

$$\begin{bmatrix} u_d \\ u_q \end{bmatrix} = \frac{2}{3} \mathbf{T} \begin{bmatrix} u_a \\ u_b \\ u_c \end{bmatrix} \tag{8.99}$$

where $\mathbf{T}$ is the transformation matrix defined as

$$\mathbf{T} = \begin{bmatrix} \cos(\omega t) & \cos(\omega t - 120°) & \cos(\omega t - 240°) \\ -\sin(\omega t) & -\sin(\omega t - 120°) & -\sin(\omega t - 240°) \end{bmatrix} \tag{8.100}$$

Now, let us transform Eq. (8.90) into rotating dq frame. By making use of (8.98), the following relation can be written

$$\frac{2}{3} \mathbf{T} L \frac{d\mathbf{i_g}}{dt} = L \frac{d}{dt} \begin{bmatrix} i_{gd} \\ i_{gq} \end{bmatrix} - \omega L \begin{bmatrix} i_{gq} \\ i_{gd} \end{bmatrix} \tag{8.101}$$

Multiplying both sides of (8.95) by $\mathbf{T}$ and making some algebraic manipulation, one can obtain the following expression

$$\mathbf{T} \mathbf{v_{in}} = \frac{1}{2} V_o \mathbf{T} \begin{bmatrix} u_a \\ u_b \\ u_c \end{bmatrix} \tag{8.102}$$

Using relation in (8.99), the transformation of (8.102) can be written as

$$\frac{2}{3} \mathbf{T} \mathbf{v_{in}} = \frac{1}{2} V_o \begin{bmatrix} u_d \\ u_q \end{bmatrix} \tag{8.103}$$

Now, the transformation of (8.90) can be written as

$$L \frac{di_{gd}}{dt} + r i_{gd} = v_{gd} + \omega L i_{gq} - \frac{1}{2} V_o u_d \tag{8.104}$$

$$L \frac{di_{gq}}{dt} + r i_{gq} = v_{gq} - \omega L i_{gd} - \frac{1}{2} V_o u_q \tag{8.105}$$

The d and q components of three-phase grid voltages in rotating dq frame is obtained as $v_{gd} = V_g$ and $v_{gq} = 0$. On the other hand, the relation between grid currents and dc current is given by

$$\frac{3}{2}[u_a \quad u_b \quad u_c] \begin{bmatrix} i_{ga} \\ i_{gb} \\ i_{gc} \end{bmatrix} = 3I_{dc} \tag{8.106}$$

Hence, the expression of dc current in terms of rotating dq control inputs can be written as

$$I_{dc} = \frac{3}{4}\left(i_{gd}u_d + i_{gq}u_q\right) \tag{8.107}$$

Now, substituting (8.107) into (8.91), we obtain

$$C\frac{dV_o}{dt} = \frac{3}{4}\left(i_{gd}u_d + i_{gq}u_q\right) - I_o \tag{8.108}$$

It can be noticed that Eqs. (8.104), (8.105) and (8.108) are nonlinear since u_d and u_q are control inputs.

8.5.2 Controller Design

Now, define the state variables and control input as follows

$$x_1 = i_{gd} - i_{gd}^* \tag{8.109}$$

$$x_2 = i_{gq} - i_{gq}^* \tag{8.110}$$

$$x_3 = V_o - V_o^* \tag{8.111}$$

$$u_d = U_{do} + \Delta u_d \tag{8.112}$$

$$u_q = U_{qo} + \Delta u_q \tag{8.113}$$

where i_{gd}^* represents the reference of i_{gd}, i_{gq}^* represents the reference of i_{gq} (usually selected as $i_{gq}^* = 0$ to achieve unity power factor), V_o^* represents the reference of V_o, Δu_d and Δu_q represent the perturbations away from the nominal control inputs (U_{do} and U_{qo}). Assuming that the actual variables track their references in the steady-state (i.e.: $i_{gd} = i_{gd}^*$, $i_{gq} = i_{gq}^* = 0$, $V_o = V_o^*$, $u_d = U_{do}$ and $u_q = U_{qo}$), Eqs. (8.104), (8.105) and (8.108) can be written as

$$ri_{gd}^* = V_g - \frac{1}{2}V_o^* U_{do} \tag{8.114}$$

$$0 = \omega L i_{gd}^* + \frac{1}{2}V_o^* U_{qo} \tag{8.115}$$

$$0 = \frac{3}{4}i_{gd}^* U_{do} - I_o \tag{8.116}$$

It is worth noting that i_{gq}^* is set to zero to achieve unity power operation. The steady-state expressions of control inputs can be solved from (8.114) and (8.115) as follows

$$U_{do} = \frac{2\left(V_g - ri_{gd}^*\right)}{V_o^*} \tag{8.117}$$

$$U_{qo} = -\frac{2\omega L i_{gd}^*}{V_o^*} \tag{8.118}$$

Now, substituting (8.117) into (8.116), one can obtain

$$\frac{3}{2V_o^*}\left(-ri_{gd}^{*\,2} + V_g i_{gd}^*\right) - I_o = 0 \tag{8.119}$$

Clearly, Eq. (8.119) is a quadratic equation whose solution is given by

$$i_{gd}^* = \frac{1}{2}\left(\frac{V_g}{r} \pm \sqrt{\frac{V_g^2}{r^2} - \frac{8V_o^* I_o}{3r}}\right) \tag{8.120}$$

Since there are two solutions for i_{gd}^*, only feasible solution should be taken into consideration. For instance, no real solution exists for i_{gd}^* if the load current satisfies the following condition

$$I_o > \frac{3V_g^2}{8rV_o^*} \tag{8.121}$$

Equation (8.121) gives the upper limit of the load current. It is worthy noting that i_{gd}^* can also be computed by using a proportional-integral (PI) regulator as follows

$$i_{gd}^* = K_p\left(V_o^* - V_o\right) + K_i \int \left(V_o^* - V_o\right) dt \tag{8.122}$$

where K_p and K_i are the proportional and integral gains, respectively. Also, substituting $i_{gd} = x_1 + i_{gd}^*$, $i_{gq} = x_2 + i_{gq}^*$, $V_o = x_3 + V_o^*$, and $u_d = U_{do} + \Delta u_d$ into (8.104) we obtain

$$L\frac{dx_1}{dt} + L\frac{di_{gd}^*}{dt} + rx_1 + ri_{gd}^* = V_g + \omega L x_2 + \omega L i_{gq}^* - \frac{1}{2}x_3 U_{do}$$
$$- \frac{1}{2}x_3 \Delta u_d - \frac{1}{2}V_o^* U_{do} - \frac{1}{2}V_o^* \Delta u_d \tag{8.123}$$

Substituting $i_{gd} = x_1 + i_{gd}^*$, $i_{gq} = x_2 + i_{gq}^*$, $V_o = x_3 + V_o^*$, and $u_q = U_{qo} + \Delta u_q$ into (8.105), we obtain

$$L\frac{dx_2}{dt} + L\frac{di_{gq}^*}{dt} + rx_2 + ri_{gq}^* = -\omega L x_1 - \omega L i_{gd}^* - \frac{1}{2}x_3 U_{qo} - \frac{1}{2}x_3 \Delta u_q - \frac{1}{2}V_o^* U_{qo} - \frac{1}{2}V_o^* \Delta u_q$$

(8.124)

Similarly, substituting $i_{gd} = x_1 + i_{gd}^*$, $i_{gq} = x_2 + i_{gq}^*$, $V_o = x_3 + V_o^*$, $u_d = U_{do} + \Delta u_d$, and $u_q = U_{qo} + \Delta u_q$ into (8.108), we obtain

$$C\frac{dx_3}{dt} + C\frac{dV_o^*}{dt}$$
$$= \frac{3}{4}\left(x_1 U_{do} + x_1 \Delta u_d + i_{gd}^* U_{do} + i_{gd}^* \Delta u_d + x_2 U_{qo} + x_2 \Delta u_q + i_{gq}^* U_{qo} + i_{gq}^* \Delta u_q\right) - I_o$$

(8.125)

Considering the steady-state Eqs. (8.114)–(8.116), the error dynamics of the rectifier can be extracted from (8.123)–(8.125) as follows

$$L\frac{dx_1}{dt} = -rx_1 + \omega L x_2 - x_3\left(\frac{V_g - ri_{gd}^*}{V_o^*}\right) - \frac{1}{2}x_3 \Delta u_d - \frac{1}{2}V_o^* \Delta u_d \qquad (8.126)$$

$$L\frac{dx_2}{dt} = -rx_2 - \omega L x_1 - x_3\frac{\omega L i_{gd}^*}{V_o^*} - \frac{1}{2}x_3 \Delta u_q - \frac{1}{2}V_o^* \Delta u_q \qquad (8.127)$$

$$C\frac{dx_3}{dt} = \frac{3}{4}\left[\frac{2\left(V_g - ri_{gd}^*\right)}{V_o^*}x_1 + x_1 \Delta u_d + i_{gd}^* \Delta u_d - \frac{2\omega L i_{gd}^*}{V_o^*}x_2 + x_2 \Delta u_q\right]$$

(8.128)

Now, let us define the following Lyapunov function

$$V(x) = \frac{3}{2}Lx_1^2 + \frac{3}{2}Lx_2^2 + Cx_3^2 \qquad (8.129)$$

The derivative of (8.129) is given by

$$\frac{dV(x)}{dt} = 3x_1 L\frac{dx_1}{dt} + 3x_2 L\frac{dx_2}{dt} + 2x_3 C\frac{dx_3}{dt} \qquad (8.130)$$

Substitution of (8.126)–(8.128) into (8.130) yields

$$\frac{dV(x)}{dt} = -3r\left(x_1^2 + x_2^2\right) - \frac{3}{2}\left(V_o^* x_1 - i_{gd}^* x_3\right)\Delta u_d - \frac{3}{2}V_o^* x_2 \Delta u_q \qquad (8.131)$$

It is clear from (8.131) that the first term is always negative. The derivative of Lyapunov function is always negative if the perturbed control inputs are selected as

$$\Delta u_d = K_d\left(V_o^* x_1 - i_{gd}^* x_3\right), \quad K_d > 0 \qquad (8.132)$$

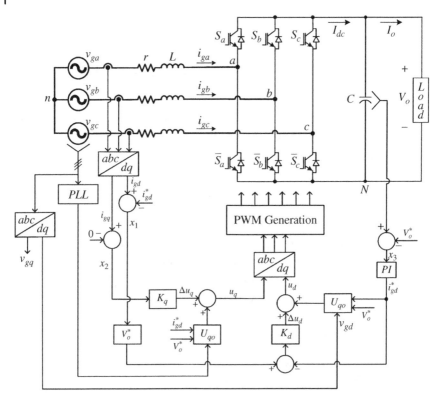

Figure 8.29 The block diagram of Lyapunov function-based control of three-phase rectifier.

$$\Delta u_q = K_q x_2, \quad K_q > 0 \tag{8.133}$$

The block diagram of Lyapunov function-based control of three-phase rectifier is shown in Figure 8.29.

8.5.3 Results

The effectiveness of the Lyapunov function-based control method has been verified by simulations. The system and control parameters are given in Table 8.5.

Figure 8.30 shows the simulated responses of the grid currents and output voltage for a step change in the load resistance from 50 to 25 Ω. It is clear from Figure 8.30a,b that the grid voltages are in phase with the grid currents before

Table 8.5 System and control parameters.

Symbol	Value
Output voltage reference, V_o^*	400 V
Inductance, L	4 mH
Capacitance, C	1000 μF
Inductor resistance, r	0.1 Ω
Load resistance, R_L	50 Ω
Grid voltage amplitude, V_g	$120\sqrt{2}$ V
Switching frequency, f_{sw}	10 kHz
Current gains, K_d and K_q	0.006 and 1.8
PI gains, K_p and K_i	1 and 100

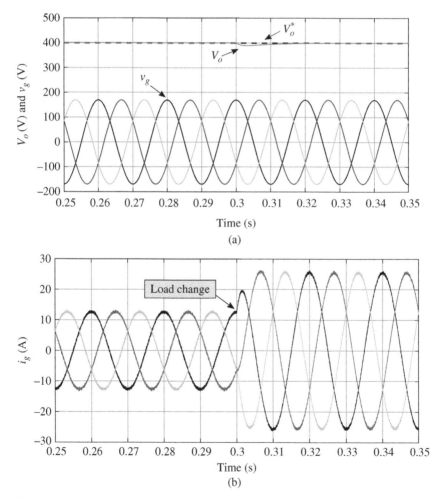

Figure 8.30 The responses of grid current and output voltage for a step change in the load resistance from 50 to 25 Ω. (a) Grid voltages, (b) Grid currents, (c) Derivative of Lyapunov function.

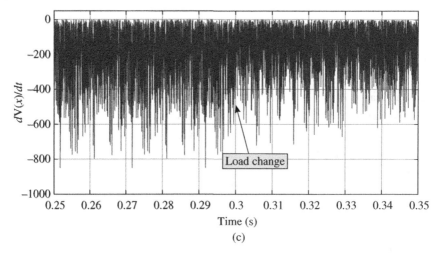

Figure 8.30 (Continued)

the load is changed (i.e.: $R_L = 50$ Ω). This result confirms the unity power factor operation of the rectifier for $i_{gq}^* = 0$ A. Also, the output voltage is regulated at its reference value which is 400 V. The negative definiteness of Lyapunov function is also guaranteed as shown in Figure 8.30c. When the load resistance is suddenly reduced to 25 Ω, the output voltage exhibits a small under-shoot and settles down to 400 V again after a short transition period. In the meantime, the amplitude of grid currents increase so as to achieve the power balance between input and output. The THD values of grid currents before and after the load change were computed as 3.09% and 1.6%, respectively. Also, it can be noticed that the derivative of Lyapunov function stays negative after the load change.

Figure 8.31 shows the response of grid currents under 50 Ω when the rectifier operates with the reactive power control. It is worthy to remark that the reactive power control can be achieved when $i_{gq}^* \neq 0$. When $i_{gq}^* = -5$ A, the grid currents lag grid voltages (see Figure 8.30) as shown in Figure 8.31a. On the other hand, when $i_{gq}^* = 5$ A, the grid currents lead grid voltages as shown in Figure 8.31b.

The results presented in this section reveal that the Lyapunov function-based control method not only guarantees the closed-loop stability, but also yields fast dynamic response for abrupt load variations. Moreover, it has the ability to control the reactive power which is essential in some industrial applications.

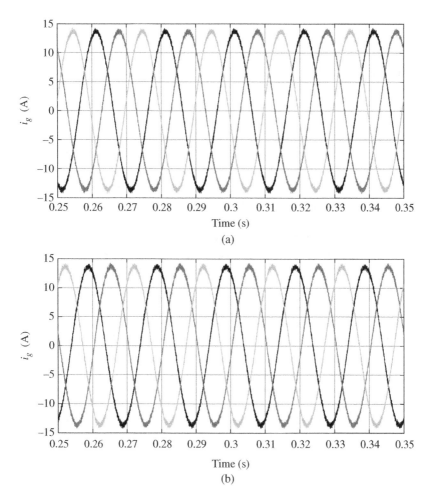

Figure 8.31 The responses of grid currents under 50 Ω resistive load when: (a) i^*_{gq} = − 5A, (b) i^*_{gq} = 5A.

References

1 H. Komurcugil, N. Altin, S. Ozdemir, and I. Sefa, "Lyapunov-function and proportional-resonant-based control strategy for single-phase grid-connected VSI with LCL filter," *IEEE Trans. Ind. Electron.*, vol. 63, no. 5, pp. 2838–2849, May 2016.

2 R. Teodorescu, F. Blaabjerg, M. Liserre, and P. Loh, "Proportional resonant controllers and filters for grid-connected voltage-source converters," *Proc. Inst. Electron. Eng. Electron. Power Appl.*, vol. 153, no. 5, pp. 750–762, Sep. 2006.

3 J. Xu, S. Xie, and T. Tang, "Active damping-based control for grid-connected LCL-filtered inverter with injected grid current only," *IEEE Trans. Ind. Electron.*, vol. 61, no. 9, pp. 4746–4758, Sep. 2014.

4 X. Hao, X. Yang, T. Liu, L. Huang, and W. Chen, "A sliding-mode controller with multiresonant sliding surface for single-phase grid-connected VSI with an LCL filter," *IEEE Trans. Power Electron.*, vol. 28, no. 5, pp. 2259–2268, May 2013.

5 S. Eren, M. Pahlevaninezhad, A. Bakhshai, and P. K. Jain, "Composite nonlinear feedback control and stability analysis of a grid-connected voltage source inverter with *LCL* filter," *IEEE Trans. Ind. Electron.*, vol. 60, no. 11, pp. 5059–5074, Nov. 2013.

6 G. Shen, X. Zhu, J. Zhang, and D. Xu, "A new feedback method for PR current control of LCL-filter-based grid-connected inverter," *IEEE Trans. Ind. Electron.*, vol. 57, no. 6, pp. 2033–2041, Jun. 2010.

7 H. Komurcugil, S. Bayhan, and H. Abu-Rub, "Lyapunov-function based control approach with cascaded PR controllers for single-phase grid-tied LCL-filtered quasi-Z-source inverters," in *IEEE CPE-POWERENG2017*, Spain, 2017, pp. 510–515, 2017.

8 Y. Liu, H. Abu-Rub, B. Ge, et al., *Impedance Source Power Electronic Converters*. Hoboken, NJ: Wiley, 2016.

9 H. Komurcugil, N. Altin, S. Ozdemir, and I. Sefa, "An extended Lyapunov-function-based control strategy for single-phase UPS inverters," *IEEE Trans. Power Electron.*, vol. 30, no. 7, pp. 3976–3983, Jul. 2015.

10 J. J. E. Slotine and W. Li, *Applied Nonlinear Control*. Englewood Cliffs, NJ, USA: Prentice-Hall, 1991.

11 H. Komurcugil and O. Kukrer, "Lyapunov-based control for three-phase PWM AC/DC voltage-source converters," *IEEE Trans. Power Electron.*, vol. 13, no. 5, pp. 801–813, Sept. 1998.

9

Model Predictive Control of Various Converters

9.1 CCS MPC Method for a Three-Phase Grid-Connected VSI

Figure 9.1 shows a three-phase grid-connected VSI. Using the Kirchhoff laws, the circuit equations can be easily obtained as follows:

$$L\frac{di_a}{dt} = \frac{V_{dc}}{2}u_a - v_{ga} \tag{9.1}$$

$$L\frac{di_b}{dt} = \frac{V_{dc}}{2}u_b - v_{gb} \tag{9.2}$$

$$L\frac{di_c}{dt} = \frac{V_{dc}}{2}u_c - v_{gc} \tag{9.3}$$

where L represents the grid impedance, i_a, i_b, i_c, are the three-phase inverter currents, v_{ga}, v_{gb}, v_{gc} are the three-phase grid voltages, and u_a, u_b and u_c are the control signals.

The aforementioned circuit equations can be transformed in the stationary reference frame or, equivalently, rotating dq-frame. The next equations of grid connected inverter were introduced in Section 5.4.12:

$$L\frac{di_d}{dt} = \frac{V_{dc}}{2}m_d - v_d - \omega i_q$$

$$L\frac{di_q}{dt} = \frac{V_{dc}}{2}m_q - v_q + \omega i_d$$

where i_d, i_q, v_d, and v_q are the inverter currents and the grid voltages in the dq-frame, respectively. The control signals are represented by m_d and m_q, and ω is the angular grid frequency.

Advanced Control of Power Converters: Techniques and MATLAB/Simulink Implementation,
First Edition. Hasan Komurcugil, Sertac Bayhan, Ramon Guzman, Mariusz Malinowski, and Haitham Abu-Rub.
© 2023 The Institute of Electrical and Electronics Engineers, Inc.
Published 2023 by John Wiley & Sons, Inc.
Companion website: www.wiley.com/go/komurcugil/advancedcontrolofpowerconverters

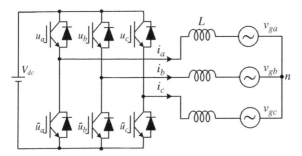

Figure 9.1 Circuit diagram of a three-phase grid-connected VSI.

9.1.1 Model Predictive Control Design

In this section, a model predictive control for the VSI of Figure 9.1 will be derived. The control design is based on the controller proposed in [1]. First, the incremental model of the converter is presented. Second, the predictive model of the converter is obtained to find the optimal control signal, according to the theory presented in the Chapter 5. Some constraints in the amplitude currents are introduced to limit the maximum grid current. Finally, some simulation results are presented to validate the controller. For the control implementation, let's assume a control horizon $N_c = 2$, a prediction horizon $N_p = 3$, and a control effort $r_w = 0.05$.

9.1.1.1 VSI Incremental Model with an Embedded Integrator

Considering the power converter having the parameters $L = 5$ mH, $V_{dc} = 400$ V, and for the grid voltages the peak voltage is $V_p = 155$ V and the angular grid frequency is $\omega = 2\pi \cdot 60$ rad/s. The sampling frequency is $f_s = 10$ kHz. According to the Eqs. (5.97), (5.98), and (5.99) presented in Chapter 5, the discrete incremental model of the converter in the dq-frame can be obtained as follows:

$$
\begin{bmatrix} \Delta i_d(k+1) \\ \Delta i_q(k+1) \\ i_d(k+1) \\ i_q(k+1) \end{bmatrix} = \begin{bmatrix} 1 & -0.038 & 0 & 0 \\ 0.038 & 1 & 0 & 0 \\ 1 & -0.038 & 1 & 0 \\ 0.038 & 1 & 0 & 1 \end{bmatrix} \begin{bmatrix} \Delta i_d(k) \\ \Delta i_q(k) \\ i_d(k) \\ i_q(k) \end{bmatrix}
$$

$$
+ \begin{bmatrix} 4 & 0 \\ 0 & 4 \\ 4 & 0 \\ 0 & 4 \end{bmatrix} \begin{bmatrix} \Delta m_d \\ \Delta m_q \end{bmatrix} + \begin{bmatrix} -0.02 & 0 \\ 0 & -0.02 \\ -0.02 & 0 \\ 0 & -0.02 \end{bmatrix} \begin{bmatrix} \Delta v_d \\ \Delta v_q \end{bmatrix}
$$

$$
\tag{9.4}
$$

$$y(k) = \begin{bmatrix} 0 & 0 & 1 & 0 \\ 0 & 0 & 0 & 1 \end{bmatrix} \begin{bmatrix} \Delta i_d(k) \\ \Delta i_q(k) \\ i_d(k) \\ i_q(k) \end{bmatrix} \tag{9.5}$$

Equations (9.4) and (9.5) can be expressed in a compact form as follows:

$$x(k + 1) = Ax(k) + B\Delta u(k) + D\Delta q(k) \tag{9.6}$$

$$y(k) = Cx(k) \tag{9.7}$$

where

$$A = \begin{bmatrix} 1 & -\omega T_s & 0 & 0 \\ \omega T_s & 1 & 0 & 0 \\ 1 & -\omega T_s & 1 & 0 \\ \omega T_s & 1 & 0 & 1 \end{bmatrix}, \quad B = \begin{bmatrix} \dfrac{T_s V_{dc}}{2L} & 0 \\ 0 & \dfrac{T_s V_{dc}}{2L} \\ \dfrac{T_s V_{dc}}{2L} & 0 \\ 0 & \dfrac{T_s V_{dc}}{2L} \end{bmatrix},$$

$$D = \begin{bmatrix} -\dfrac{T_s}{L} & 0 \\ 0 & -\dfrac{T_s}{L} \\ -\dfrac{T_s}{L} & 0 \\ 0 & -\dfrac{T_s}{L} \end{bmatrix}, \quad C = \begin{bmatrix} 0 & 0 & 1 & 0 \\ 0 & 0 & 0 & 1 \end{bmatrix}$$

9.1.1.2 Predictive Model of the Converter

The incremental model defined in the previous section is used to obtain the predictive model of the system. According to (5.35), the predictive model is defined as

$$Y = Fx(k_i) + G_b \Delta u + G_d \Delta Q$$

The incremental disturbance vector is defined as $\Delta q = \begin{bmatrix} \Delta v_d & \Delta v_q \end{bmatrix}^T$. Since v_d and v_q are more or less constant, the quantities Δv_d and Δv_q are approximately zero. Then, the last term of the predictive model can be removed from the equation, yielding:

$$Y = Fx(k_i) + G_b \Delta u$$

The matrices F and G_b can be obtained from matrices A, B, and C using the expressions defined in (5.36) and (5.37), respectively, yielding:

$$F = \begin{bmatrix} CA \\ CA^2 \\ CA^3 \end{bmatrix} = \begin{bmatrix} 1 & -0.0002 & 1 & 0 \\ 0.0002 & 1 & 0 & 1 \\ 2 & -0.0006 & 1 & 0 \\ 0.0006 & 2 & 0 & 1 \\ 3 & -0.0011 & 1 & 0 \\ 0.0011 & 3 & 0 & 1 \end{bmatrix} \tag{9.8}$$

$$G_b = \begin{bmatrix} CB & 0 \\ CAB & CB \\ CA^2B & CAB \end{bmatrix} = \begin{bmatrix} 4 & 0 & 0 & 0 \\ 0 & 0.0001 & 0 & 0 \\ 8 & 0 & 4 & 0 \\ 0.0008 & 0.0002 & 0 & 0.0001 \\ 12 & 0 & 8 & 0 \\ 0.0023 & 0.0003 & 0.0008 & 0.0002 \end{bmatrix} \tag{9.9}$$

9.1.1.3 Cost Function Minimization

The next step in a predictive control system is to use the predictive model in a cost function to be minimized. The cost function has been introduced in Chapter 5 by Eq. (5.40). By using the predictive model in cost function equation, one has

$$J = \left(Y_{ref} - Fx(k_i) + G_b\Delta u\right)^T \left(Y_{ref} - Fx(k_i) + G_b\Delta u\right) + \Delta U^T R\Delta U \tag{9.10}$$

where $Y_{ref}{}^T = R_r y^*(k_i) \underbrace{\begin{bmatrix} I_{2\times2} & I_{2\times2} & I_{2\times2} \end{bmatrix}}_{N_p = 3} = \begin{bmatrix} i_d^* \\ i_q^* \end{bmatrix} = \begin{bmatrix} i_d^* & i_d^* & i_d^* \\ i_q^* & i_q^* & i_q^* \end{bmatrix}$ being $I_{2\times2}$ the

two-dimension identity matrix.

As shown in Chapter 5, by taking the time derivative of Eq. (9.10), the optimal control signal is obtained as shown in (5.45), but eliminating the perturbation term:

$$\Delta U = \left(G_b^T G_b + R\right)^{-1} G_b^T \left(R_r y^*(k_i) - F\begin{bmatrix} \Delta x_m(k_i) \\ y(k_i) \end{bmatrix}\right) \tag{9.11}$$

where the matrix R is a diagonal matrix with the control effort $r_\omega = 0.05$ in the main diagonal:

$$R = \begin{pmatrix} 0.05 & 0 & 0 & 0 \\ 0 & 0.05 & 0 & 0 \\ 0 & 0 & 0.05 & 0 \\ 0 & 0 & 0 & 0.05 \end{pmatrix}$$

Note that the dimension of the matrix R coincides with the dimension of the matrix $G_b^T G_b$.

As it was explained in Chapter 5, Eq. (9.11) can be expressed as a function of the gains K_r and K_c, yielding:

$$\Delta U = -K_r(y(k_i) - y^*(k_i)) - K_c \Delta x_m(k_i) \tag{9.12}$$

where the gains K_r and K_c can be obtained from the result of computing the matrix $\left(G_b^T G_b + R\right)^{-1} G_b^T F$. In our case

$$\left(G_b^T G_b + R\right)^{-1} G_b^T F = \begin{pmatrix} 0.2494 & 0.0002 & 0.2483 & 0.0001 \\ 0 & 0.0280 & 0 & 0.0120 \\ 0.0010 & -0.0004 & -0.2472 & -0.0001 \\ 0 & 0.0160 & 0 & 0.0060 \end{pmatrix}$$

$$\tag{9.13}$$

Then, according to Section 5.4.11, the gains K_r and K_c are defined as follows:

$$K_c = \begin{pmatrix} 0.2494 & 0.0002 \\ 0 & 0.0280 \\ 0.0010 & -0.0004 \\ 0 & 0.0160 \end{pmatrix} \qquad K_r = \begin{pmatrix} 0.2483 & 0.0001 \\ 0 & 0.0120 \\ -0.2472 & -0.0001 \\ 0 & 0.0060 \end{pmatrix}$$

9.1.1.4 Inclusion of Constraints

This section is dedicated to the inclusion of constraints in the MPC design for the VSI. Concretely, the constraints will be applied to limit the VSI current and also to the maximum value of control signal, m_d. Since the constraints are applied to a continuous signal, the control design must be designed in the dq-frame as shown in Section 9.1.1.

Inclusion of Output Constraints

First, the output constraints are formulated as shown in Eqs. (5.67)–(5.69). As explained before, this constraint will be introduced to the cost function to limit the maximum value of the VSI current. Then, according to (5.69), the output constraints are formulated as follows:

$$G_b \Delta U \le Y_{\max} - Fx(k_i) \tag{9.14}$$

or equivalently

$$M \Delta U \le \delta$$

where the matrices M and δ are defined as:

$$M = G_b \quad \delta = Y_{\max} - Fx(k_i)$$

Since the constraint will be only applied to the first value of the vector Y (i.e. $y(k+1)$), only the first row of G_b and F are necessary. The it can be written as

$$(4 \quad 0 \quad 0 \quad 0) \begin{pmatrix} \Delta m_d(k_i) \\ \Delta m_q(k_i) \\ \Delta m_d(k_i + 1) \\ \Delta m_q(k_i + 1) \end{pmatrix} \leq (I_{dmax} - (1 \quad -0.0002 \quad 1 \quad 0)x(k_i))$$

$$(9.15)$$

where $x(k_i) = \begin{pmatrix} \Delta i_d(k_i) \\ \Delta i_q(k_i) \\ i_d(k_i) \\ i_q(k_i) \end{pmatrix}$

Then, according to (9.15), M and δ can be defined as:

$$M = (4 \quad 0 \quad 0 \quad 0) \quad \delta = (I_{dmax} - (1 \quad -0.0002 \quad 1 \quad 0)x(k_i))$$

Constraints on the Control Signal

According to (5.62), the constraints applied to the control signal can be formulated as follows:

$$u_{\min} \leq u \leq u_{\min}$$

If it is desired to limit the maximum value of the control signal m_d, one can write that:

$$m_d \leq m_{d,\max} \rightarrow m_d(k-1) + \Delta m_d \leq m_{d,\max} \rightarrow \Delta m_d \leq m_{d,\max} - m_d(k-1)$$

where according to the general formulation $M\Delta u \leq \delta$, it can be written as

$$M = 1 \quad \delta = m_{d,\max} - m_d(k-1)$$

Finally, to include both constraints, the output constraint and the control signal constraint, Eq. (9.15) can be rewritten as

$$\begin{pmatrix} 4 & 0 & 0 & 0 \\ 1 & 0 & 0 & 0 \end{pmatrix} \begin{pmatrix} \Delta m_d(k_i) \\ \Delta m_q(k_i) \\ \Delta m_d(k_i + 1) \\ \Delta m_q(k_i + 1) \end{pmatrix} \leq \begin{pmatrix} I_{d,\max} - (1 \quad -0.0002 \quad 1 \quad 0)x(k_i) \\ m_{d,\max} - m_d(k-1) \end{pmatrix}$$

$$(9.16)$$

where $M = \begin{pmatrix} 4 & 0 & 0 & 0 \\ 1 & 0 & 0 & 0 \end{pmatrix}$ and $\delta = \begin{pmatrix} I_{d\max} - (1 \quad -0.0002 \quad 1 \quad 0)x(k_i) \\ m_{d,\max} - m_d(k-1) \end{pmatrix}$.

9.1.2 MATLAB®/Simulink® Implementation

This section deals with the implementation of the control system using MATLAB® and Simulink®. The simulation is performed using a Simulink model of the converter and a MATLAB function for the controller implementation. Figure 9.2 shows the Simulink blocks used for the simulation. As shown in the figure, the power converter is composed by a three-phase branch of IGBT switches connected to the grid through three inductors. Besides, a controller block is also depicted, and the content of this block is shown in Figure 9.3. As shown in the figure, the controller block is developed using a MATLAB function, and the space vector modulator (SVM) block used as a modulator. The inputs to the MATLAB function are the three-phase voltages and currents, and the reference currents in dq-frame.

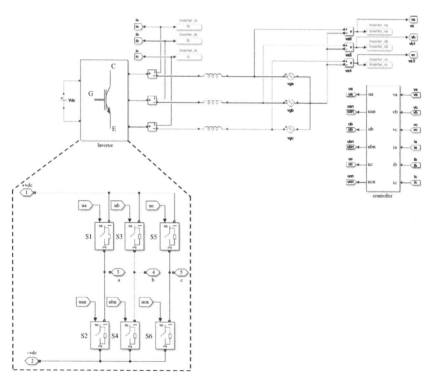

Figure 9.2 Simulink Block diagram of the MPC simulation.

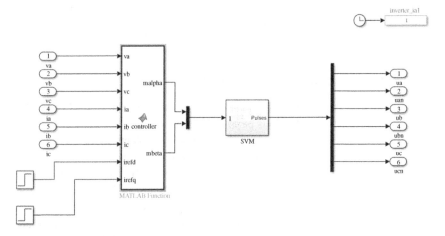

Figure 9.3 Simulink Block diagram of the MPC controller.

The output of the controller are the control signals in the alpha-beta frame, m_α and m_β. These control signals in the alpha-beta frame are the inputs of the SVM block, which generates the control actions u_a, u_b, and u_c used to control the switches.

The MATLAB function used to implement the controller is listed as follows:

```
function [malpha, mbeta] = controller(va, vb, vc, ia, ib,
ic, irefd, irefq, data)

persistent md mq w w_1 w_2 vq vq_1 vq_2 Xf xm xm_old

if isempty(md), md=0; end
if isempty(mq,  mq=0; end
if isempty(w_1), w_1=0; end
if isempty(w_2), w_2=0; end
if isempty(vq), vq=0; end
if isempty(vq_1), vq_1=0; end
if isempty(vq_2), vq_2=0; end
if isempty(xm_old), xm_old=[0;0]; end
if isempty(xm), xm=[0;0]; end
if isempty(Xf ), Xf=[0;0;0;0]; end


%%% 3-phase PLL

   w=data.num(1)*vq+data.num(2)*vq_1+data.num(3)*vq_2-
data.den(2)*w_1-data.den(3)*w_2;
```

```
    w_2=w_1;
    w_1=w;
    vq_2=vq_1;
    vq_1=vq;
  if(w>=2*pi)
     w_1=w_1-2*pi;
     w_2=w_2-2*pi;
  end;
```

%%% abc to dq transformation

```
vd=(2/3)*(cos(w)*va+cos(w-2*pi/3)*vb+cos(w-4*pi/3)*vc);
vq=(2/3)*(-sin(w)*va-sin(w-2*pi/3)*vb-sin(w-4*pi/3)*vc);

id=(2/3)*(cos(w)*ia+cos(w-2*pi/3)*ib+cos(w-4*pi/3)*ic);
iq=(2/3)*(-sin(w)*ia-sin(w-2*pi/3)*ib-sin(w-4*pi/3)*ic);
```

%%% Control design with MPC (J=u'Eu+u'P)

```
  data.P=-(data.Gb'*data.F(:,[3,4])*[irefd;irefq]-data.
  Gb'*data.F*Xf);
  data.Deltau=-data.Kr*[id-irefd;iq-irefq]-data.Kc*
  (xm-xm_old);
```

%%% optimal solution with constraints in idmax

```
data.gamma=[data.idmax-data.F(1,:)*Xf;1-md];
data.Deltau=optim_mimo(data.E,data.P,data.M,data.gamma,
data.Deltau);
```

%%% Obtaining the dq control signals

```
    md=md+data.Deltau(1);
    mq=mq+data.Deltau(2);
    u=[md;mq];
    xm_old=xm;
```

%%% State observer to estimate the state variables. The
use of the observer is optional. You can comment %%these
lines if you do not want to use the observer.

```
    xm=data.Am*xm+data.Bm*u+data.Dm*[vd;vq]+data.Lk*([id;
    iq]-data.Cm*xm);
    y=data.Cm*xm;
```

```
%%% With no observer, the id and iq currents are measured
y=[id;iq]. You  must uncomment this line if you %%%want to
use measured currents.
```

```
%y=[id;iq];
```

```
%%% Incremental state-space vector
```

```
Xf=[xm-xm_old;y];
```

```
%%% Obyaining the duty cicles, malpha and mbeta. dq to
alpha-beta transformation
```

```
malpha=md*cos(w)-mq*sin(w);
mbeta=md*sin(w)+mq*cos(w);
```

As it can be seen from the code, the controller is a function called controller where the inputs to this function are the three-phase voltages and currents, the reference currents in *dq* frame and the global variable data. The function provides as an output the control signals in the alpha-beta frame. In the function, first the typical three-phase phase-locked loop (PLL) is used. The PLL is based on a finite difference equation obtained from the continuous transfer function:

$$\frac{\Omega(s)}{V_q(s)} = \frac{100s + 50000}{s^2} \tag{9.17}$$

The PLL forces the value $v_q = 0$, and when this condition holds, the PLL is synchronized with the three-phase grid voltages [2]. The next step is to use the phase value obtained from the PLL to apply the Park transformation. This transformation is applied to obtain the *dq* components of the three-phase grid voltages and currents. Once the currents and voltages are transformed, the control action can be computed using (9.12). Note that the values of K_c and K_r can be obtained off-line and how to obtain their values will be discussed later. At this point, the control signals are obtained without constraints. Then, in order to obtain the constrained solution, the quadratic programing function qp () is used [3]. This function modifies the optimum control signal obtained with (9.12) according to the constraints defined by the matrices M and δ. Note that the value of M can be defined as a constant matrix, while the value of δ needs to be computed every sampling period, since its value has a dependence of the state-space vector at time k_i. The MATLAB code of this quadratic programing function is discussed in detail in [1] and is listed below:

```
function x=qp(E,P,M,gamma,DeltaU);
global data
[n1,m1]=size(M);
x=DeltaU;
j=0;

for i=1:n1
if (M(i,:)*x>gamma(i))
    k=k+1;
else
j=j+0;
end
end
if (j==0)
    return;
end
H=M*inv(E)*M';
K=(M*inv(E)*P+gamma);
[n,m]=size(K);
x_ini=zeros(n,m);
lambda=x_ini;
al=10;
for km=1:50
lambda_p=lambda;
for i=1:n
w= H(i,:)*lambda-H(i,i)*lambda(i,1);
w=w+K(i,1);
la=-w/H(i,i);
lambda(i,1)=max(0,la);
end
al=(lambda-lambda_p)'*(lambda-lambda_p);
if (al<10e-8);
    break;
end
end
x=DeltaU-inv(E)*M'*lambda;
```

Finally, a MATLAB function, which contains the global variables and constants, the computation of the MPC gains, and starts the Simulink model, is necessary.

This function defines the values of the system parameters and contains the discrete model of the converter. The matrices of the discrete state-space model are used in a function called MPC_matrices(), which computes the matrices G_b and F of the predictive model and also provides the matrices for the incremental model with the embedded integrator. Besides, the discretization of the transfer function defined in (9.17) for the PLL is also computed using the MATLAB function cd2m(). This function also provides a plot of the three-phase grid currents. The code of this function and also the code for the MPC_matrices() are listed as follows:

```
clear all
clc
disp('Beginning Predictive Control simulation...')

global data

%%% Parameters of the VSI Simulink model

L=5e-3;   %Inverter input inductor
R=0.1;
Vdc=400;

%Simulation parameters:

fs=10e3;
data.h=1/fs;
tstop=0.2;
max_step_size=1e-6;
min_step_size=0.1e-6;
warning('off');

% Continuous state space model

data.L1 = 5e-3; % Inductance [H]
data.Vdc = Vdc; % DC-link voltage [V]

data.w=2*pi*60;

 data.Ac=[0 -data.w*data.L1;data.w*data.L1 0];
 data.Bc=[data.Vdc/(2*data.L1) 0;0 data.Vdc/2*(data.L1)];
 data.Dc=[-1/data.L1 0;0 -1/data.L1];
 data.Am=eye(2)+data.Ac*data.h;
```

```
data.Bm=data.h*data.Bc;
data.Dm=data.h*data.Dc;
data.Cm=[1 0;0 1];

%%% d-q PLL filter

[data.num,data.den]=c2dm([100 50000],[1 0 0],data.
h,'tustin');

%% Predictive model matrices and gains

data.Nc=2; %Prediction Horizon
data.Np=3; %Control horizon

[data.A,data.B,data.C,data.F,data.Gb] = MPC_matrices(data.
Am,data.Bm,data.Cm,data.Nc,data.Np);
[n,m]=size(data.Gb'*data.Gb);
 R_w=0.05*eye(m,m);
 data.E=(data.Gb'*data.Gb+R_w);
 K=inv(data.E)*data.Gb'*data.F;
 data.Kc=K(:,[1,2]);
 data.Kr=K(:,[3,4]);
 data.M=[data.Gb(1,:);[1 zeros(1,2*data.Nc-1)]];
 data.gamma=zeros(2,1);
 data.P=zeros(2*data.Nc,1);
 data.idmax=10;
 data.iqmax=0;
 data.Deltau=zeros(2*data.Nc,1);

%%% Observer gain
data.Lk=0.7656*eye(2);
sim('VSI_model_L.mdl');
figure;
grid on;
box on;
hold on;

plot(t,ia);
plot(t,ib);
plot(t,ic);
```

```
function [A,B,C,F,Gb] = MPC_matrices(Am,Bm,Cm,Nc,Np)

[m1,n1]=size(Cm);
[nb,n_in]=size(Bm);
[y1 y2]=size(Cm*Bm);
[q,q1]=size(Am);
A=zeros(q+m1,q+m1);
A(1:q,1:q)=Am;
A(q+1:q+m1,1:n1)=Cm*Am;
A(q+1:m1+q,q+1:q+m1)=eye(m1);
B=Bm;
B(nb+1:nb+y1,1:y2)=Cm*Bm;
C=zeros(m1,n1);
C(1:m1,n1+1:n1+m1)=eye(m1);
[x1,x2]=size(C*A);
for kk=1:Np
    nn=kk-1;
F(x1*nn+1:x1*nn+x1,1:x2)=C*A^kk;
end
[x3,x4]=size(C*B);
for i=1:Nc
   mm=i-1;
    for j=1:Np
        nn=j-1;
        if j<i
        Gb(x3*nn+1:x3*nn+x3,x4*mm+1:x4*mm+x4)=zeros(x3,x4);
        else
        Gb(x3*nn+1:x3*nn+x3,x4*mm+1:x4*mm+x4)=C*A^(j-i)*B;
        end
    end
end

[n,m]=size(C);
end
```

9.1.3 Simulation Studies

In this section, the simulation results are presented. The MPC controller is tested under two different situations. First, a reference current of $I_d = 15$ A and $I_q = 0$ without any constraint is used. In the second simulation, constraints in the current

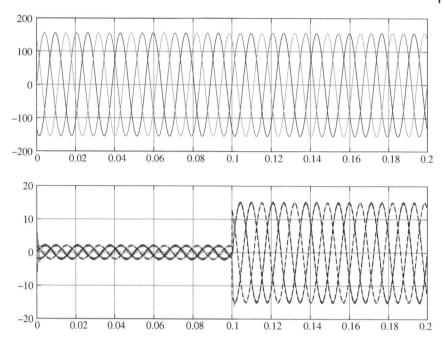

Figure 9.4 From top to bottom: three-phase grid voltages and three-phase grid currents and references without any constraint in the current when a load step change from 2 to 15 A is done.

and in the control signal $I_d \leq 10$ A and $m_d \leq 1$ will be presented. In both cases, a current step change will be included.

Figure 9.4 shows from top to bottom the three-phase grid voltages and the three-phase grid currents and their references (dashed line) when a reference current step change is done from 2 to 15 A. From $t = 0$ to $t = 0.1$ s, the reference currents are $I_{d, \ ref} = 2$ A and $I_{q, \ ref} = 0$, and from $t = 0.1$ s to $t = 0.2$ s, the reference currents are $I_{d, \ ref} = 15$ A and $I_{q, \ ref} = 0$. With this reference values, assuming that the currents track their references, the peak current value in natural frame is computed as

$$I = \sqrt{I_{d,ref}^2 + I_{q,ref}^2} = I_{d,ref} \tag{9.18}$$

Since in this case the MPC does not consider any constraint, the peak values of the three-phase grid currents will coincide with the value of $I_{d, \ ref}$, i.e. 2 or 15 A.

Figure 9.5 shows the control signals m_d and m_q. As it can be seen from the figure, at $t = 0.1$ s the value of $m_d = 3$. This change in the control signal occurs while the current step change happens. Since there is not any constraint in the control signals, the value of m_d has not any limitation.

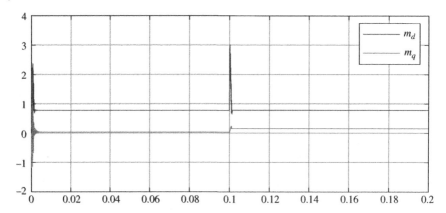

Figure 9.5 Control signals without any constraint.

In Figure 9.6, a constraint in the current I_d is considered. In this case, its value is limited to $I_d \leq 10$ A, while the reference current is maintained at the same value of the previous simulation. As shown in the figure, the three-phase currents are limited to 10 A even in the case that the reference current, in dashed line, has a step change from 2 to 15 A.

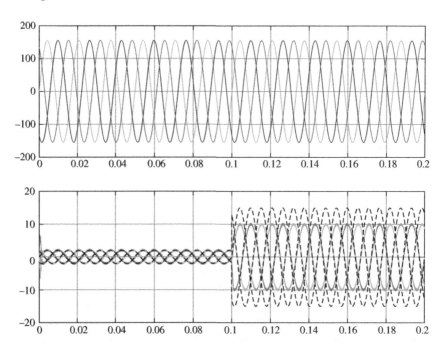

Figure 9.6 From top to bottom: three-phase grid voltages and three-phase grid currents and references with a constraint in the current when a load step change from 2 to 15 A is done.

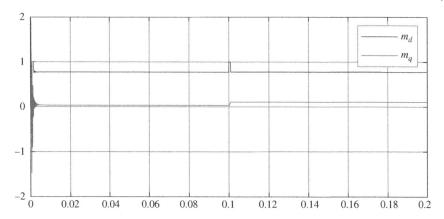

Figure 9.7 Control signals m_d and m_q with the constraint $m_d \leq 1$.

Figure 9.7 shows the control signals m_d and m_q. As it can be seen from the figure, the value of m_d is limited to 1 due to the inclusion of the constraint in the control signal.

9.2 Model Predictive Control Method for Single-Phase Three-Level Shunt Active Filter

An effective model predictive control (MPC) method for single-phase three-level T-type inverter-based shunt active power filters (SAPFs) is presented in this sub chapter. The presented control technique is based on the controller proposed in [4]. Contrary to most of the existing MPC methods, the presented MPC method eliminates the need for using weighting factor and additional constraints required for balancing DC capacitor voltages in the cost function. The design of cost function is based on the energy function. Since the factor used in the formulation of the energy function does not have any adverse influence on the performance of the system, the cost function becomes weighting factor free. The weighting factor free-based MPC brings simplicity in the practical implementation. The effectiveness of the presented MPC method has been investigated in steady state as well as dynamic transients caused by load changes. The theoretical considerations are verified through experimental studies performed on a 3 kVA system.

9.2.1 Modeling of Shunt Active Filter (SAPF)

The single-phase three-level SAPF using a T-type inverter is depicted in Figure 9.8. Clearly, the SAPF has four switches per each leg. The state of each switch is defined as

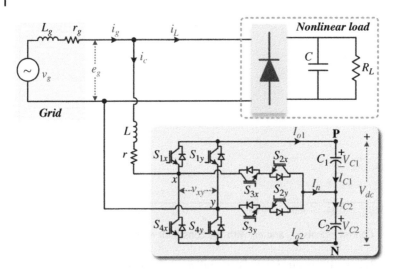

Figure 9.8 Single-phase three-level T-type inverter based SAPF.

$$S_{ij} = \begin{cases} 1 & \text{closed} \\ 0 & \text{open} \end{cases} \tag{9.19}$$

where $i = 1, 2, 3, 4$ and $j = x, y$. The SAPF produces three different pole voltages with respect to the neutral point O. These voltages exist when the midpoint of each rectifier leg is connected to positive (P), neutral (O), and negative (N) points via proper switching. The operating states, states of switching devices, and produced pole voltages are shown in Table 9.1. Apparently, the SAPF operates in the P state producing a pole voltage $v_{jo} = +V_{dc}/2$ when S_{1j} and S_{2j} are closed (ON) and S_{3j} and S_{4j} are open (OFF). On the other hand, the rectifier operates in the O state, which produces a pole voltage $v_{jo} = 0V$ when S_{2j} and S_{3j} are ON and S_{1j} and S_{4j} are OFF. Finally, the SAPF operates in the N state producing pole voltage $\pm V_{dc}/2$ when S_{1j} and S_{2j} are OFF and S_{3j} and S_{4j} are ON.

Table 9.1 Operating states, switching states and pole voltages.

Operating state	S_{1j}	S_{2j}	S_{3j}	S_{4j}	v_{jo}
P	ON	ON	OFF	OFF	$+V_{dc}/2$
O	OFF	ON	ON	OFF	0
N	OFF	OFF	ON	ON	$-V_{dc}/2$

The control input switching functions of the rectifier can be defined as

$$S_1 = S_{1x} - S_{1y} \tag{9.20}$$

$$S_2 = S_{2x} - S_{2y} \tag{9.21}$$

The three-level input voltage and neutral current can be defined in terms of the switching functions as follows:

$$v_{xy} = S_1 V_{C1} + S_2 V_{C2} \tag{9.22}$$

$$I_n = I_{C2} - I_{C1} = (S_2 - S_1)i_c \tag{9.23}$$

By making use of different switching combinations, v_{xy} can be generated as a five-level voltage with levels 0, $\pm V_{dc}/2$, and $\pm V_{dc}$. The differential equations of the SAPF can be written as

$$\frac{di_c}{dt} = \frac{1}{L}\left(e_g - v_{xy} - ri_c\right) \tag{9.24}$$

$$\frac{dV_{dc}}{dt} = \frac{2I_{o1}}{C_1} \tag{9.25}$$

where $v_{xy} = uV_{dc}$, $I_{o1} = ui_c$, u is the switching function. In the derivation of (9.25), it is assumed that $V_{C1} = V_{dc}/2$. On the other hand, the capacitor currents in terms of the switching functions (S_1 and S_2) and grid current are obtained as

$$I_{C1} = \frac{1}{2}(S_1 i_c - S_2 i_c) \tag{9.26}$$

$$I_{C2} = \frac{1}{2}(S_2 i_c - S_1 i_c) \tag{9.27}$$

The design of the passive components (filter inductance and DC capacitors) is calculated as follows [5], [6]:

$$L_{\max} = \frac{v_g}{2\sqrt{2}f_s \Delta i_c} \tag{9.28}$$

$$C_1 = C_2 = \frac{v_g I_{c\max}}{2f\left[V_{dc}^2 - V_{dc\min}^2\right]} \tag{9.29}$$

where $L_{\max}$ is the maximum value of filter inductance, Δi_c is the high-frequency ripple in compensation current (i_c), f_s is switching frequency, $I_{C\max}$ is the maximum compensation current, f is the grid frequency, and V_{dcmin} is the minimum DC side voltage. Using parameters listed in Table 9.2, the maximum filter inductance is calculated as 2.62 mH for 9 kHz average switching frequency and 15% ripple ratio in i_c. It is worth noting that the filter inductance (L) used in the experimental setup was selected to be 2 mH. It should be noted that the minimum V_{dc} should be larger than the peak value of grid voltage. Since the experimental

Table 9.2 System and control parameters.

Description and symbol	Value
Grid voltage amplitude: E_m	$120\sqrt{2}$ V
DC-link voltage reference: V_{dc}^*	250 V
Inductance: L	2 mH
Inductor resistance: r	0.1 Ω
Grid inductance: L_g	2 mH
Inductor resistance: r_g	0.1 Ω
DC capacitors: $C_1 = C_2$	470 μF
Nonlinear load (R_L, C)	25 Ω, 470 μF
PI gains: K_p, K_i	0.3, 5
Controller gain: β_2	1
Sampling time: T_s	50 μs

tests are performed using 120 V rms grid voltage, V_{dcmin} is selected to be 170 V. Substituting $v_g = 120$ V, $I_{Cmax} = 12$A, $f = 50$ Hz, $V_{dc} = 250$ V, and $V_{dcmin} = 170$ V into (9.29), we obtain the values of C_1 and C_2 as 428 μF. Since these values are not available, 470 μF capacitors are used in the experimental setup.

9.2.2 The Energy-Function-Based MPC

The block diagram of the energy function based MPC technique is depicted in Figure 9.9 [4]. The presented approach consists of the following two steps.

9.2.2.1 Design of Energy-Function-Based MPC

The main control objectives of the three-level SAPF are to regulate V_{dc} to its reference, regulate capacitor voltages, obtain sinusoidal grid current at unity power factor, and guarantee stable operation under various load types. It is worth to note that when the regulation of V_{dc} is accomplished, the control of capacitor voltages is achieved automatically. Also, the unity power factor operation occurs when the grid current is in phase with the grid voltage. In order to meet the latter objective, the grid voltage and reference grid current should be in the form of

$$e_g = E_m \sin(\omega t), i_g^* = I_m^* \sin(\omega t) \tag{9.30}$$

The sine wave template in i_g^* is obtained via a classical phase-locked loop (PLL). The amplitude of the reference grid current is produced by a proportional-integral (PI) regulator as follows:

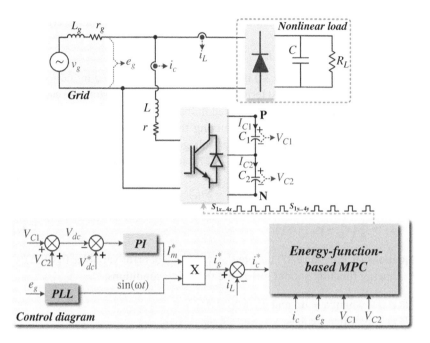

Figure 9.9 Block diagram of the energy function based MPC with three-level T-type inverter-based SAPF. *Source:* Komurcugil et al. [4]/IEEE.

$$I_m^* = K_p\left(V_{dc}^* - V_{dc}\right) + K_i \int \left(V_{dc}^* - V_{dc}\right) dt \tag{9.31}$$

where K_p and K_i are proportional and integral gains, respectively, and V_{dc}^* is reference of V_{dc}. From (9.24), it follows that the derivative of i_c^* can be written as

$$\frac{di_c^*}{dt} = \frac{1}{L}\left(e_g - v_{xy}^* - ri_c^*\right) \tag{9.32}$$

where i_c^* denotes the reference of i_c, which is defined as

$$i_c^* = i_g^* - i_L \tag{9.33}$$

In (9.33), the measurement of i_L is required. Unlike the classical MPC involving a cost function, the presented MPC employs an energy function. It can be seen from Figure 9.9 that there are two DC-side capacitors and one AC-side inductor, which have the ability to store energy in the SAPF. Therefore, the energy function in the continuous time can be expressed in terms of capacitor voltage error (x_1) and inductor current error (x_2) as follows:

$$E(x) = \frac{1}{2}\beta_1 x_1^2 + \frac{1}{2}\beta_2 x_2^2 \tag{9.34}$$

where the constants β_1 and β_2 should be positive and x_1 and x_2 are the error variables defined as follows:

$$x_1 = V_{C1} - V_{C2}, \quad x_2 = i_c - i_c^* \tag{9.35}$$

It can be noticed that Eq. (9.34) is in the form of energy stored in inductor and capacitor. For instance, the first term in (9.34) gives the energy stored in capacitor, while the second term is the energy stored in inductor. The details of formulating $E(x)$ can be found in [7]. It is worth noting that inclusion of x_1 in $E(x)$ has a self-balanced effect on the DC capacitor voltages. Now, taking the derivative of (9.35) and making use of (9.24), (9.26), (9.27), and (9.32) yields

$$\dot{x}_1 = \frac{1}{2C_1}(S_1 i_c - S_2 i_c) - \frac{1}{2C_2}(S_2 i_c - S_1 i_c) \tag{9.36}$$

$$\dot{x}_2 = \frac{1}{L}\left(v_{xy}^* - v_{xy} - r x_2\right) \tag{9.37}$$

Equations (9.36) and (9.37) denote the dynamics of error variables in (9.35). The energy function should satisfy the following conditions:

i) $E(x) > 0$ when $x_1 \neq 0$ and $x_2 \neq 0$
ii) $E(x) \to \infty$ when $\|x_1\| \to \infty$ and $\|x_2\| \to \infty$
iii) $\dot{E}(x) < 0$

Clearly, the first two conditions are satisfied. The third condition (negative definiteness of $E(x)$) which guarantees the stability of the system should also be satisfied. It is worth to note that $E(x)$ increases or decreases depending on the values of x_1 and x_2. When the error variables converge to zero ($x_1 = 0$ and $x_2 = 0$), $E(x)$ also tends to zero. Therefore, the main goal is to minimize $E(x)$ and at the same time achieve $\dot{E}(x) < 0$.

Taking the derivative of (9.34) yields

$$\dot{E}(x) = \beta_1 \dot{x}_1 x_1 + \beta_2 \dot{x}_2 x_2 \tag{9.38}$$

Now, substituting (9.22), (9.36), and (9.37) into (9.38) gives

$$\dot{E}(x) = \frac{\beta_1}{2C_1}(S_1 i_c - S_2 i_c)x_1 - \frac{\beta_1}{2C_2}(S_2 i_c - S_1 i_c)x_1 + \frac{\beta_2}{L}\left(v_{xy}^* - S_1 V_{C1} - S_2 V_{C2} - r x_2\right)x_2 \tag{9.39}$$

Substituting $i_c = x_2 + i_c^*$ into (9.39) and assuming that $C_1 = C_2 = C$, one can obtain

$$\dot{E}(x) = S_1 x_1 x_2 \left(\frac{\beta_1}{C} - \frac{\beta_2}{L}\right) + S_2 x_1 x_2 \left(\frac{\beta_2}{L} - \frac{\beta_1}{C}\right)$$
$$+ \frac{\beta_1}{C}(S_1 - S_2)i_c^* x_1 + \frac{\beta_2}{L}\left(v_{xy}^* x_2 - S_1 V_{C1} x_2 - S_2 V_{C2} x_2 - r x_2^2\right) \tag{9.40}$$

Selecting $\beta_1 = C\beta_2/L$ eliminates the terms $S_1 x_1 x_2$ and $S_2 x_1 x_2$, which also reduces Eq. (9.40) to

$$\dot{E}(x) = \frac{\beta_2}{L}\left((S_1 - S_2)i_c^* x_1 + v_{xy}^* x_2 - S_1 V_{C1} x_2 - S_2 V_{C2} x_2 - r x_2^2\right) \tag{9.41}$$

The stability of the closed-loop system is assured if $\dot{E}(x) < 0$. The main aim in this study is to design MPC strategy by using (9.41). Since Eq. (9.41) is in continuous time, it should be expressed in discrete time to design the MPC.

9.2.2.2 Discrete-Time Model

Now, $\dot{E}(x)$ in (9.41) can be expressed in discrete time at $(k + 1)^{\text{th}}$ sampling instant as follows:

$$\begin{aligned}
\dot{E}_x^{(n)}(k + 1) = {} & \frac{\beta_2}{L}\left(\left(S_1^{(n)}(k) - S_2^{(n)}(k)\right)i_c^*(k + 1)x_1(k + 1)\right. \\
& + v_{xy}^*(k + 1)x_2(k + 1) - S_1^{(n)}(k)V_{C1}(k + 1)x_2(k + 1) \\
& \left. - S_2^{(n)}(k)V_{C2}(k + 1)x_2(k + 1) - r x_2^2(k + 1)\right)
\end{aligned} \tag{9.42}$$

Equation (9.42) can be used to select the optimum switching functions $S_1(k)$ and $S_2(k)$ such that $\dot{E}_x(k + 1)$ is negative. It is worth to note that the choice of β_2/L has no effect on the performance of the controller provided that $\beta_2 > 0$. In order to verify this claim, the performance of the energy function-based MPC is investigated by using various β_2 values (see Figure 9.18b). It can be observed that once $\dot{E}_x(k + 1) < 0$ is achieved, the system works successfully regardless of β_2 values. Therefore, unlike the classical MPC method whose performance depends on WF, the presented MPC method is WF free that yields a simplification in the design of the controller. The error variables in (9.42) can be written as

$$x_1(k + 1) = V_{C1}(k + 1) - V_{C2}(k + 1) \tag{9.43}$$

$$x_2(k + 1) = i_c(k + 1) - i_c^*(k + 1) \tag{9.44}$$

Applying the first-order forward Euler approximation to (9.24), (9.26), and (9.27), the future values of $i_c(k)$, $V_{C1}(k)$, and $V_{C2}(k)$ at $(k + 1)^{\text{th}}$ sampling instant can be obtained as follows:

$$i_c(k + 1) = \left(1 - \frac{r}{L}T_s\right)i_c(k) + \frac{T_s}{L}\left(e_g(k) - v_{xy}(k)\right) \tag{9.45}$$

$$V_{C1}(k + 1) = V_{C1}(k) + \frac{T_s}{2C_1}\left(S_1(k)i_c(k) - S_2(k)i_c(k)\right) \tag{9.46}$$

$$V_{C2}(k + 1) = V_{C2}(k) + \frac{T_s}{2C_2}\left(S_2(k)i_c(k) + S_1(k)i_c(k)\right) \tag{9.47}$$

where T_s is the sampling period and $v_{xy}(k)$ is given by

$$v_{xy}(k) = S_1(k)V_{C1}(k) + S_2(k)V_{C2}(k) \tag{9.48}$$

On the other hand, the grid voltage, reference grid current, and input voltage of rectifier at $(k + 1)^{th}$ sampling instant, which are needed in (9.42) can be predicted as

$$e_g(k + 1) = \frac{3}{2}e_g(k) - \frac{1}{2}e_g(k - 1) \tag{9.49}$$

$$i_c^*(k + 1) = \frac{3}{2}i_c^*(k) - \frac{1}{2}i_c^*(k - 1) \tag{9.50}$$

$$v_{xy}^*(k + 1) = e_g(k + 1) - \frac{L}{T_s}\left(i_c^*(k + 1) - i_c^*(k)\right) - ri_c^*(k + 1) \tag{9.51}$$

Algorithm 9.1   Energy-Function-Based MPC

```
1: Measure e_g(k), i_c(k), V_C1(k), and V_C2(k).
2: Compute i_c*(k).
3: Compute (31)-(33).
4:       for n=1,....., 9 do
5:           Compute (25) and (26).
6:           Evaluate (24).
7:       end for
8: return minimum Ėx(k + 1).
9: Choose switching states which yield minimum Ėx(k + 1)
```

The steps of the energy-function-based MPC method are given in the algorithm. First, the system variables are measured to be used in the control algorithm. Then, the compensation current reference (i_c^*) is computed. The predictions of $e_g(k + 1)$, $i_c^*(k + 1)$ and $v_{xy}^*(k + 1)$ are calculated to be used in the energy function. Thereafter, the predicted values of the error variables $(x_1(k + 1)$ and $x_2(k + 1))$ are calculated for each possible switching state. The derivative of energy function $(\dot{E}_x(k + 1))$ is evaluated using the predicted values in the previous steps. In the last step, the optimal switching vector is determined by minimizing $\dot{E}_x(k + 1)$. The flowchart of the MPC algorithm is given in Figure 9.10 [4].

9.2.3   Experimental Studies

The effectiveness of the presented MPC method is verified by experimental studies by implementing the block diagram shown in Figure 9.9 [4]. In the experimental studies, a full-wave diode-bridge rectifier is used as the nonlinear load. The experimental studies have been carried out on the prototype shown in Figure 9.11 where the power grid is emulated via a regenerative grid simulator (Chroma 61860). The MPC method was implemented by using OPAL-RT OP5600. The pulse width modulation (PWM) signals generated by OPAL-RT are applied to drive the switches in the T-type inverter. The system and control parameters used in the experimental studies are listed in Table 9.2.

9.2.3.1 Steady-State and Dynamic Response Tests

Figure 9.12 shows the waveforms of grid voltage (e_g), grid current (i_g), load current (i_L), filter current (i_c), filter current reference (i_c^*), DC-link voltage (V_{dc}), DC-link voltage reference $\left(V_{dc}^*\right)$, and capacitor voltages (V_{C1}, V_{C2}). Despite the highly distorted load current, the grid current is sinusoidal and in phase with the grid voltage. Hence, the unity power factor operation is achieved. The filter current and its reference are overlapped, which implies that the controller has almost zero tracking error. It is obvious that the control of the DC-link and capacitor voltages is achieved at 250 and 125 V, respectively.

Figure 9.13 shows the spectrums of load and grid currents corresponding to Figure 9.12. It can be seen from Figure 9.13a that the load current contains significant odd harmonic components. The measured THD of load current is 67.6%. On the other hand, most odd harmonic components in the grid current are suppressed as shown in Figure 9.13b. Clearly, only the 3rd, 5th, and 7th harmonic components are discernible with small magnitudes. Majority of odd harmonic components such as 9th, 11th, 13th, 15th, and other components are suppressed effectively. As a consequence of this fact, the THD of grid current is measured to be 2.7%. The effectiveness of the controller on the operation of SAPF can be understood if the measured THD values are considered. Comparing the

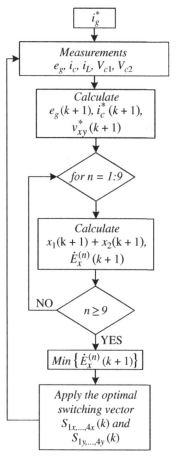

Figure 9.10 Flowchart of the energy-function based FCS-MPC. *Source:* Komurcugil et al. [4]/IEEE.

Oscilloscope

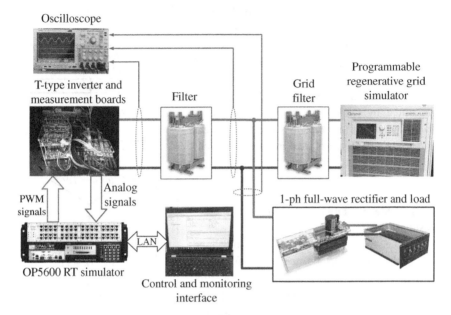

PWM signals

Analog signals

OP5600 RT simulator

Control and monitoring interface

Figure 9.11 Experimental setup for the SAPF. *Source:* Komurcugil et al. [4], from IEEE.

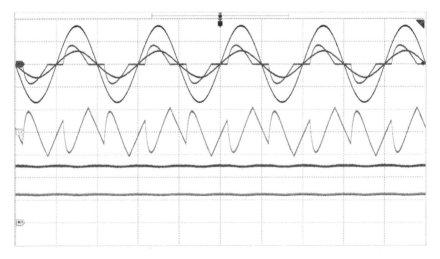

Figure 9.12 Waveforms of grid-voltage (e_g), grid current (i_g), load current (i_L), filter current (i_c), filter current reference (i_{c^*}), DC-link voltage (V_{dc}), DC-link voltage reference (V_{dc^*}), and capacitor voltages (V_{C1} and V_{C2}).

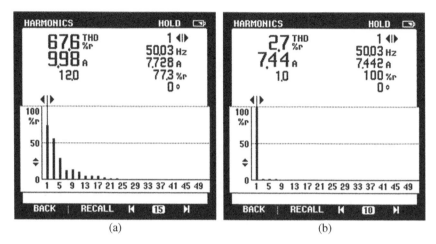

Figure 9.13 Measured spectrums of load and grid currents corresponding to Figure 9.12. (a) Load current spectrum, (b) Grid current spectrum.

measured THD values of load and grid currents, one can see that the presented SAPF operates with high performance.

Figure 9.14 shows operation of the SAPF for a step change in the nonlinear load current. It is worth noting that all variables are regulated before the step change occurs. When the load current increases (DC load resistance has been changed from 25 to 12.5 Ω), the DC-link voltage exhibits small undershoot and tracks its reference successfully as shown in Figure 9.14a. The capacitor voltages behave similar to the DC-link voltage. The grid current is also increased in response to the load current increment. On the other hand, when the load current is decreased (DC load resistance has been changed from 12.5 to 25 Ω), the DC-link voltage gives rise to small overshoot and tracks its reference successfully after the transient is over as shown in Figure 9.14b. Similarly, the capacitor voltages are also regulated in this case. Finally, it can be seen that the grid current is decreased in response to the decrement in load current.

Figure 9.15 shows the waveforms of grid current, PI output, capacitor voltage error, and filter current error corresponding to the load current changes in Figure 9.15. Apparently, the grid current increases to track its reference when the load current is increased as shown in Figure 9.15a. Similarly, as a result of decrement in the load current, the grid current decreases as well and tracks its reference as shown in Figure 9.15b. In both cases, the error variables are completely zero, which indicate that the controller acts very fast in tracking the references.

Figure 9.16 shows the waveforms of grid voltage, grid current, load current (i_L), filter current, DC-link voltage, and capacitor voltages for start-up. Initially, the system is started with SAPF disabled. In this case, the grid current is equal to the load current and, eventually, the filter current is zero. The DC-link and capacitor

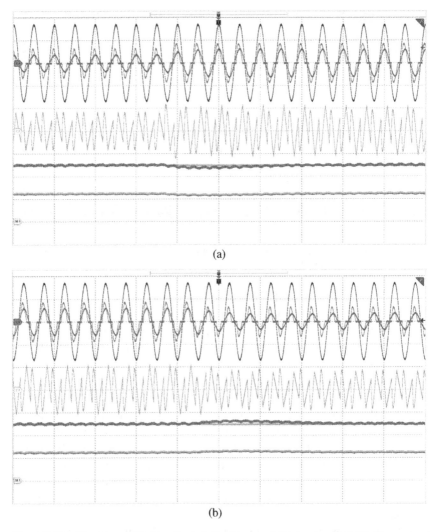

(a)

(b)

Figure 9.14 Waveforms of grid-voltage (e_g), grid current (i_g), load current (i_L), filter current (i_c), filter current reference (i_c^*), DC-link voltage (V_{dc}), DC-link voltage reference (V_{dc}^*), and capacitor voltages (V_{C1} and V_{C2}) for step change in load current. (a) When the load resistance is changed from 25 to 12.5 Ω; (b) when the load resistance is changed from 12.5 to 25 Ω.

voltages are also zero. When the SAPF is enabled, the filter current is not zero anymore, which implies that the grid current and load current are not equal. In this case, the filter current is added with the load current at the point of common coupling such that the grid current becomes sinusoidal and in phase with grid voltage. On the other hand, DC-link and capacitor voltages gradually converge to 250 and 125 V, respectively.

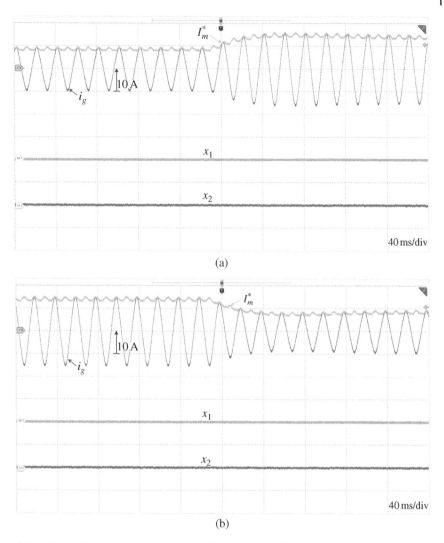

Figure 9.15 Waveforms of grid current (i_g), PI output (i_m^*), capacitor voltage error (x_1), and filter current error (x_2) corresponding to the load current changes in Figure 9.14. (a) Load current is increased, (b) Load current is decreased.

9.2.3.2 Comparison with Classical MPC Method

Figure 9.17 shows operation of SAPF under classical MPC that employs a WF. Even though the control of DC-link and capacitor voltages is accomplished, and the grid current is sinusoidal, the unity power factor is not satisfied completely as shown in Figure 9.17a. The spectrum of grid current is presented in Figure 9.17b. The THD of grid current is measured as 3.80%. Comparing the THDs

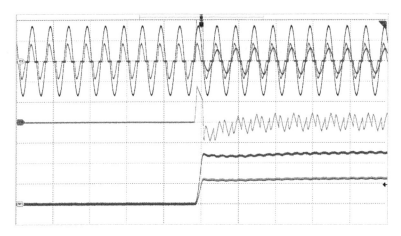

Figure 9.16 Waveforms of grid-voltage (e_g), grid current (i_g), load current (i_L), filter current (i_c), DC-link voltage (V_{dc}), and capacitor voltages (V_{C1} and V_{C2}) for start-up.

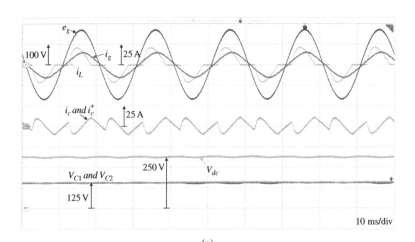

(a)

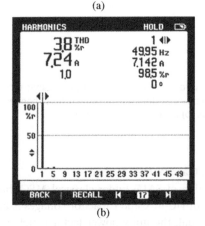

(b)

Figure 9.17 Operation of SAPF under classical MPC; (a) Waveforms of grid-voltage (e_g), grid current (i_g), load current (i_L), filter current (i_c), filter current reference (i_c^*), DC-link voltage (V_{dc}), and capacitor voltages (V_{C1} and V_{C2}), (b) Spectrum of grid current (i_4).

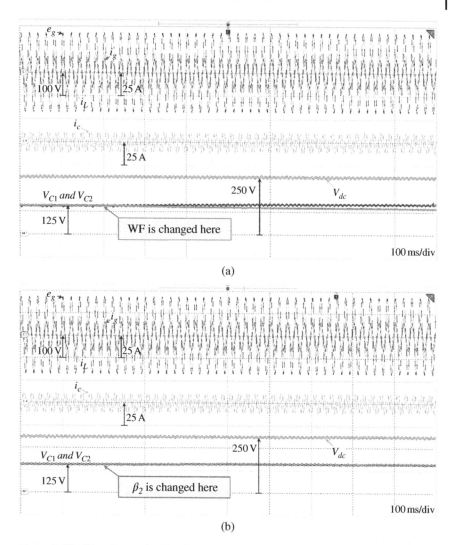

Figure 9.18 Waveforms of grid-voltage (e_g), grid current (i_g), load current (i_L), filter current (i_c), DC-link voltage (V_{dc}), and capacitor voltages (V_{C1} and V_{C2}) obtained by: (a) Classical MPC for a step change in WF from 1 to 10, (b) presented MPC for a step change in β_2 from 1 to 10.

of grid current obtained by presented MPC and classical MPC, one can see that the presented MPC method yields smaller THD value as shown in Figure 9.13b.

Figure 9.18 shows the waveforms obtained by the classical and presented MPC methods for a step variation in WF and β_2, respectively. It is worth

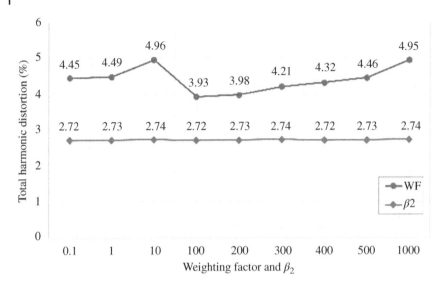

Figure 9.19 THD comparison of classical and presented MPC methods under weighting factor and β_2 variations.

noting that the control of DC-link and capacitor voltages and reactive power compensation together with the unity power factor are satisfied before the step change in WF as shown in Figure 9.18a. However, when the WF value is changed from 1 to 10, the capacitor voltages deviate from 125 V and become unbalanced. If the system is operated with WF = 10 for a long time, it would adversely affect the operation of SAPF since the control of DC-link voltage will be lost. This can be considered as the main disadvantage of the classical MPC that is dependent on the WF value. In literature, no procedure for tuning the WF for optimum performance is reported. The tuning of WF is usually achieved by trial-and-error method. On the other hand, changing β_2 value from 1 to 10 has no effect on the operation of the system as shown in Figure 9.18b. Since the presented MPC does not require the WF, it brings a simplification in the design of the controller.

Figure 9.19 shows the influence of varying WF and β_2 on the THD of grid current. It is evident that the THD is subject to the WF value variations. This result clearly shows the disadvantage of classical MPC in which WF tuning is essential to obtain good performance. Since there is no preset rule for tuning WF, the design of WF becomes challenging in some applications. However, the THD is not affected when the value of β_2 is increased. Contrary to the classical MPC, the coefficient (β_2) employed in the cost function of the presented MPC has no effect on the THD 4

9.3 Model Predictive Control of Quasi-Z Source Three-Phase Four-Leg Inverter

This section presents a model predictive control (MPC) scheme for quasi-Z source (qZS) three-phase four-leg inverters. To cope with drawbacks of the traditional voltage source inverters (VSI)s, the qZS three-phase four-leg inverter topology is explained in this study. More information about the qZS inverters can be found in [5]. The control technique in this subsection is based on the controller proposed in [8] and [9]. The presented controller handles each phase current independently. As a result of this, the qZS four-leg inverter has fault-tolerant capability, for example if one leg fails, the others can work normally. Simulation studies were performed to verify the steady-state and transient-state performances of the control strategy under balanced/unbalanced reference currents and load conditions.

9.3.1 qZS Four-Leg Inverter Model

The qZS four-leg inverter topology with R–L output filter is shown in Figure 9.20. This topology can be investigated as two stages: (i) the qZS network and (ii) four-leg inverter with R–L output filter and load. In the first stage of this topology, the

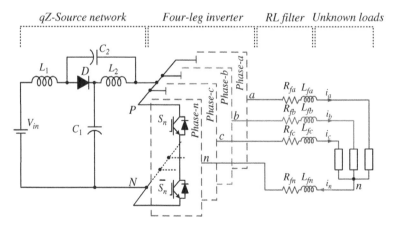

Figure 9.20 qZS three-phase four-leg inverter topology. *Source:* Bayhan et al. [9]/IEEE.

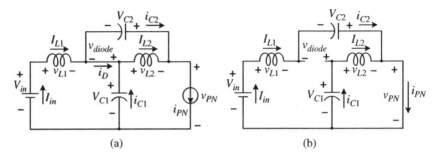

Figure 9.21 Equivalent circuit of the qZS network. (a) In non-shoot-through state. (b) In shoot-through state.

qZS is made of an L–C impedance network, which can boost the DC voltage in response to the so-called shoot-through zero state of the inverter switching cycle. In the shoot-through zero state, two semiconductor switches in the same leg are simultaneously switched on to create short circuit across the DC link. During this state, energy is transferred in the qZS network from the capacitors to the inductors, and this state is used to boost the DC voltage.

In the second stage of this topology, the four-leg inverter is used. As shown in Figure 9.20, the load neutral point is connected to the mid-point of the inverter's fourth phase leg to allow for zero sequence current/voltage. However, the addition of an extra leg makes the switching schemes more complicated compared to a three-leg VSI. Nevertheless, using the extra phase leg improves inverter capability and reliability. The four-leg inverter can be used under balanced/unbalanced and/ or linear/nonlinear load conditions.

The equivalent circuits of the qZS network in non-shoot-through and shoot-through states are illustrated in Figure 9.21a,b, respectively [10]. All voltages and currents are defined in this figure and the polarities are shown with arrows.

1) *Non-shoot-Through State:* During the non-shoot-through state, four-leg inverter model is represented by a constant current source; it can be seen from Figure 9.21a. By applying Kirchhoff's voltage law to Figure 9.21a, inductor voltages (v_{L1} and v_{L2}), DC-link voltage (v_{PN}), and diode voltage (v_{diode}) are written as

$$v_{L1} = V_{in} - V_{C1}, v_{L2} = -V_{C2} \tag{9.52}$$

$$v_{PN} = V_{C1} - v_{L2} = V_{C1} + V_{C2}, v_{diode} = 0 \tag{9.53}$$

2) *Shoot-Through State:* During the shoot-through state, four-leg inverter model is represented by short-circuit, it can be seen from Figure 9.21b. By applying Kirchhoff's voltage law to Figure 9.21b, inductors voltages (v_{L1} and v_{L2}), DC-link voltage (v_{PN}), and diode voltage (v_{diode}) are written as

$$v_{L1} = V_{C2} + V_{in}, v_{L2} = V_{C1} \tag{9.54}$$

$$v_{PN} = 0, v_{diode} = V_{C1} + V_{C2} \tag{9.55}$$

At steady state, the average voltage of the capacitors over one switching cycle are

$$\left. \begin{aligned} V_{C1} &= V_{in} \frac{T_1}{T_1 - T_0} \\ V_{C2} &= V_{in} \frac{T_0}{T_1 - T_0} \end{aligned} \right\} \tag{9.56}$$

where T_0 is the duration of the shoot-through state, T_1 is the duration of the non-shoot-through state, and V_{in} is the input DC voltage.

From (9.53), (9.55), and (9.56), the peak DC-link voltage across the inverter bridge in Figure 9.20 is

$$v_{PN} = V_{C1} + V_{C2} = V_{in} \frac{T}{T_1 - T_0} = BV_{in} \tag{9.57}$$

where T is the switching cycle ($T_0 + T_1$) and B is the boost factor of the qZSI. The average current of the inductors L_1 and L_2 can be calculated from the system power P:

$$I_{L1} = I_{L2} = I_{in} = P/V_{in} \tag{9.58}$$

Applying Kirchhoff's current law and (9.58) results in

$$i_{C1} = i_{C2} = i_{PN} - I_{L1} \tag{9.59}$$

The voltage gain (G) of the qZSI can be expressed as

$$G = \hat{v}_{ln}/0.5 v_{PN} = MB \tag{9.60}$$

where M is the modulation index and $\hat{v}_{ln}$ is the peak AC-phase voltage.

The equivalent circuit of the four-leg inverter with the output R–L filter is shown in Figure 9.22, where the L_{fj} is the filter inductance, R_{fj} is the filter resistance, and R_j is the load resistance for each of the phase $j = a, b, c$.

For a three-phase four-leg inverter, the addition of the fourth leg makes the switching states 16 (2^4). The valid switching states with the corresponding phase and line voltages for the traditional four-leg inverter are presented in [10]. In addition to these switching states, for this application, one extra switching state is required to ensure shoot-through state. Therefore, a total of 17 switching states are used in this application. The voltage in each leg of the four-leg inverter can be expressed as

$$\left. \begin{aligned} v_{aN} &= S_a v_{dc} \\ v_{bN} &= S_b v_{dc} \\ v_{cN} &= S_c v_{dc} \\ v_{nN} &= S_n v_{dc} \end{aligned} \right\} \tag{9.61}$$

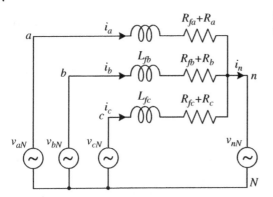

Figure 9.22 Equivalent circuit of the three-phase four-leg inverter.

where S_a, S_b, S_c, and S_n are the switching states, and v_{dc} and v_{nN} are DC link and load neutral voltages, respectively.

The output voltage of this inverter can be written in terms of the previous inverter voltages:

$$
\left.
\begin{aligned}
v_{an} &= (S_a - S_n)v_{dc} \\
v_{bn} &= (S_b - S_n)v_{dc} \\
v_{cn} &= (S_c - S_n)v_{dc}
\end{aligned}
\right\}
\tag{9.62}
$$

By applying Kirchhoff's voltage law to Figure 9.22, the inverter voltages can be expressed in terms of load-neutral voltages and load currents as follows:

$$
\left.
\begin{aligned}
v_{aN} &= \left(R_{fa} + R_a\right)i_a + L_{fa}\frac{di_a}{dt} + v_{nN} \\
v_{bN} &= \left(R_{fb} + R_b\right)i_b + L_{fb}\frac{di_b}{dt} + v_{nN} \\
v_{cN} &= \left(R_{fc} + R_c\right)i_c + L_{fc}\frac{di_c}{dt} + v_{nN}
\end{aligned}
\right\}
\tag{9.63}
$$

From (9.62) and (9.63), the output voltages can be expressed as

$$
\left.
\begin{aligned}
v_{an} &= \left(R_{fa} + R_a\right)i_a + L_{fa}\frac{di_a}{dt} \\
v_{bn} &= \left(R_{fb} + R_b\right)i_b + L_{fb}\frac{di_b}{dt} \\
v_{cn} &= \left(R_{fc} + R_c\right)i_c + L_{fc}\frac{di_c}{dt}
\end{aligned}
\right\}
\tag{9.64}
$$

which is simplified to

$$
v_j = \left(R_{fj} + R_j\right)i_j + L_{fj}\frac{di_j}{dt}, j = a, b, c
\tag{9.65}
$$

and neutral current in can be written as

$$i_n = i_a + i_b + i_c \tag{9.66}$$

The expression for output current, derived from (9.65), is

$$\frac{di_j}{dt} = \frac{1}{L_f}\left[v_j - \left(R_{fj} + R_j\right)i_j\right], j = a, b, c \tag{9.67}$$

9.3.2 MPC Algorithm

The MPC scheme is shown in Figure 9.23. It has two main layers: (i) a predictive model and (ii) cost function optimization. The discrete-time model of the system is used to predict future behavior of the control variables. The cost function is used to minimize the error between the reference and the predicted control variables in the next sampling time. This control technique has several advantages as follows: easy to implement in both linear and nonlinear systems, it shows high accuracy and fast dynamic response, and it has very small steady-state error throughout different operating points. Here, the MPC scheme is described in the following steps:

1) determination of references.
2) build discrete-time models of the system.
3) define a cost function g.
4) prepare control algorithm.

9.3.2.1 Determination of References

DC-link voltage and output currents references are normally obtained through maximum power point tracking algorithm for RESs. This control technique, however, aims to improve the control capability of the qZS four-leg inverter. For this reason, without loss of generality, these references are left to be defined by the user.

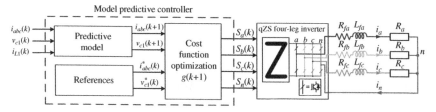

Figure 9.23 Block diagram of the MPC scheme proposed in [8]. *Source:* Bayhan et al. [8]/IEEE.

9.3.2.2 Discrete-Time Models of the System

The control of the qZS four-leg inverter output currents (i_a, i_b, i_c) and capacitor voltage (V_{C1}) required two discrete-time models be created from the continuous-time equations. To do that, the general structure of the forward-difference Euler Eq. (9.68) is used to compute the differential equations of the output current and the capacitor voltage:

$$\frac{df}{dt} \approx \frac{f(x_0 + h) - f(x_0)}{h} \tag{9.68}$$

To estimate the value in the next sample time, for a suitably small-time step, (9.68) becomes the discretization equation:

$$\frac{\Delta f(k)}{\Delta t} \approx \frac{f(k + 1) - f(k)}{T_s} \tag{9.69}$$

where T_s is the sampling time.

1) *Predictive Model I:* This model is used to predict future behavior of each of the output currents (i_a, i_b, i_c). The continuous-time expression for each phase current is given in (9.67). By substituting (9.69) into (9.67), the discrete-time model for each output phase current is

$$i_j(k + 1) = A_v v_j(k + 1) + A_i i_j(k), j = a, b, c \tag{9.70}$$

where $i_j(k + 1)$ is the predicted output current vector at the next sampling time and A_v and A_i are constants as defined by

$$\left.\begin{aligned} A_v &= \frac{T_s}{L_f + (R + R_f)T_s} \\ A_i &= \frac{L_f}{L_f + (R + R_f)T_s} \end{aligned}\right\} \tag{9.71}$$

2) *Predictive Model II:* This model is used to predict future behavior of the capacitor voltage (V_{C1}). The continuous time model of the capacitor current can be expressed as

$$i_{C1} = C_1 \frac{d(V_{C1} - i_{C1}r_c)}{dt} \tag{9.72}$$

where C_1 and r_c are the capacitance and the equivalent series resistance (ESR) of the capacitor, respectively. Based on (9.72), the capacitor voltage is derived as

$$\frac{dV_{C1}}{dt} = r_c \frac{di_{C1}}{dt} + \frac{1}{C_1} i_{C1} \tag{9.73}$$

By substituting (9.69) into (9.73), the discrete-time model of the V_{C1} can be obtained as

$$V_{C1}(k + 1) = V_{C1}(k) + i_{C1}(k + 1)r_c + i_{C1}(k)\left(\frac{T_s}{C_1} - r_C\right) \tag{9.74}$$

where $V_{C1}(k + 1)$ is the predicted capacitor voltage at the next sampling time and $i_{C1}(k)$ is capacitor current that depends on the states of the qZSI topology. According to the operational principle of qZS network explained in Section II-B, for non-shoot-through and shoot-through states, capacitor current can be defined as follows:

1) During non-shoot-through state

$$i_{C1} = I_{L1} - [S_a i_a + S_b i_b + S_c i_c] \tag{9.75}$$

2) During shoot-through state

$$i_{C1} = -I_{L1} \tag{9.76}$$

9.3.2.3 Cost Function Optimization

The selection of the cost function is a key part of the MPC scheme. The presented MPC scheme has two cost functions, which are used to minimize output current and capacitor voltage errors in the next sampling time. The output current cost function is defined as

$$
\begin{aligned}
g_i = & \left\| i_j^*(k + 1) - i_j(k + 1) \right\| \left[i_a^*(k + 1) - i_a(k + 1) \right]^2 \\
& + \left[i_b^*(k + 1) - i_b(k + 1) \right]^2 + \left[i_c^*(k + 1) - i_c(k + 1) \right]^2
\end{aligned}
\tag{9.77}
$$

where $i_j^*(k + 1)$ is the reference output current vector and $i_j(k + 1)$ is the predicted output current vector in the next step ($j = a, b, c$).

The cost function of capacitor voltage can also be defined as

$$g_v = \lambda^* \left| v_{C1}^*(k + 1) - v_{C1}(k + 1) \right| \tag{9.78}$$

where $v_{C1}^*(k + 1)$ and $v_{C1}(k + 1)$ are the reference and predicted capacitor voltages, respectively. The weighting factor (λ) was determined by using cost function classification technique that was detailed in Chapter 5. The complete cost function is

$$g(k + 1) = g_i(k + 1) + g_v(k + 1) \tag{9.79}$$

9.3.2.4 Control Algorithm

The flowchart for the control algorithm is given in Figure 9.24. Cost function minimization is implemented as a repeated loop for each voltage vector to predict the values, evaluate the cost function, and store the minimum value and the index value of the corresponding switching state. The control algorithm can be summarized in the next steps.

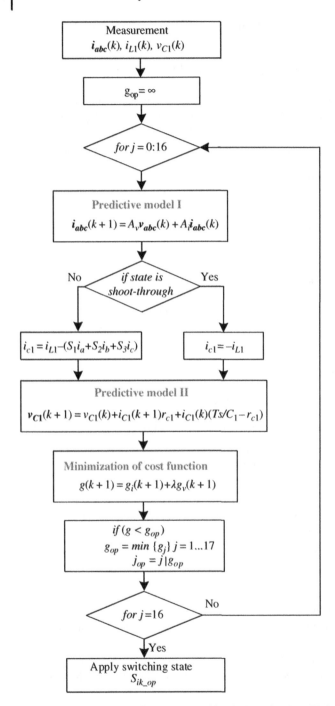

Figure 9.24 Flowchart of the presented MPC algorithm for qZS four-leg inverters.

1) Sampling the output phase currents (i_{abc}), inductor current (i_{L1}), and capacitor voltage (v_{C1}).
2) These are used to predict output currents and capacitor voltage using the predictive model I, and II, respectively.
3) All predictions are evaluated using the cost function.
4) The optimal switching state that corresponds to the optimal voltage vector that minimizes the cost function is selected to be applied at the next sampling time.

9.3.3 Simulation Results

To verify the presented MPC scheme for qZS four-leg inverter, a simulation model was carried out using MATLAB&Simulink software. The parameters used in the simulations are given in Table 9.3.

Steady-state results of the MPC scheme under balanced reference output currents and balanced loads are shown in Figure 9.25. During the test, the reference currents (i_a^*, i_b^*, i_c^*), and reference capacitor voltage (V_{C1}^*) are set to 10 A, and 300 V, respectively. It can be observed from Figure 9.25 that the output current tracks its reference value very well while the capacitor and DC-link voltages are kept constant. Furthermore, neutral current is zero since the reference output currents are balanced as shown in Figure 9.25b.

The MPC scheme is also tested to verify the robustness of the technique against unbalanced reference currents ($i_a^* = 10$ A, $i_b^* = i_c^* = 5$ A) and unbalanced loads ($R_a = 10\,\Omega$, $R_b = 7.5\,\Omega$, $R_c = 5\,\Omega$). This is the typical application for three-phase four-wire systems, where the load demand on each phase is different. Steady-state results of the MPC scheme under unbalanced reference output currents and unbalanced loads are shown in Figure 9.26. It can be observer from results, the controller is able to control each phase current independently while the capacitor and DC-link voltages are kept constant. The neutral current, which is sum of the

Table 9.3 qZS four-leg inverter and load parameters.

Parameter	Value
Input DC voltage (V_{in})	200 V
qZS network inductances (L_1, L_2)	1.2 mH
qZS network capacitors (C_1, C_2)	1000 μF
Load resistance (R)	10 Ω
Filter inductance (L_f)	10 mH
Filter resistance (R_f)	0.05 Ω
Nominal frequency (f_o)	50 Hz
Sampling time (T_s)	20 Ω

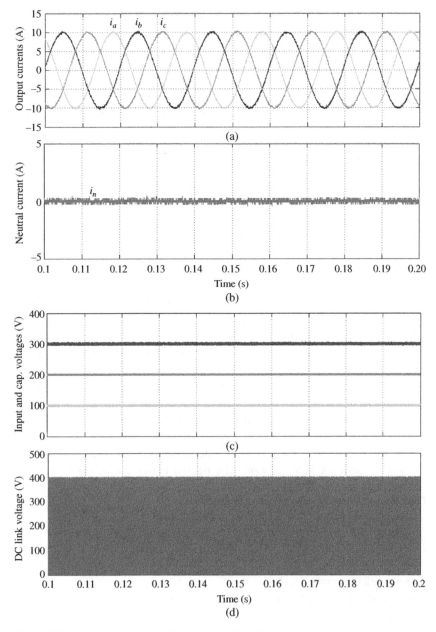

Figure 9.25 Steady-state simulation results with balanced reference currents and balanced loads. (a) Three-phase output currents (i_a, i_b, i_c); (b) neutral current (i_n); (c) input voltage (V_{in}), and capacitor voltages (V_{C1}), and (V_{C2}); (d) DC link voltage (v_{PN}).

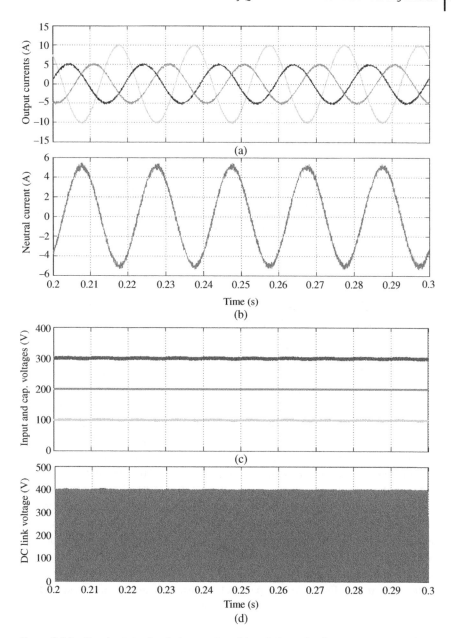

Figure 9.26 Steady-state simulation results with unbalanced reference currents and unbalanced loads. (a) Three-phase output currents (i_a, i_b, i_c); (b) neutral current (in); (c) input voltage (V_{in}), and capacitor voltages (V_{C1}), and (V_{C2}); (d) DC link voltage (v_{PN}).

three-phase currents, flows through the fourth leg because of the unbalance reference currents.

The transient-state performance of the MPC scheme for reference output currents step from 5 to 15 A are shown in Figure 9.27. For this test, reference currents and loads are balanced. It can be observed from these results that the transient time is very short, and the output current tracks its reference while the capacitor and DC-link voltages are kept constant. The neutral current is zero because of balanced reference load currents.

In Figure 9.28, the results are presented with unbalanced reference current step changed and balanced loads. For this test, all reference output currents set to 5 A at the beginning. Then, reference currents are set to $i_a^* = 7$ A, $i_b^* = 0$ A, and $i_c^* = 15$ A. Results of this test show that the controller handles each phase current independently while the capacitor and DC-link voltages are kept constant, even under unbalanced reference currents step.

9.4 Weighting Factorless Model Predictive Control for DC–DC SEPIC Converters

This section presents a weighting factorless model predictive current control method for DC–DC single-ended primary-inductor converters (SEPICs). The control technique presented in this subsection is based on the controller proposed in [11]. In the conventional model predictive control (MPC), the weighting factor should be returned when the operating condition of the converter changes. The weighting factor avoids excessive switching frequency in the DC–DC converters. Based on the relation between the inductor current ripple and switching frequency, a weighting factorless MPC is proposed in [11]. The effect of the weighting factor on the switching frequency is investigated. The MPC eliminates the need for employing a weighting factor, which is usually determined by trial-and-error in the conventional MPC. The presented control strategy is verified experimentally under input voltage, load, and parameter variations. The results obtained from traditional MPC and MPC in [11] are compared. It is shown that the presented MPC regulates the average switching frequency when the operating mode of the converter is changed. Furthermore, parameter mismatch results are presented for both conventional and presented MPC methods.

9.4.1 Principle of Control Strategy

The main objective of MPC is to generate the switching signals by minimizing the cost function. The cost function calculates the error between the predicted value of the desired control variable and the measured (actual) control variable. The

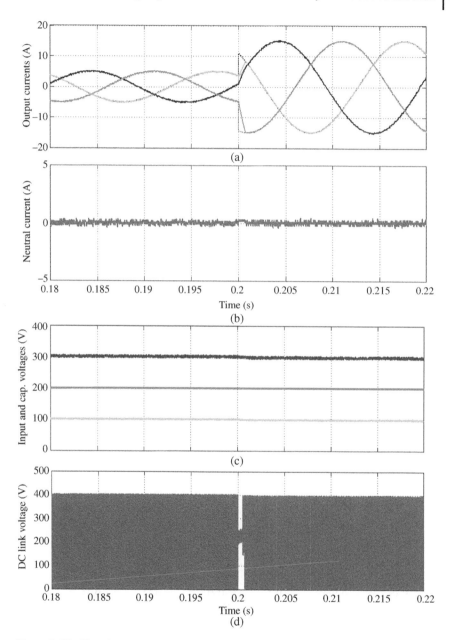

Figure 9.27 Transient-state simulation results with balanced reference currents and balanced loads. (a) Three-phase output currents (i_a, i_b, i_c); (b) neutral current (i_n); (c) input voltage (V_{in}), and capacitor voltages (V_{C1}), and (V_{C2}); (d) DC link voltage (v_{PN}).

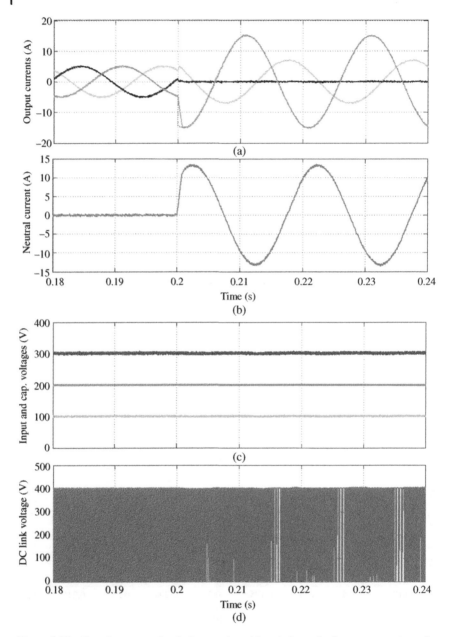

Figure 9.28 Transient-state simulation results with unbalanced reference currents and balanced loads. (a) Three-phase output currents $(i_a,\ i_b,\ i_c)$; (b) neutral current (i_n); (c) input voltage (V_{in}), and capacitor voltages (V_{C1}), and (V_{C2}); (d) DC link voltage (v_{PN}).

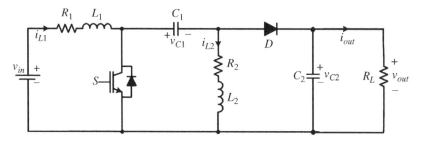

Figure 9.29 Circuit diagram of DC–DC SEPIC converter.

prediction step can be defined as another critical point for the MPC. The details of the design procedures are explained step by step in the following sub-sections.

9.4.1.1 Conventional Model Predictive Current Control

Figure 9.29 shows the circuit of a DC–DC SEPIC converter. The differential equation of i_{L1} can be written as

$$\frac{di_{L1}}{dt} = \frac{1}{L_1}\left[v_{in} - (R_1 i_{L1} S) - (v_{C1} + v_{out})(1 - S)\right] \tag{9.80}$$

where S denotes the switching state, which is 1 for ON state and 0 for OFF state. The continuous-time derivative of i_{L1} can be approximated in discrete-time by using Euler's forward method as follows:

$$\frac{di_{L1}}{dt} \approx \frac{i_{L1}(k + 1) - i_{L1}(k)}{T_s} \tag{9.81}$$

where T_s is the sampling period. By making use of (9.80) and (9.81), the future value of i_{L1} at $(k + 1)^{\text{th}}$ sampling instant can be obtained as

$$i_{L1}(k + 1) = \frac{T_s}{L_1}\left[v_{in}(k) - (SR_1 i_{L1}(k)) - (v_{C1}(k) + v_{out}(k))(1 - S)\right] + i_{L1}(k) \tag{9.82}$$

Equation (9.82) predicts the value of i_{L1} in the next sampling interval. The objective of the MPC method is to minimize the error between the predicted and reference values. The current error in the next sampling interval can be obtained by the following cost function:

$$g_i(k + 1) = \left|i_{L1}^*(k) - i_{L1}(k + 1)\right| \tag{9.83}$$

As mentioned before, the main objective of MPC method is to determine the optimum control action in each sampling time for all possible switching states such that the cost function in (9.83) is minimum. In [11] and [12], it is pointed out that the cost function requires a second term to penalize the difference between two consecutive switching states as follows:

$$g_s(k) = |S(k) - S(k-1)| \tag{9.84}$$

Hence, the total cost function can be written by combining (9.83) and (9.84) as follows:

$$g(k+1) = g_i(k+1) + \lambda g_s(k) \tag{9.85}$$

Since the SEPIC converter has two possible switching states, the total cost function in (9.85) is calculated for both ON and OFF states. Eq. (9.85) can also be defined as

$$g(k+1) = \begin{cases} g_i(k+1), & g_s(k) = 0 \\ g_i(k+1) + \lambda, & g_s(k) = 1 \end{cases} \tag{9.86}$$

The first equation in (9.86) refers to the case where $S(k) = S(k-1)$ implying $g_s(k) = 0$. In this case, the total cost function is equal to the inductor current error. On the other hand, the second equation in (9.86) refers to the case where $S(k) \neq S(k-1)$, which implies that $g_s(k) = 1$. Hence, it is worth noting that the value of total cost function increases due to the inclusion of weighting factor (λ) in (9.86). In the case of $S(k) \neq S(k-1)$, the switch position used in evaluating the algorithm steps at k^{th} instant differs from that of used in the previous sampling period at $(k-1)^{th}$ instant. Therefore, $S(k)$ is referred to as the complementary switch position. For instance, if the position of the switch in the $(k-1)^{th}$ interval is ON, the complementary switch position is OFF and vice versa. Therefore, the total cost function computed when $S(k) \neq S(k-1)$ is referred to as the complementary cost function.

9.4.1.2 Cost Function Analysis of Conventional MPC

The value of weighting factor can be chosen for a desired switching frequency. However, since the value of cost function in (9.85) is dependent on the operating point, the switching frequency changes when the operating point of the converter is changed. Therefore, the cost function should be designed by taking into account the measured and actual system parameters. For the sake of analyzing the cost function, an evolution of the inductor current, inductor current error and switch states are given in Figure 9.30 2. The error of inductor current in Figure 9.30 is obtained by comparing the inductor current with a constant reference. It is apparent from Figure 9.30 that the typical inductor current contains a ripple (Δi_{L1}). It is well known that the current ripple in DC–DC converters is inevitable. However, the cost function in (9.83) aims to obtain zero current error in each sampling interval.

Figure 9.30 Evolution of inductor current, inductor current error, and switch states.

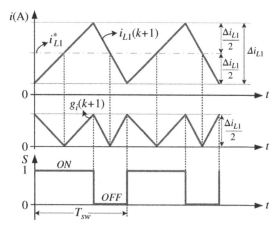

Due to this fact, the switch position is changed in each consecutive sampling step, which causes the excessive switching frequency to be half of the sampling frequency mentioned in [13]. It is evident in Figure 9.30 that an error band always exists, which avoids the excessive switching frequency. In conventional MPC, the value of the complementary cost function is increased by the weighting factor and current error. Thus, the switch position is not changed while the cost value of the actual switch position is lower than the value of complementary cost function. Therefore, an error band occurs for the current error. Since the error band depends on the current error and manually tuned the weighting factor, the inclusion of the weighting factor reduces the switching frequency only, in the conventional method. For a desired fixed switching frequency, the error band should be $\Delta i_{L1}/2$ as shown in Figure 9.30. Also, the error band is dependent on the operating parameters, such as input voltage and duty ratio. In the conventional MPC employing a constant weighting factor, the switching frequency varies when the operating point of the converter is changed.

Assuming that the reference current is constant, the absolute error of Δi_{L1} reaches zero axes two times in a switching period as seen in Figure 9.30. In addition, the error is equal to half of the Δi_{L1} when the switch position is changed from ON to OFF and vice versa. Therefore, Δi_{L1} is dependent on the switching frequency. The ripple of inductor current can be written as follows [14]:

$$\Delta i_{L1} = \frac{v_{in}d}{L_1 f_{sw}} \tag{9.87}$$

where d is duty ratio and f_{sw} is the switching frequency. By considering (9.87), the error band can be obtained for the desired switching frequency. It should be taken into account that the resolution of the switching signal is directly proportional to the sampling frequency as follows:

$$S_{res} = \frac{f_s}{f_{sw}} \tag{9.88}$$

where f_s and f_{sw} denote the sampling frequency and switching frequency, respectively. Eq. (9.88) can be considered to determine the switching frequency.

It is well known that the finite control set MPC method generates the switching signals without using a comparator. Hence, the measurement or estimation of the duty ratio is essential. For the sake of reducing the implementation complexity, the duty ratio is estimated as follows:

$$\frac{d}{1-d} = \frac{v_{out}}{v_{in}} = \frac{i_{in}}{i_{out}} \tag{9.89}$$

where i_{in} is the average of i_{L1}. Assuming that v_{out} is equal to its reference (v_{out}^*) in the steady state, the duty ratio can be written as

$$d = \frac{v_{out}^*}{v_{out}^* + v_{in}} \tag{9.90}$$

Now, substituting (9.90) into (9.87) yields the expression for the inductor current ripple in terms of switching frequency, input voltage, reference output voltage, and inductance as follows:

$$\Delta i_{L1} = \frac{v_{in}\, v_{out}^*}{L_1 f_{sw}(v_{in} + v_{out}^*)} \tag{9.91}$$

Figure 9.31 shows Δi_{L1} calculated by varying v_{in} from 30 to 60 V when $v_{out}^* = 48$ V. To investigate the influence of various load conditions on Δi_{L1}, the load resistance is varied from 10 to 40 Ω for each input voltage level. It should be noted that i_{in} is calculated by using (9.89). As can be seen from Figure 9.31, the ripple is highly dependent on the input voltage. For example, the ripple increases when the input voltage is increased and vice versa. However, it is worth noting that some deviations in Δi_{L1} may occur due to the variations of converter efficiency.

9.4.1.3 Cost Function Design of Presented MPC in [11]

The conventional cost function is restructured to reduce the switching frequency variations and eliminate the necessity of using weighting factor. As mentioned for the conventional cost function in (9.85), the weighting factor increases the cost value of the complimentary switch position. This means that until the calculated current error reaches the cost value, the position of the switch is not changed. In

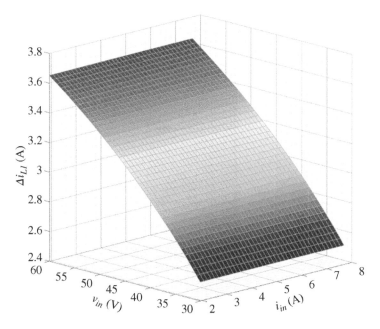

Figure 9.31 Effect of varying v_{in} and i_{in} on the inductor current ripple.

other words, the limitation can be defined as the allowed error band. To identify the allowed error band with the complementary switch position, the current error should be eliminated. Hence, the conventional cost function in (9.85) is configured to eliminate the current error for complementary switching state as follows:

$$g(k + 1) = g_i(k + 1)(1 - g_s(k)) + \lambda g_s(k) \tag{9.92}$$

As clearly seen in Eq. (9.92), only the weighting factor λ is effective for the complementary cost function. Thus, the error band can be easily determined by the weighting factor. By considering Figure 9.30, the relation between weighting factor and allowed error band can be defined as

$$\lambda = \frac{\Delta i_{L1}}{2} \tag{9.93}$$

Finally, the cost function in (9.92) can be modified by replacing λ with (9.93) as follows:

$$g(k + 1) = g_i(k + 1)(1 - g_s(k)) + \frac{\Delta i_{L1}}{2} g_s(k) \tag{9.94}$$

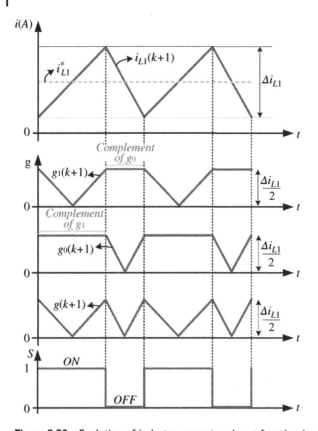

Figure 9.32 Evolution of inductor current and cost function in one switching cycle.

Figure 9.32 shows the evolution of the inductor current and cost function for ON and OFF states of the switch. While $g_1(k+1)$ denotes the cost function in the ON state, $g_0(k+1)$ is the cost function in the OFF state.

The optimum cost function is denoted by $g(k+1)$. As clearly seen from Figure 9.32, the value of complementary cost function is fixed to the allowed ripple value ($\Delta_{iL1}/2$). Until the calculated cost function reaches the error limit, the complementary cost function is not selected, and the switch position is not changed. Such an operation not only guarantees the average switching frequency control but also provides average current control with a constant current reference. Furthermore, the necessity of using reference generation or average calculation is eliminated.

Algorithm 9.2   MPC Algorithm

```
function S(k) = MPC (i_L1(k), v_in(k), v_c1(k), v_out(k), i*_L1(k))
    g_OP = ∞; S_P = [0; 1]; S_OP= Ø;
    Δi_L1 = Eq. (12)
    for j=1 to length(s_p) do
        i_L1(k+1) = Eq. (3)
        g_i(k+1) = Eq. (4)
        g_s(k) = Eq. (5)
        g(k+1) = Eq. (15)
        if g(k+1) < g_OP then
        g_OP =g(k+1);
            S_OP = j;
        end if
    end for
    S(k) = S_P(S_OP);
end function
```

The model predictive current control strategy is summarized in Algorithm 9.2. $S(k)$ is the control output, which is the switching signal of the converter. The state variables are measured at k^{th} sampling instant, and the inductor current ripple is estimated according to (9.91). The current error in (9.83) is calculated using the predicted current (9.82) for the next sampling interval. The complementary switch position is determined by Eq. (9.84), and it is used in the total cost function. The minimization part selects the minimum cost function, and it is assigned as the new optimal cost value. Furthermore, the optimum switching state is also determined by this step. The procedure is repeated for each possible switching state, and the appropriate switch position is determined by S_{OP}.

9.4.1.4   Output Voltage Control
The output voltage of the converter is controlled by using a proportional-integral (PI) regulator, which generates the current reference for i_{L1}. The voltage error can be written as follows:

$$x = v_{out}^* - v_{out} \tag{9.95}$$

The reference current is then determined for the desired reference voltage by the PI controller as follows:

$$i_{L1}^* = k_p x + k_i \int x \, dt \tag{9.96}$$

where i_{L1}^* is the reference inductor current, which is used in the MPC algorithm.

9.4.2 Experimental Results

The effectiveness of the control strategy has been investigated experimentally on a DC-DC SEPIC converter prototype. The block diagram of the control strategy is given in Figure 9.33. Experimental setup parameters are given in Table 9.4. The control algorithm is embedded into a TMS320F28379D 32-bit MCU. Current and voltages are measured by using transducers. According to the control algorithm, the control signal is generated by the MCU, and it is applied to the gate terminal of IGBT via gate driver. It is worth noting that the average switching frequency has been calculated by another microcontroller.

9.4.2.1 Switching Frequency Control Test

The performance of the control approach in controlling the average switching frequency is tested. Figure 9.34 shows the responses of v_{out}, v_{in}, i_{L1}, and f_{sw} for abrupt changes in the reference switching frequency from 10 to 5 kHz and from 5 to 10 kHz. Initially, the system is operated with 10 kHz average switching frequency.

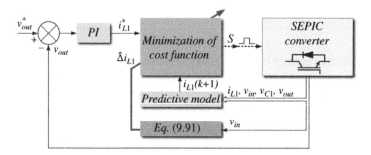

Figure 9.33 Block diagram of the control strategy. *Source:* Adapted from Guler et al. [11].

Table 9.4 System and control parameters.

Description and Symbol	Value
Input voltage: v_{in}	30 V and 60 V
Capacitors: C_1, C_2	300 μF, 1000 μF
Inductors: L_1, L_2	730 μH
Reference output voltage: v_{out}^*	48 V
Proportional and integral gains: k_p and k_i	0.25 and 100
Sampling time: T_s	10 μs
Switching frequency: f_{sw}	10 kHz
Loads: R_{L1}, R_{L2}, R_{L3}	40 Ω, 20 Ω, 10 Ω

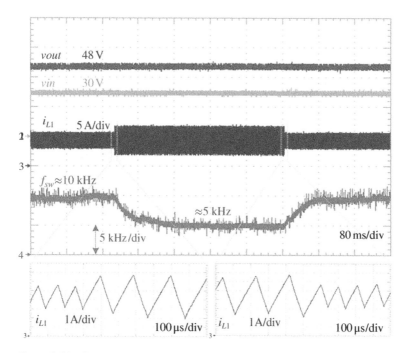

Figure 9.34 Responses of v_{out}, v_{in}, i_{L1}, and f_{sw} for abrupt changes in the reference switching frequency from 10 to 5 kHz and from 5 to 10 kHz.

When the reference frequency is changed from 10 to 5 kHz, the ripple of inductor current is increased (almost doubled) considerably as clearly discernible in Figure 9.34. Also, the average switching frequency gradually converges to 5 kHz. Similarly, when the reference switching frequency is changed from 5 to 10 kHz, the ripple of inductor current is reduced, and the average switching frequency tracks its reference after the transient period is over. It is worth noting that neither v_{out} nor v_{in} are affected by the abrupt change in the reference switching frequency.

9.4.2.2 Dynamic Response Test Under Input Voltage Variation

Figure 9.35 shows the dynamic responses of v_{out}, v_{in}, i_{L1}, and f_{sw} for changing v_{in} from 60 to 30 V and from 30 to 60 V again. The results are obtained from MPC employing a constant weighting factor (conventional MPC) and presented MPC (weighting factorless control approach). In both cases, the system is operated with $v_{in} = 60$ V and $v_{out}^* = 48$ V, which implies that the converter operates in the buck mode initially.

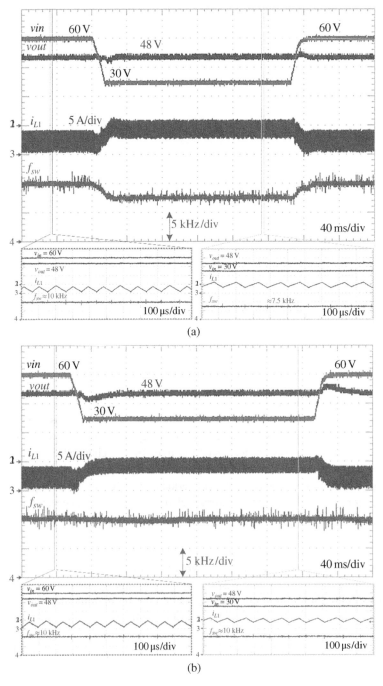

Figure 9.35 Dynamic responses of v_{out}, v_{in}, i_{L1}, and f_{sw} for changing vin from 60 to 30 V and from 30 to 60 V: (a) conventional MPC with constant weighting factor and (b) presented weighting factorless MPC.

Figure 9.35a shows the responses obtained by the conventional MPC. Initially, the average switching frequency is fixed to 10 kHz by the manually tuned constant weighting factor. When v_{in} is changed from 60 to 30 V, the operation mode of the converter is changed from buck mode to boost mode. Except for the small transition period, the output voltage is successfully regulated at its reference value. However, the switching frequency reduces to 7.5 kHz, which clearly shows that the conventional MPC with the weighting factor is not able to maintain the switching frequency at 10 kHz. When v_{in} is changed from 30 to 60 V, the average switching frequency returns to its initial value.

Figure 9.35b shows the responses obtained by the presented weighting factorless MPC. As seen from the figure, the average switching frequency at the start-up is approximately fixed to 10 kHz by the presented MPC. When v_{in} is changed from 60 to 30 V, the output voltage is regulated at its reference after 60 ms with 8% undershoot. It is obvious that the average switching frequency is maintained at 10 kHz when the converter changes its operating mode from buck mode to boost mode. Similarly, when v_{in} is changed from 30 to 60 V, the output voltage is regulated at its reference value and average switching frequency is kept at 10 kHz again. These results show that the presented weighting factorless MPC control approach not only regulates the output voltage but also controls the average switching frequency under input voltage variations.

Figure 9.36 shows responses of v_{out}, v_{in}, i_{L1}, and $\Delta i_{L1}/2$ for changing v_{in} from 60 to 30 V and from 30 to 60 V. The value of $\Delta i_{L1}/2$, which determines the allowed error band for 10 kHz switching frequency, is approximately equal to half of the

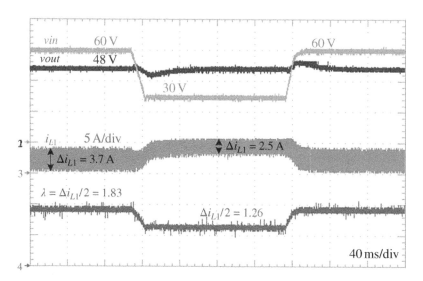

Figure 9.36 Responses of v_{out}, v_{in}, i_{L1}, and $\Delta i_{L1}/2$ for changing vin from 60 to 30 V and from 30 to 60 V.

inductor current ripple given in (9.93). These results verify the correctness of the theoretical considerations.

9.4.2.3 Dynamic Response Test Under Load Change

Figure 9.37 shows the dynamic responses of v_{out}, v_{in}, i_{L1}, and i_{out} against abrupt load changes in the boost and buck modes. In Figure 9.37a, the converter is

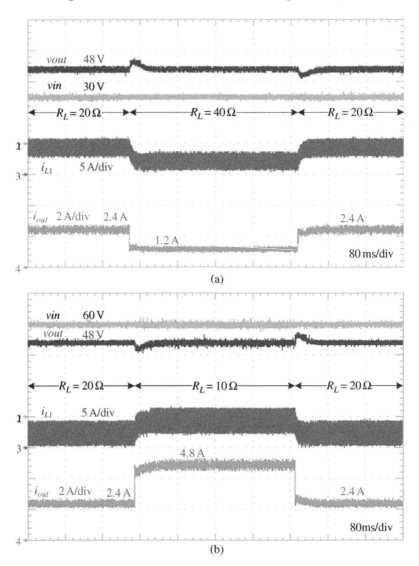

(a)

(b)

Figure 9.37 Dynamic responses of v_{out}, v_{in}, i_{L1}, and i_{out} against abrupt load changes: (a) boost mode and (b) buck mode.

operated with $v_{in} = 30$ V and $v_{out}^* = 48$ V in the boost mode. Initially, the load resistance is 20 Ω and i_{out} is 2.4 A. When the load resistance is changed from 20 to 40 Ω, the output voltage exhibits a small overshoot and is regulated at 48 V in 40 ms. During this interval, the output current is decreased from 2.4 to 1.2 A. Then, the load resistance is changed from 40 to 20 Ω and i_{out} returns to the initial condition. Then, the operation of system is also tested in the buck mode with $v_{in} = 60$ V, $v_{out}^* = 48$ V and the results are given in Figure 9.37b. In this case, the output current is maintained at 2.4 A. The dynamic response test is realized by changing the load resistance from 20 to 10 Ω. Similar to the boost mode, the output voltage exhibits a small overshoot and is regulated at 48 V. However, the output current changes its value from 2.4 to 4.8 A during this load change. Finally, the load resistance is changed from 10 to 20 Ω again.

For the sake of demonstrating the response of average switching frequency, the dynamic response test in Figure 9.37 is repeated under the same operating conditions, and obtained results are presented in Figure 9.38. Initially, the converter operates in the boost mode under 20 Ω load resistance, with an average switching frequency equal to 10 kHz. When the load resistance is changed from 20 to 40 Ω, the switching frequency increases to 10.3 kHz. Then the load resistance is changed from 40 to 20 Ω, and the switching frequency returned to 10 kHz. Figure 9.38b shows the response of average switching frequency in the buck mode. Initially, $v_{in} = 60$ V, the load resistance is 20 Ω, and the average switching frequency is 10 kHz. Next, the load resistance is reduced from 20to 10 Ω, and the average switching frequency is reduced to 9 kHz. Finally, the load resistance is changed from 10 to 20 Ω, which results in a change in the switching frequency from 9 to 10 kHz. As mentioned in the theoretical analysis, Δi_{L1} is estimated based on input and reference voltages with ideal power conditions. However, efficiency may change depending on the operating point of the converter. Therefore, the averaged switching frequency has changed in the load step test.

9.4.2.4 Influence of Parameter Mismatch

In general, the MPC method selects the appropriate switch position as result of the model-based prediction. Therefore, the system parameters are the most important factor in determining the accuracy of the controller [14]. The system parameters are subject to variations due to the operating conditions, temperature, age, and tolerances. Since the prediction of i_{L1} at $(k + 1)^{th}$ sampling instant depends on L_1, the influence of variations in L_1 on the performance of the controller should be investigated.

Figure 9.39 shows the steady-state responses of v_{out}, v_{in}, i_{L1}, and f_{sw} obtained by the conventional MPC and presented MPC methods in the boost mode when the value of L_1 in the prediction equation deviates from its nominal value by +10%. It is

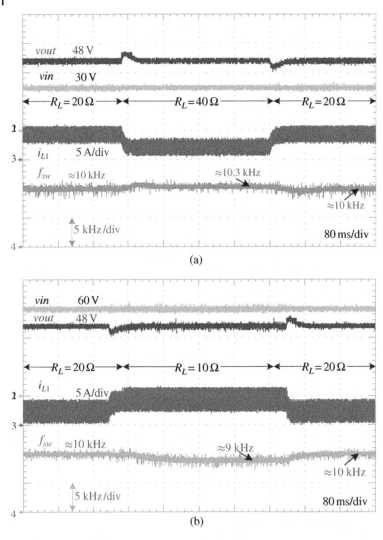

Figure 9.38 Switching frequency results for load change test: (a) boost mode and (b) buck mode.

evident that the output voltage is not affected from this parameter mismatch in both methods. However, while the average switching frequency is decreased from 10 to 9.3 kHz in the conventional MPC, it is slightly increased from 10 to 10.5 kHz in the presented MPC method. This opposite behavior is mainly caused by the design of cost function in these methods. In the conventional MPC, the predicted current takes larger values than its actual values, and it causes to expand the error

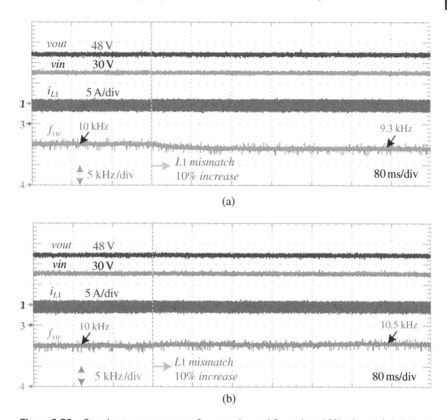

Figure 9.39 Steady-state responses of v_{out}, v_{in}, i_{L1}, and f_{sw} under +10% mismatch in L_1 in the boost mode obtained by: (a) conventional MPC, (b) presented MPC.

band, which is mentioned in Figure 9.30. Therefore, the average switching frequency is reduced in the conventional MPC method.

In the presented MPC method, L_1 is used in predicting i_{L1} and estimating Δi_{L1}. Like the traditional MPC, the parameter mismatch causes an increase of predicted current value. However, the estimated Δi_{L1} is lower than its actual value, which narrows the allowed error leading to a rise in the average switching frequency.

The influence of variation in L_1 is also investigated in the buck mode. Figure 9.40 shows the steady-state responses under −10% mismatch in L_1 obtained by conventional and presented MPC methods. While the average switching frequency is increased to 10.8 kHz in the conventional MPC, it is decreased to 9.8 kHz in the presented MPC. As mentioned before, the reason for such behavior comes from the design of cost functions.

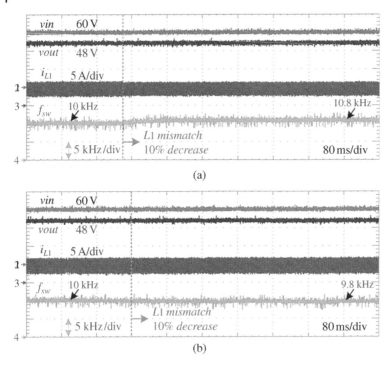

Figure 9.40 Steady-state responses of v_{out}, v_{in}, i_{L1}, and f_{sw} under −10% mismatch in L_1 in the buck mode obtained by: (a) conventional MPC, (b) presented MPC.

9.5 Model Predictive Droop Control of Distributed Generation Inverters in Islanded AC Microgrid

In general, the distributed generation (DG) inverters can operate in parallel in islanded microgrids through droop characteristic-based control technique. This control technique is suitable for long-distance DG inverters to achieve proper power sharing while ensuring good voltage regulation. On the other hand, the traditional droop control shows some disadvantages, such as the requirement for multi-loop feedback control, which results in complex controller design and slow dynamic response. To overcome this drawback, model predictive-based droop control technique is presented in this section. To verify the control technique performance, simulation studies are performed with MATLAB/Simulink. The control technique presented in this subsection is based on the controller proposed in [14].

9.5.1 Conventional Droop Control

The equivalent circuit of n voltage source inverters (VSI)s connected to common AC microgrid bus is shown in Figure 9.41. It can be seen that each DG inverter can

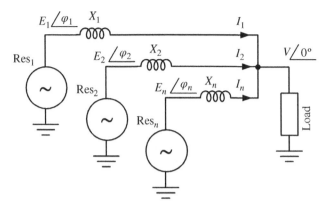

Figure 9.41 Equivalent circuit of DG inverters in Islanding AC microgrid.

be represented by synchronous generator equivalent circuit. The basic idea of the droop control technique is to imitate the behavior of a synchronous generator whose frequency is reduced as the active power increases. When the inverter output impedance is highly inductive, the active and reactive power of nth converter connected to the AC microgrid can be defined as

$$\left.\begin{aligned} P_n &= \frac{VE_n}{X_n} \sin \varphi_n \\ Q_n &= \frac{VE_n \cos \varphi_n - V^2}{X_n} \end{aligned}\right\} \tag{9.97}$$

where E_n is the inverter output voltage, V is the point of common coupling (PCC) voltage, X_n is the inverter output reactance, ϕ_n is the phase angle between the inverter output voltage and PCC voltage. It is clear that the reactive power depends on the E_n, whereas the active power is dependent on the ϕ_n. Using this information, the P/ω and Q/E droop characteristics can be drawn, as shown in Figure 9.42

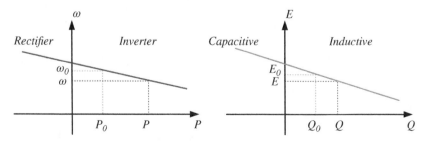

Figure 9.42 P/ω and Q/E droop characteristics.

[15]. These characteristics can be explicated as follows: when the frequency decreases from ω_0 to ω, the DG is allowed to increase its active power from P_0 to P. A falling frequency is evidence of an increase in the active power demand of the system. In other words, parallel connected units with the same droop characteristic increase their active power outputs to handle the fall in frequency. Increasing active power of the parallel units will prevent the fall in frequency. Thus, the units will settle at active power outputs and frequency at a steady-state point on the droop characteristic. To sum up, the droop characteristic-based control technique allows parallel connected units to share load without the units affecting each other. This idea can also be extended to explain the voltage droop characteristic.

Although this principle is common in synchronous generator control, it can be integrated in voltage source inverters (VSIs) by using the well-known P/Q droop method, which can be expressed as

$$\left.\begin{aligned} \omega_n &= \omega_0 - k_P(P_n - P_0) \\ E_n &= E_0 - k_Q(Q_n - Q_0) \end{aligned}\right\} \tag{9.98}$$

where n is the index representing each VSIs, ω_0 and E_0 are the base frequency and voltage, respectively, P_n and Q_n are the active and reactive power of the units, P_0 and Q_0 are the base active and reactive power, respectively, and k_P and k_Q are the active and reactive proportional droop coefficients, respectively.

The choice of k_P and k_Q impacts the network stability, so they must be carefully and appropriately designed. Usually, the droops are coordinated to make each DG system supply apparent power proportional to its capacity. The active and reactive proportional droop coefficients (k_P and k_Q) can be expressed by

$$\left.\begin{aligned} k_P &= \frac{\Delta\omega_{\max}}{P_{\max}} \\ k_Q &= \frac{\Delta E_{\max}}{Q_{\max}} \end{aligned}\right\} \tag{9.99}$$

where, $\Delta\omega_{\max}$ is the maximum allowed voltage frequency droop; $\Delta E_{\max}$ is the maximum allowed voltage amplitude droop; $P_{\max}$ is the maximum allowed active power; and $Q_{\max}$ is the maximum allowed reactive power. The control algorithm with conventional droop control is illustrated in Figure 9.43. The power stage consists of input power source such as PV panel, DC to AC converter, and LC filter with line inductor. The control structure consists of three control loops: (i) power sharing loop; (ii) voltage control loop; and (iii) current control loop.

- To obtain the amplitude and frequency of the reference voltage according to the droop characteristic, a power sharing loop is needed to be used.

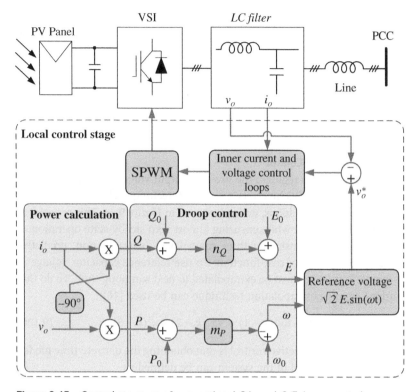

Figure 9.43 Control structure of conventional P/ω and Q/E droop control.

- To control output voltage of the LC filter, a voltage controller is needed to be used.
- To control output current, a voltage controller is needed to be used.

9.5.2 Control Technique

Figure 9.44 depicts the block diagram of the predictive droop control technique for DG inverter. It consists of two main layer: droop control and MPC.

9.5.2.1 Reference Voltage Generation Through Droop Control

To ensure power sharing in islanded AC microgrid, the capacitor reference voltage $(v_c(k))$ for the individual DG inverter must be defined. To do that, traditional droop control, which is given in Section 9.5.1, is used.

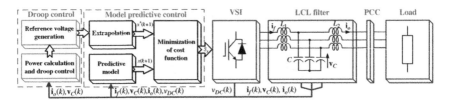

Figure 9.44 The MPC based droop control algorithm.

9.5.2.2 Model Predictive Control

MPC technique is used to regulate the inverter output voltage in order to ensure power sharing among the DG inverters. Using MPC technique eliminates cascaded current and voltage control loops, which is shown in Figure 9.43. This results in a very fast dynamic response while ensuring almost zero steady-state operation. It can be seen that MPC consists of three main layers: extrapolation, predictive model, and minimization of cost function. To use reference capacitor voltage in the cost function, the $V*c$ must be extrapolated to next sampling time. To do that fourth-order Lagrange extrapolation technique can be used [16]:

$$x^*(k+1) = 4x^*(k) - 6x^*(k-1) + 4x^*(k-2) - x^*(k-3) \tag{9.100}$$

The next step in the predictive control is that obtaining the discrete-time model. The continuous-time model of the system can be defined as

$$\mathbf{i}_f = \frac{2}{3}\left(i_{fa} + ai_{fb} + a^2 i_{fc}\right) \tag{9.101}$$

$$\mathbf{v}_C = \frac{2}{3}\left(v_{Ca} + av_{Cb} + a^2 v_{Cc}\right) \tag{9.102}$$

$$\mathbf{i}_o = \frac{2}{3}\left(i_{oa} + ai_{ob} + a^2 i_{oc}\right) \tag{9.103}$$

where $\mathbf{i}_f$ is the filter current, $\mathbf{v}_C$ is the output voltage, and $\mathbf{i}_o$ is the output current. The continuous-time model of LC filter voltage and current can be expressed by

$$\left.\begin{array}{l} L\dfrac{d\mathbf{i}_f}{dt} = \mathbf{v}_i - \mathbf{v}_C \\[2mm] C\dfrac{d\mathbf{v}_C}{dt} = \mathbf{i}_f - \mathbf{i}_o \end{array}\right\} \tag{9.104}$$

where L is the inductance and C is the capacitance of the LC filter.

These equations can be represented as state-space system as

$$\frac{d\mathbf{x}}{dt} = \mathbf{A}\mathbf{x} + \mathbf{B}\mathbf{v}_i + \mathbf{B_d}\mathbf{i}_o \tag{9.105}$$

where

$$
\left.\begin{array}{ll}
\mathbf{x} = \begin{bmatrix} \mathbf{i}_f \\ \mathbf{v}_C \end{bmatrix} & \mathbf{A} = \begin{bmatrix} 0 & -1/L \\ 1/C & 0 \end{bmatrix} \\[18pt]
\mathbf{B} = \begin{bmatrix} 1/L \\ 0 \end{bmatrix} & \mathbf{B}_d = \begin{bmatrix} 0 \\ -1/C \end{bmatrix}
\end{array}\right\}
\tag{9.106}
$$

To use mathematical equations in the prediction model, discrete-time model of the system must be obtained from continuous-time equations. To do that, the general structure of forward-difference Euler equation can be used in order to compute the differential equations [16]. The system discrete-time model can be derived from (9.105) and it can be expressed as

$$
\mathbf{x}(k+1) = \mathbf{A}_q \mathbf{x}(k) + \mathbf{B}_q \mathbf{v}_i(k) + \mathbf{B}_{dq} \mathbf{i}_o(k)
\tag{9.107}
$$

where

$$
\left.\begin{array}{l}
\mathbf{A}_q = e^{\mathbf{A}T_s} \\[10pt]
\mathbf{B}_q = \displaystyle\int_0^{T_s} e^{\mathbf{A}\tau} \mathbf{B} d\tau \\[16pt]
\mathbf{B}_{dq} = \displaystyle\int_0^{T_s} e^{\mathbf{A}\tau} \mathbf{B}_d d\tau
\end{array}\right\}
\tag{9.108}
$$

where T_s is the sampling time.

These discrete-time equations are utilized in the prediction model to predict future behavior of the system. In the prediction model, the actual value of $\mathbf{v}_C(k)$, $\mathbf{i}_f(k)$, and $\mathbf{i}_o(k)$ are used to obtain the output voltage at the next sampling time $\mathbf{v}_C(k+1)$.

Since traditional three-phase VSI has seven different voltage vector, the cost function g compares the seven predicted voltage vectors $\mathbf{v}_C(k+1)$ with the extrapolated reference voltage vectors $\mathbf{v}_C(k+1)$ to obtain the optimal value vector $\mathbf{v}_i$, which should be applied to the inverter. In other words, the $\mathbf{v}_i$ is the optimal voltage vector that minimizes the error between the reference and predicted output voltage at the next sampling time.

The cost function is defined as

$$
g = \left(v_{C\alpha}^* - v_{C\alpha}\right)^2 + \left(v_{C\beta}^* - v_{C\beta}\right)^2
\tag{9.109}
$$

where $v_{C\alpha}^*$ and $v_{C\beta}^*$ are the real and imaginary parts of the output-voltage reference vector $\mathbf{v}_C^*$, while $v_{C\alpha}$ and $v_{C\beta}$ are the real and imaginary parts of the predicted output-voltage vector $v_c(k + 1)$. This cost function has been chosen to obtain the lowest voltage error.

9.5.3 Simulation Results

Figure 9.45 shows a single-line diagram of the microgrid system model, which is connected to a 380 V, 50-Hz point of common coupling (PCC). The test model contains two DGs with voltage ratings of 380 V, and maximum power generation limits (arbitrarily chosen to be 5 and 7.5 kVA, respectively) are included in the simulations. Table 9.5 summarizes the parameters used in the simulation model.

Firstly, the power sharing ratio between VSI1 and VSI2 is set to 1 : 1. The load that is connected to the PCC is 10 kW and 1 KVAR. Figure 9.46a–f show the simulation results of this case. The currents provided by the inverters are equal, as shown in Figure 9.46a,b, while the load current is the sum of the DG1 and

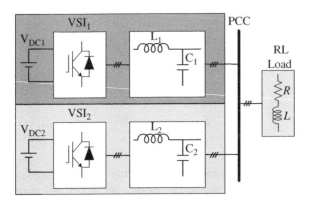

Figure 9.45 Block diagram of the simulated model.

Table 9.5 Simulation model parameters.

DG 1	
Nominal power, S_1	5 kVA
Inductance, L_1	10 mH
Capacitance, C_1	20 μF
DG 2	
Nominal power, S_1	7.5 kVA
Inductance, L_1	10 mH
Capacitance, C_1	20 μF

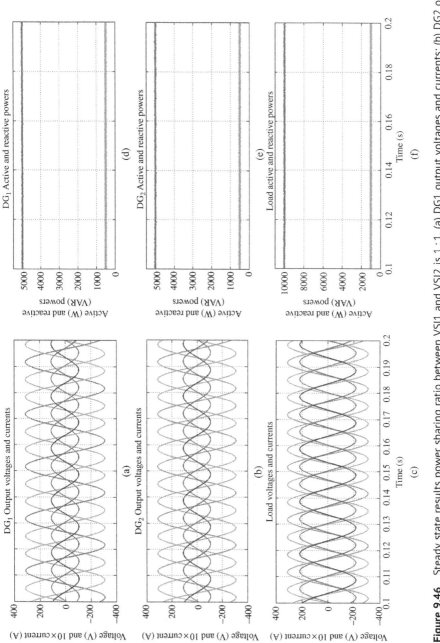

Figure 9.46 Steady state results power sharing ratio between VSI1 and VSI2 is 1 : 1. (a) DG1 output voltages and currents; (b) DG2 output voltages and currents; (c) Load output voltages and currents; (d) DG1 active and reactive powers; (e) DG2 active and reactive powers. (f) Load active and reactive powers.

DG2 output currents, which is depicted in Figure 9.46c. Figure 9.46d–f show the active and reactive powers of DG1, DG2, and load, respectively. It can be seen that the generated active and reactive powers of the DGs are equal and the load active and reactive powers are the sum of the DG1 and DG2 active and reactive powers.

Secondly, the power sharing ratio between VSI1 and VSI2 is set to 2 : 3. The load which is connected to the PCC is 12.5 kW and 0 KVAR in this case. Figure 9.47a–f show the simulation results of this case. The ratio of inverter currents is 2 : 3, as shown in Figure 9.47a,b, while the load current is the sum of the DG1 and DG2 output currents, as shown in Figure 9.47c. Furthermore, the generated active power in DG1 is 5 kW, whereas in DG2 is 7.5 kW, which is depicted in Figure 9.47d and e, respectively. Please note that the load active power is the sum of the DG1 and DG2 generated active powers.

To test the load impact on the MPC based droop control technique, the load connected to the PCC is increased from 5 to 10 kW. The power sharing ratio between VSI1 and VSI2 is set to 1 : 1. Figure 9.48a–f show the simulation results of this case. At the beginning, DGs provide the same amount of active power (2.5 kW) to the load. The load is stepped up from 5 to 10 kW at 0.15 seconds. After this point, the DGs share the total load (10 kW) and take 5 kW each. Negligible small deterioration is notice on the voltage and current signals at the load change instant. The result shows that the technique performs well against load change.

9.6 FCS-MPC for a Three-Phase Shunt Active Power Filter

In this chapter, the finite control set model predictive control is combined with the vector operation technique to be applied in the control of a three-phase active power filter. The control technique presented in this subsection is based on control technique proposed in [17]. Typically, in the finite control set technique applied to three-phase power converters, eight different vectors are considered in order to obtain the optimum control signal by minimizing a cost function. On the other hand, the vector operation technique is based on dividing the grid voltage period into six different regions. The main advantage of combining both techniques is that for each region the number of possible voltage vectors to be considered can be reduced to a half, thus reducing the computational load employed by the control algorithm. Besides, in each region, only two phase-legs are switching at high frequency while the remaining phase-leg is maintained to a constant DC-voltage value during this interval. Accordingly, a reduction of the switching losses is obtained. Unlike the typical model predictive control methods which make use of the discrete differential equations of the converter, this method considers a

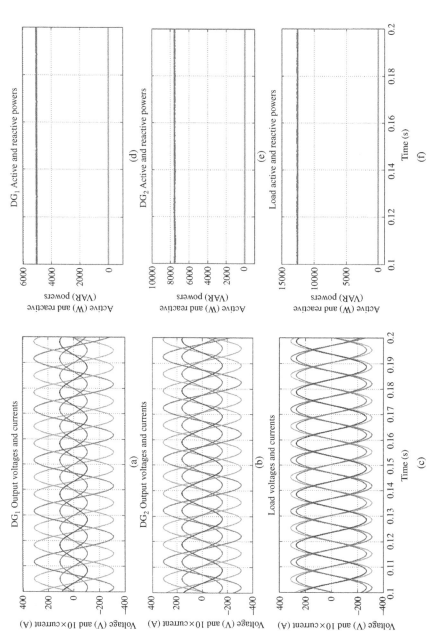

Figure 9.47 Steady state results power sharing ratio between VSI1 and VSI2 is 2 : 3. (a) DG1 output voltages and currents; (b) DG2 output voltages and currents; (c) Load output voltages and currents; (d) DG1 active and reactive power; (e) DG2 active and reactive powers; (f) Load active and reactive powers.

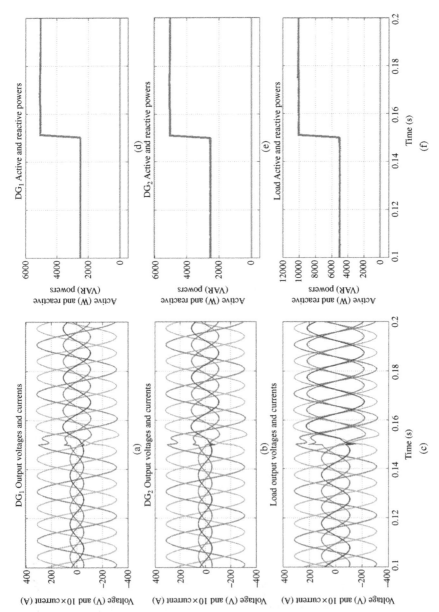

Figure 9.48 Load impact with power sharing ratio between VSI1 and VSI2 is 1 : 1. (a) DG1 output voltages and currents; (b) DG2 output voltages and currents; (c) Load output voltages and currents; (d) DG1 active and reactive powers; (e) DG2 active and reactive power; (f) Load active and reactive powers.

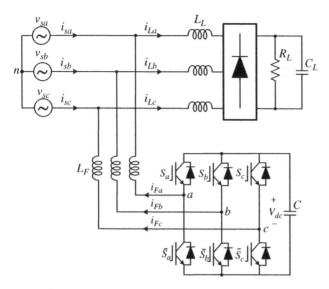

Figure 9.49 Three-phase shunt active power filter.

Kalman filter (KF) in order to improve the behavior of the closed-loop system in noisy environments. Selected experimental results are exposed in order the demonstrate the validity of the control proposal.

9.6.1 System Modeling

The three-phase SAPF circuit is shown in Figure 9.49. From this circuit, the following equations for each phase-leg can be defined as follows:

$$L_F \frac{di_{Fa}}{dt} = v_{aN} - v_{nN} - v_{sa} \tag{9.110}$$

$$L_F \frac{di_{Fb}}{dt} = v_{bN} - v_{nN} - v_{sb} \tag{9.111}$$

$$L_F \frac{di_{Fc}}{dt} = v_{cN} - v_{nN} - v_{sc} \tag{9.112}$$

A space phasor can be defined as $\vec{f}(t) = \frac{2}{3}\left(f_a(t) + af_b(t) + a^2f_c(t)\right)$ where $f_a(t)$, $f_b(t)$ and $f_c(t)$ can be three-phase voltages or currents indistinctly and $a = e^{j2\pi/3}$. On applying this concept to Eqs. (9.110)–(9.112), one has

$$L_F \frac{d}{dt} \frac{2}{3}\left(i_{Fa} + ai_{Fb} + a^2i_{Fc}\right) = \frac{2}{3}\left(v_{aN} + av_{bN} + a^2v_{cN}\right) - \frac{2}{3}\left(1 + a + a^2\right)$$
$$+ \frac{2}{3}\left(v_{sa} + av_{sb} + a^2v_{sc}\right)$$

$$\tag{9.113}$$

It is easy to find that $1 + a + a^2 = 0$, then the aforementioned equation can be rewritten as follows:

$$L_F \frac{d\mathbf{i}_F}{dt} = \mathbf{v}_{apf} - \mathbf{v}_s \tag{9.114}$$

where

$$\mathbf{i}_F = \frac{2}{3}\left(i_{Fa} + ai_{Fb} + a^2 i_{Fc}\right) \tag{9.115}$$

$$\mathbf{v}_s = \frac{2}{3}\left(v_{sa} + av_{sb} + a^2 v_{sc}\right) \tag{9.116}$$

$$\mathbf{v}_{apf} = \frac{2}{3}v_{dc}\left(S_a + aS_b + a^2 S_c\right) \tag{9.117}$$

being S_i the switching states such that $S_i \in \{0, 1\}$.

Equation (9.56) can be separated into the real and imaginary parts corresponding to the α and β components:

$$L_F \frac{di_{Fa}}{dt} = v_{apf\alpha} - v_{s\alpha} \tag{9.118}$$

$$L_F \frac{di_{F\beta}}{dt} = v_{apf\beta} - v_{s\beta} \tag{9.119}$$

In the FCS-MPC, a discrete converter model is required in order to predict the N_p current samples ahead. Usually, the predicted current is calculated from the discrete differential equation. Unlike other works of the same topic, in this chapter a KF will be used to estimate the current and the voltages at the point of common coupling (PCC). Then, to achieve this objective, the PCC voltage is added to the model as a state variable. As $v_{s\alpha}$ and $v_{s\beta}$ are signals in quadrature, the following equations can be written:

$$\frac{dv_{s\alpha}}{dt} = -\omega_0 v_{s\beta} \tag{9.120}$$

$$\frac{dv_{s\beta}}{dt} = \omega_0 v_{s\alpha} \tag{9.121}$$

where ω_o is the grid angular frequency.

From (9.118)–(9.121) the following state-space model is obtained:

$$\frac{d\mathbf{x}}{dt} = \mathbf{A}_c\mathbf{x} + \mathbf{B}_c\mathbf{u} \tag{9.122}$$

$$y(t) = \mathbf{C}\mathbf{x}(t) \tag{9.123}$$

where

$$\mathbf{A}_c = \begin{pmatrix} 0 & 0 & 1/L_F & 0 \\ 0 & 0 & 0 & 1/L_F \\ 0 & 0 & 0 & -\omega_o \\ 0 & 0 & \omega_o & 0 \end{pmatrix} \tag{9.124}$$

$$\mathbf{B}_c = \begin{pmatrix} -1/L_F & 0 \\ 0 & -1/L_F \\ 0 & 0 \\ 0 & 0 \end{pmatrix} \tag{9.125}$$

$$\mathbf{x} = \begin{pmatrix} i_{F\alpha} & i_{F\beta} & v_{s\alpha} & v_{s\beta} \end{pmatrix}^T \tag{9.126}$$

$$\mathbf{u} = \begin{pmatrix} v_{apf\alpha} & v_{apf\beta} \end{pmatrix}^T \tag{9.127}$$

$$\mathbf{C} = \begin{pmatrix} 1 & 0 & 0 & 0 \\ 0 & 1 & 0 & 0 \end{pmatrix} \tag{9.128}$$

This state-space model can be discretized using the first order approximation. Then, considering a sampling frequency T_s the following discrete state-space model is obtained:

$$\mathbf{x}(k+1) = \mathbf{A}\mathbf{x}(k) + \mathbf{B}\mathbf{u}(k) + \mathbf{\eta}(k) \tag{9.129}$$

$$y(k) = \mathbf{C}\mathbf{x}(k) + \mathbf{w}(k) \tag{9.130}$$

where according to the first order approximation it holds that

$$\mathbf{A} = \mathbf{I} + \mathbf{A}_c T_s \tag{9.131}$$

$$\mathbf{B} = \mathbf{B}_c T_s \tag{9.132}$$

Being $\mathbf{I}$ the identity matrix.

In the discrete model it has been considered the process and measurement noise vectors defined by $\mathbf{\eta}$ and $\mathbf{w}$, respectively. According to these noise vectors, the process and measurement noise covariances matrices can be defined:

$$\mathbf{R}(k) = E\{\mathbf{w}(k)\mathbf{w}^T(k)\} \tag{9.133}$$

$$\mathbf{Q}(k) = E\{\mathbf{\eta}(k)\mathbf{\eta}^T(k)\} \tag{9.134}$$

Those quantities will be used in a KF to estimate the states as will be shown later.

9.6.2 Control Technique

The control diagram is depicted in Figure 9.50. A KF is used in order to estimate the states variables of one sample in advance. The PI controller regulates the filter output voltage and also is used to obtain the value of the gain k in order to calculate the reference currents. This gain and the state estimation are the inputs of a cost

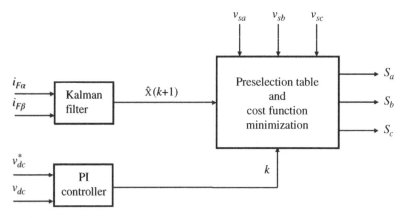

Figure 9.50 The block diagram of the control technique.

function. First, a previous vector pre-selection task is done and only four possible vectors will be considered. This preselection is done according to Table 9.7. Then, for each one of these selected vectors, the error between the predicted grid current, and its reference is computed in order to find the optimum one that minimizes the error. Once the optimum is obtained, the switching state associated to this vector is applied to the converter. In the next sections, this control algorithm is explained in detail.

9.6.3 FCS-MPC with Reduced States

The FCS-MPC is based on the predictive control technique where the system under control has a finite number of switching states. The goal is to obtain the optimum switching state from a cost function, which minimizes the error between a predicted state variable and its reference. According to the power converter shown in Figure 9.49, eight switching states should be considered. However, if the grid voltage is divided in six regions of 60°, the number of possible switching states can be reduced to four, as it will be shown later. As a consequence, the computational load of the control algorithm will be clearly reduced.

9.6.3.1 Vector Selection Based on Vector Operation

Equation 9.117 shows that different control switching states result in eight different voltage vectors which are represented in Figure 9.51. Accordingly, the voltage vector $\mathbf{v}_{apf}$ can take eight different values for each switching state.

Each vector has a real and imaginary part such as $\mathrm{Re}\{\mathbf{v}_{apf}\} = v_{apf\alpha}$, and $\mathrm{Im}\{\mathbf{v}_{apf}\} = v_{apf\beta}$. In Table 9.6, the SAPF voltage $\mathbf{v}_{apf}$ is represented according to each

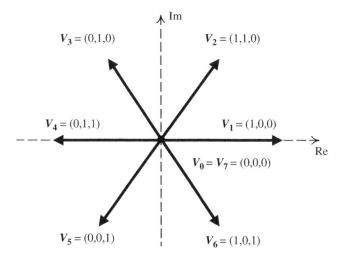

Figure 9.51 Voltage vectors.

Table 9.6 Switching states and voltage vectors.

S_a	S_b	S_c	v_{apf}
0	0	0	$\mathbf{V}_0 = 0$
1	0	0	$\mathbf{V}_1 = \dfrac{2}{3}v_{dc}$
1	1	0	$\mathbf{V}_2 = \dfrac{1}{3}v_{dc} + j\dfrac{\sqrt{3}}{3}v_{dc}$
0	1	0	$\mathbf{V}_3 = -\dfrac{1}{3}v_{dc} + j\dfrac{\sqrt{3}}{3}v_{dc}$
0	1	1	$\mathbf{V}_4 = -\dfrac{3}{3}v_{dc}$
0	0	1	$\mathbf{V}_5 = -\dfrac{1}{3}v_{dc} - j\dfrac{\sqrt{3}}{3}v_{dc}$
1	0	1	$\mathbf{V}_6 = \dfrac{1}{3}v_{dc} - j\dfrac{\sqrt{3}}{3}v_{dc}$
1	1	1	$\mathbf{V}_7 = 0$

switching state. According to the sign of the grid voltages, the grid period can be divided into six 60° regions. Figure 9.52 shows one period of the three-phase grid voltages. As it can be seen, only two grid voltages have the same sign (positive or negative) in each region, and their corresponding phase-legs are selected as high

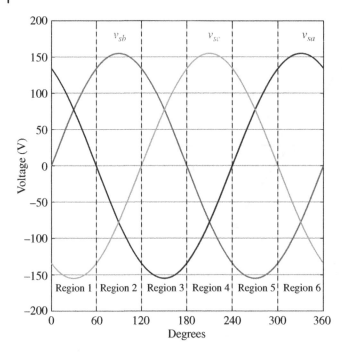

Figure 9.52 Three-phase grid voltages divided into six regions of 60°.

frequency legs. Note that, the remaining phase-leg is maintained to a constant DC voltage, vdc or −vdc, which is determined according to the vector operation technique [18], as follows:

- When the sign of two grid voltages is positive, the top switch of the remaining phase-leg is set to OFF state while the bottom switch is in ON state during this interval.
- When the sign of two grid voltages is negative, the top switch of the remaining phase-leg is set to ON state while the bottom switch is in OFF state during this interval.

With this algorithm, the eight possible vectors are reduced to only four, in each 60° region (Figure 9.52). Therefore, only four vectors must be considered in the computation of the optimum voltage vector which will be applied to the SAPF. Then, looking at Figure 9.52, and according to Table 9.6, the algorithm can be summarized as shown in Table 9.7, where 1 is related to v_{dc} and 0 is related to $-v_{dc}$ and the signs + and − are related to the positive or negative sign of the grid voltages and the symbol × indicates the high frequency phase-legs. As shown in the Table 9.7, the combination of the FCSMPC with the vector operation technique

Table 9.7 Pre-selection of voltage vectors for each region.

Region	v_{sa}	v_{sb}	v_{sc}	S_a	S_b	S_c	v_{apf}
0°–60°	+	+	−	×	×	0	$\mathbf{V}_0, \mathbf{V}_1, \mathbf{V}_2, \mathbf{V}_3$
60°–120°	−	+	−	×	1	×	$\mathbf{V}_2, \mathbf{V}_3, \mathbf{V}_4, \mathbf{V}_7$
120°–180°	−	+	+	0	×	×	$\mathbf{V}_0, \mathbf{V}_3, \mathbf{V}_4, \mathbf{V}_5$
180°–240°	−	−	+	×	×	1	$\mathbf{V}_4, \mathbf{V}_5, \mathbf{V}_6, \mathbf{V}_7$
240°–300°	+	−	+	×	0	×	$\mathbf{V}_0, \mathbf{V}_1, \mathbf{V}_5, \mathbf{V}_7$
300°-360°	+	−	−	1	×	×	$\mathbf{V}_1, \mathbf{V}_2, \mathbf{V}_6, \mathbf{V}_7$

leads to a reduction of a half of the number of voltage vectors to be considered in the optimization stage, and as a consequence, to a reduction of the computational time.

9.6.3.2 Cost Function Minimization Procedure

As mentioned before, a cost function is used to select the adequate converter voltage vector which satisfies the minimum error between the grid current and its reference. Then, the switching state which provides the optimal voltage vector is applied to control the SAPF. Let us define the following cost function in ab frame:

$$g = \left| \hat{i}^*_{s\alpha} - \hat{i}_{s\alpha} \right| + \left| \hat{i}^*_{s\beta} - \hat{i}_{s\beta} \right| \tag{9.135}$$

where $\hat{i}^*_{s\alpha} = k\hat{v}_{s\alpha}$ and $\hat{i}^*_{s\beta} = k\hat{v}_{s\beta}$ are the reference currents obtained from the estimated PCC voltages, $\hat{i}_{s\alpha} = \hat{i}_{F\alpha} + i_{L\alpha}$ and $\hat{i}_{s\beta} = \hat{i}_{F\beta} + i_{L\beta}$ are the estimated grid currents and k is a gain obtained from a Proportional-Integra (PI) controller expressed as

$$k = k_p \left(v^*_{dc} - v_{dc} \right) + k_i \int_0^t \left(v^*_{dc} - v_{dc} \right) d\tau \tag{9.136}$$

with k_p and k_i the proportional and integral gains, respectively.

The objective of the controller is to find the voltage vector that minimizes the cost function (9.135). Then, for each region, the cost function is evaluated using the four possible control vectors. The vector which minimizes the function is the vector that must applied to the SAPF.

9.6.3.3 Kalman Filter

In order to reduce the noise of the close loop system, Kalman filter is considered in this section as an alternative to other prediction methods [18]. A prediction of the filter currents and the PCC voltages are obtained in order to be used in a cost function presented in the next section. A modification in the KF algorithm is adopted in order to compensate the delay between the control process and the sampling instant [19]. The main idea consist on substituting the measurement y (k) by $y(k+1)$ in (9.130). For a better understanding, a brief explanation of the KF algorithm is presented below. However, in [19], a block diagram of the modified KF algorithm is presented in order to understand clearly its implementation.

The KF can be considered as an optimum state observer in presence of noise. Then, the main idea consists of obtaining the best estimation of the states but eliminating the effect of the noise. For this purpose, the Kalman gain is computed according to the noise covariance matrix in order to minimize the mean square error between the measured values of the states and the predicted ones. The Kalman gain expression is as follows:

$$\mathbf{L}(k) = \mathbf{P}^-(k)\mathbf{C}^T\left(\mathbf{C}\mathbf{P}^-(k)\mathbf{C}^T + \mathbf{R}(k)\right)^{-1} \tag{9.137}$$

where $\mathbf{P}^-(k)$ is the a priori error covariance and $\mathbf{R}(k)$ is given by (9.133). The implementation of the traditional KF algorithm has two parts: (i) measurement update and (ii) time update. The recursive steps are defined as follows:

First, in the measurement update step, the measurements and the error covariance matrix are computed as follows:

$$\hat{\mathbf{x}}(k) = \hat{\mathbf{x}}^-(k) + \mathbf{L}(k)(y(k) - \mathbf{C}\hat{\mathbf{x}}^-(k)) \tag{9.138}$$

$$\mathbf{P}(k) = (\mathbf{I} - \mathbf{L}(k)\mathbf{C})\mathbf{P}^-(k) \tag{9.139}$$

where $\mathbf{I}$ is the identity matrix and $\mathbf{L}(k)$ is the Kalman gain defined in (9.137).

In the next step, the time update step, a new prediction of the sates and the error covariance matrix is done, that is

$$\hat{\mathbf{x}}^-(k+1) = \mathbf{A}\hat{\mathbf{x}}(k) + \mathbf{B}u(k) \tag{9.140}$$

$$\mathbf{P}^-(k+1) = \mathbf{A}\mathbf{P}(k)\mathbf{A}^T + \mathbf{Q}(k) \tag{9.141}$$

Note that in this procedure, the algorithm is estimating the states at time k, since the measurement, $y(k)$, is obtained also at time k. Therefore, in order to overcome the system delay between the sampling time and the control process, a modification in the traditional KF is adopted. As mentioned before, the modified KF algorithm substitutes the value of $y(k)$ by $y(k+1)$ in order to estimate one step in advance, thus compensating any delay in the system. Then, the modified algorithm is obtained by substituting (9.138) in (9.140) but using $y(k+1)$ instead of $y(k)$ obtaining:

$$\hat{\mathbf{x}}^-(k+1) = \mathbf{A}\hat{\mathbf{x}}(k) + \mathbf{B}u(k) = \mathbf{A}(\hat{\mathbf{x}}^-(k) + \mathbf{L}(k)(y(k+1) - \mathbf{C}\hat{\mathbf{x}}^-(k))) + \mathbf{B}u(k)$$
$$= \mathbf{A}(\mathbf{I} - \mathbf{L}(k)\mathbf{C})\hat{\mathbf{x}}^-(k) + \mathbf{B}u(k) + \mathbf{A}\mathbf{L}(k)y(k+1)$$

$$\tag{9.142}$$

Since the value of the sample $y(k+1)$ is unknown, this term must be removed from the observer equation. For this purpose, an auxiliary variable $\mathbf{F}(k)$ is used:

$$\hat{\mathbf{x}}^-(k) = \mathbf{F}(k) + \mathbf{L}(k-1)y(k) \tag{9.143}$$

then

$$\mathbf{F}(k) = \hat{\mathbf{x}}^-(k) - \mathbf{AL}(k-1)y(k) \tag{9.144}$$

The auxiliary variable $\mathbf{F}(k)$ can be computed at time $k+1$ as follows:

$$\mathbf{F}(k+1) = \hat{\mathbf{x}}^-(k+1) - \mathbf{AL}(k)y(k+1) \tag{9.145}$$

Now, using (9.142) in (9.145), the output $y(k+1)$ is canceled in the expression of $\mathbf{F}(k)$, yielding

$$\mathbf{F}(k+1) = \mathbf{A}(\mathbf{I} - \mathbf{L}(k)\mathbf{C})\hat{\mathbf{x}}^-(k) + \mathbf{B}\mathbf{u}(k) \tag{9.146}$$

and replacing (9.143) in (9.146) a recursive equation for $\mathbf{F}(k)$ can be found:

$$\mathbf{F}(k+1) = \mathbf{A}(\mathbf{I} - \mathbf{L}(k)\mathbf{C})\hat{\mathbf{x}}^-(k)(\mathbf{F}(k) + \mathbf{L}(k-1)y(k)) + \mathbf{B}\mathbf{u}(k) \tag{9.147}$$

Equation (9.147) is computed recursively to obtain the value of $\mathbf{F}(k)$. Once, $\mathbf{F}(k)$ is found it is used in (9.143), which allows to obtain the estimated states one sample in advance. The final step is to update this estimation using Eq. (9.138).

9.6.4 Experimental Results

The prototype of the SAPF is shown if Figure 9.53. The prototype has been built using a 4.5-kVA SEMICRON full bridge as the power converter and a TMS320F28M36 floating point DSP as the control platform. The grid is generated using a PACIFIC 360-AMX source. The system parameters are listed in Table 9.8.

i) *Response of the SAPF to Load Variations*

Figure 9.54 shows the main waveforms of the SAPF in case of a sudden load step change. The changes in the load are from full load to half load and half load to full load. The figure shows from top to bottom the grid currents, the nonlinear load currents, the filter currents, and the output voltage. It can be observed from the figure that the distortion caused by the load is perfectly compensated by the filter, providing sinusoidal grid currents. In order to validate the performances of the SAPF, the harmonic spectrum of the grid current for phase-leg a is shown. Figure 9.55 compares the THD after and before the compensation. Figure 9.55a shows the spectrum before compensation, in which THD is 24.08%, while Figure 9.55b shows the spectrum of the grid currents after the filter compensation, in which THD is 2.04%. An important reduction of the THD is achieved showing the good performances of the control algorithm.

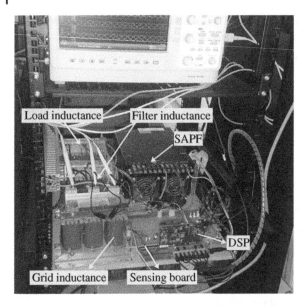

Figure 9.53 SAPF prototype.

Table 9.8 System parameters.

Symbol	Description	Value
LF	Filter input inductance	5 mH
C	Output capacitor	1500 μF
v_{dc}	dc voltage	400 V
fs	Sampling frequency	40 kHz
fsw	Switching frequency	4 kHz
f_{grid}	Grid frequency	60 Hz
v_{grid}	Grid voltage	110 V_{rms}
Lg	Grid inductance	0.5 mH
kp	Proportional gain	0.03
ki	Integral gain	0.5
RL	Load resistor	48 Ω–24 Ω
LL	Load inductance	5 mH
CL	Load capacitor	100 μF
$Ri\,(k)$	Single phase system noise power	0.24 V^2
$Qi\,(k)$	Process covariance matrix	$0.005I_3$

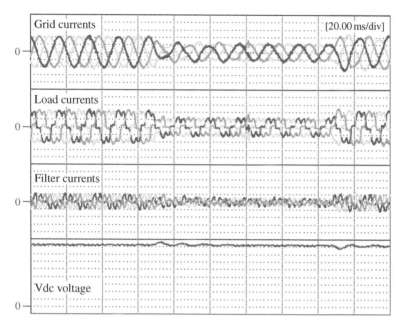

Figure 9.54 A sudden step change in the load from full load to half load and half load to full load. From top to bottom grid currents (5 A/div), load currents (5 A/div), filter currents (5 A/div), output voltage (50 V/div).

ii) *SAPF Performances under a Distorted Grid*

In Figure 9.56 the performances of the controller in the case of a distorted grid are presented. This figure illustrates from top to bottom: the grid currents, the load currents, and the distorted grid voltages, in which THD is around 14%. Even in case of grid harmonics, the grid currents are practically sinusoidal thanks to the good estimation of the grid voltages, which allows to generate the reference current only with the fundamental component of the grid voltage. This is an interesting property of the method since any synchronization system as for instance a PLL is not needed. Note that by augmenting the number of harmonics in system matrix, an estimation of the n-harmonics can be performed.

iii) *SAPF Performances under Grid Voltage Sags*

Figure 9.57 shows the SAPF performances when a grid voltage sag occurs. In this experiment, the sag has been characterized by a positive and negative sequences $V^+ = 0.8$ p.u. and $V^- = 0.4$ p.u., respectively. The phase angle between the sequences is around $\varphi = -\pi/6$. The reference currents have been obtained using the positive sequences of the grid voltage according to $\hat{i}_{abc}^* = \dfrac{P^*}{\left|\hat{v}^+\right|^2}\hat{v}_{abc}^+$. The positive sequence can be obtained by using the expressions

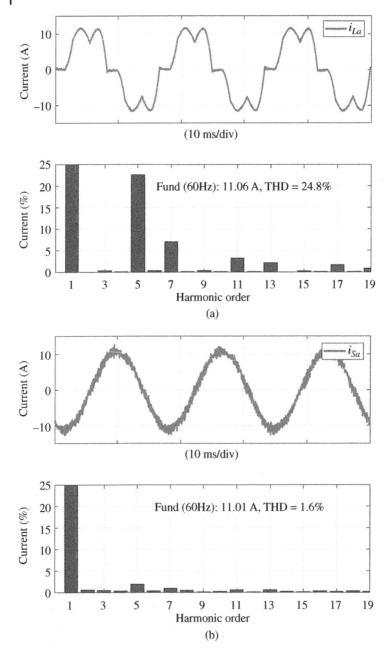

Figure 9.55 Grid current harmonics for phase-leg *a*: (a) Before compensation and (b) after compensation.

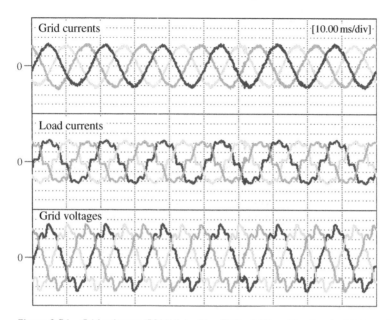

Figure 9.56 Grid voltages (50 V/div) with a THD = 14% and load and grid currents (5 A/div).

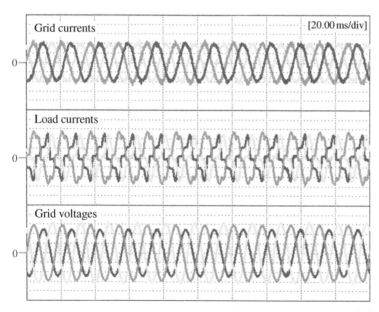

Figure 9.57 From top to bottom: grid currents (5 A/div), load currents (5 A/div) and grid voltages (50 V/div) under unbalanced grid fault.

Table 9.9 Comparative analysis.

Algorithm	f_s	t_{ex}	Load(%)	Averaged f_{sw}	THD
Conventional MPC	40 kHz	23 µs	92	4 kHz	2.1%
Presented MPC	60 kHz	15 µs	90	6 kHz	1,6%

presented in [20]. Since the grid current tracks only the positive sequence of the PCC voltage, the current amplitude is kept constant during the voltage sag.

iv) *Comparative Analysis*

In this section a comparison between the presented control algorithm with the conventional FCS-MPC is performed. Table 9.9 gives a comparison based on the execution time t_{ex} of the controller, computational load and memory usage in bytes is performed. The computational load can be computed according to

$$Load(\%) = f_s \times t_{ex} \times 100 \tag{9.148}$$

As shown in Table 9.9, the execution time for the complete control algorithm is around 15 µs (this is the case of considering only four vectors). In case of considering eight vectors (traditional FCS-MPC), the execution time is increased up to 23 µs, as expected. The KF is designed according to [21], where only the Kalman gain for one phase leg is computed and considered equal for the remaining phases. This is because the system noise is similar in the three-phase legs. With this assumption, the time employed by the KF in the traditional FCS-MPC (eight vectors) is around 18 µs, while in case of the presented method (four vectors) it is reduced to 10 µs. Note that this reduction is not exactly proportional to the number of voltage vectors. The remaining time to complete the values presented in Table 9.9, is due to the remaining parts of the control algorithm. In order to obtain the execution time, one timer of the DSP is used to measure the time of the controller task. In the conventional FCS-MPC the prediction of the grid voltages and currents is done eight times, one time for each vector, while with the presented control algorithm this computation is reduced to only four vectors in each sampling period. As it can be seen, the total time employed by the algorithm is noticeably smaller. Thanks to this, the sampling frequency can be increased from 40 kHz in the conventional FCS-MPC up to 60 kHz for the presented control algorithm. In the same way as in SMC happens, the relation between sampling frequency and switching frequency should be around 10. Then, the increment of the switching frequency yields to an improvement of the THD of the grid currents. One advantage of the presented controller is that the algorithm can be implemented in the traditional signal processor without the necessity of using faster processor such as Field Programmable Gate Arrays (FPGAs).

9.7 FCS-MPC for a Single-Phase T-Type Rectifier

This section presents grid voltage sensorless model predictive control for a single-phase T-type rectifier with an active power decoupling circuit. The presented control technique in this section is based on the controller proposed in [22]. The presented sensorless technique is based on a model reference adaptive system (MRAS) and tested under distorted grid conditions. The developed control technique is to ensure the following objectives: (i) sensorless grid voltage estimation; (ii) the second-order ripple power elimination; (iii) reference current generation based on power equilibrium; (iv) ensuring unity power factor under all operating conditions; and (v) capacitor voltage balance. The developed control structure offers simplicity, and it is cost-effective due to the absence of a grid voltage sensor. An experimental prototype is established, and the main results, including the steady-state and dynamic performances, are presented to validate the effectiveness of the presented control.

9.7.1 Modeling of Single-Phase T-Type Rectifier

Figure 9.58 shows the single-phase T-type rectifier with the 2ω ripple reduction auxiliary circuit. In this topology, the output voltage (V_{dc}) must be higher than the peak value of the input voltage source (v_s) to ensure a proper control of the input current. Therefore, the system works like a boost converter.

It is clearly seen that the rectifier part consists of eight active power switches (four switches per-leg). The state of each switch is designed as

$$S_{ij} = \begin{cases} 1 & \text{ON} \\ 0 & \text{OFF} \end{cases} \tag{9.149}$$

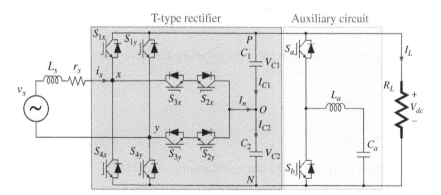

Figure 9.58 Single-phase T-type rectifier with an active power decoupling circuit (auxiliary circuit) [22].

Table 9.10 Relationship among the pole voltage and switching states.

S_{1j}	S_{2j}	S_{3j}	S_{4j}	Pole voltage	Operating state
ON	ON	OFF	OFF	$+0.5\,V_{dc}$	P
OFF	ON	ON	OFF	0	O
OFF	OFF	ON	ON	$-0.5\,V_{dc}$	N

where $i = 1, 2, 3, 4$ and $j = x, y$. The rectifier produces three various pole voltages $(+V_{dc}/2, 0, \text{and} -V_{dc}/2)$ with respect to the switching states as shown in Table 9.10 [22]. The current that flows from the T-type rectifier to the load is defined as the positive current. The switching functions of the rectifier can be defined as

$$\left.\begin{aligned} S_1 = S_{1x} - S_{1y} \\ S_2 = S_{2x} - S_{2y} \end{aligned}\right\} \tag{9.150}$$

The input voltage and neutral current can be defined in terms of the switching functions as follows:

$$v_{xy} = S_1 V_{C1} + S_2 V_{C2} \tag{9.151}$$

$$I_n = I_{C2} - I_{C1} = (S_2 - S_1) i_s \tag{9.152}$$

From Figure 9.58, the equations of the T-type PWM rectifier can be written as

$$v_s = L_s \frac{di_s}{dt} + r_s i_s + v_{xy} \tag{9.153}$$

$$I_{C1} = C_1 \frac{dV_{C1}}{dt}, \quad I_{C2} = C_2 \frac{dV_{C2}}{dt} \tag{9.154}$$

Since $V_{dc} = V_{C1} + V_{C2}$ and assuming that both capacitors on the DC-link have the same capacitance and are exposed to the same voltage ($V_{C1} = V_{C2} = V_{dc}/2$), the capacitor currents are equal but opposite as

$$I_{C2} = -I_{C1} \tag{9.155}$$

The capacitor currents in terms of the switching functions and grid current are obtained as

$$I_{C1} = \frac{1}{2}(S_1 i_s - S_2 i_s) \tag{9.156}$$

$$I_{C2} = \frac{1}{2}(S_2 i_s - S_1 i_s) \tag{9.157}$$

The derivative of grid current can be obtained from (9.153) as

$$\frac{di_s}{dt} = \frac{1}{L_s}\left(v_s - v_{xy} - r_s i_s\right) \tag{9.158}$$

where r_s is the internal resistance of the L_s.

The auxiliary (power decoupling) circuit is connected at the DC-link of the single-phase T-type rectifier, as depicted in Figure 9.58. The auxiliary circuit is composed of two power switches, the auxiliary capacitor C_a, and the auxiliary inductor L_a that transfers the ripple energy between the C_a and the DC-link. It is worth mentioning that the main converter (T-type rectifier) regulates the DC-link voltage, whereas the auxiliary circuit controls the ripple power.

The operation mode of the auxiliary circuit depends on the energy transfer between the DC-link and the C_a. In buck mode, the switch S_a is controlled to transfer the ripple energy from the DC-link to the C_a. The DC-link charges both the L_a and C_a during the turn-on interval of S_a, whereas the L_a release its energy to the C_a during the turn-off interval of S_a. In boost mode, the switch S_b is controlled to release the energy from the C_a to the DC-link. The L_a is charged by the C_a during the turn-on interval of S_b, whereas the both the C_a and L_a release energy back to the DC-link during the turn-off interval of S_b. More information can be found in [22].

9.7.2 Model Predictive Control

Figure 9.59 depicts the block diagram of the control technique. The lack of a grid voltage sensor and the fulfillment of many objectives reveals the value of the presented method. The control technique consists of four main parts, including sensorless grid voltage estimation, reference current generation based on power equilibrium, MPC for the main converter, and MPC for the power decoupling circuit based on instantaneous ripple power control.

9.7.2.1 Sensorless Grid Voltage Estimation
The grid voltage estimation is based on MRAS that is a very attractive solution for sensorless motor drives. The main idea behind MRAS is to design a closed-loop controller with parameters that can be updated to change the response of the system [23]. In MRAS, two independent models, namely, reference and adaptive models, are used to compute the same variable. First, the output of the adaptive model is compared to the response of the reference model. Then, the suitable controller is used to minimize the error between the adaptive model output and the reference model output. Finally, the controller output drives the parameter in the adaptive model that leads to the adaptive model response to match the response of the reference model.

The estimation technique uses active and reactive power exchange between the grid and the rectifier. The reference model for active P_1 and reactive Q_1 powers can be obtained by using a fictitious two-phase (α-β) reference frame.

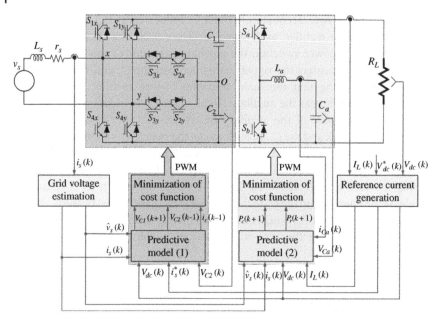

Figure 9.59 Control diagram of the grid-voltage sensorless MPC technique.
Source: Bayhan et al. [22]/IEEE.

$$
\left.
\begin{aligned}
P_1 &= \frac{1}{2}\left(v_{xy\alpha}i_{s\alpha} + v_{xy\beta}i_{s\beta}\right) \\
Q_1 &= \frac{1}{2}\left(v_{xy\beta}i_{s\alpha} - v_{xy\alpha}i_{s\beta}\right)
\end{aligned}
\right\}
\tag{9.159}
$$

where v_{xy} is the rectifier input voltage and it is defined in (9.151) and i_s is the grid current.

The T-type rectifier is connected to the grid through the input inductance L_s, as shown in Figure 9.59. In such a system, by neglecting the internal resistance r_s, the active P_2 and reactive Q_2 powers can be derived as

$$
\left.
\begin{aligned}
P_2 &= \frac{V_{xy}V_s \sin\delta}{2X_{L_s}} \\
Q_2 &= \frac{V_{xy}^2 - V_{xy}V_s \cos\delta}{2X_{L_s}}
\end{aligned}
\right\}
\tag{9.160}
$$

It can be seen that the reference model for active and reactive powers can be computed by (9.159) where all the variables can be measured or calculated. It is also clear that (9.159) is independent of the grid voltage. On the other hand, although, (9.160) can be used to calculate the same active and reactive powers,

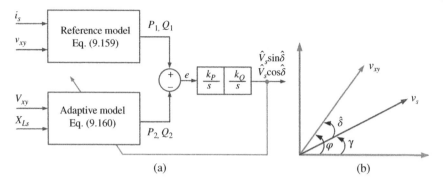

Figure 9.60 (a) Block diagram of MRAS based grid voltage observer; (b) Phasor representation of the grid voltage (v_s) and the rectifier. *Source:* Bayhan et al. [22]/IEEE.

it depends on the grid voltage amplitude Vs and phase angle δ. For that reason, (9.160) can be used as the adaptive model. The computed results of (9.159) and (9.160) should be identical under steady-state conditions. By using this knowledge, the grid voltage can be estimated.

The block diagram of the MRAS-based grid voltage observer is given in Figure 9.60a. The error between the reference model (9.159) and the adaptive model (9.160) is processed using a suitable controller to get an accurate estimation of the grid voltage. The estimated $\hat{V}_s \sin \hat{\delta}$ and $\hat{V}_s \cos \hat{\delta}$ can now be used to compute the unit vectors $\cos \gamma$ and $\sin \gamma$. $\hat{\delta}$ is the angle between the V_s and V_{xy} and the angle γ is equal to $\varphi - \hat{\delta}$ as shown in Figure 9.60b. Trigonometric identities can be used to calculate the unit vectors as

$$\left.\begin{array}{l} \cos \gamma = \cos \varphi \cos \hat{\delta} + \sin \varphi \sin \hat{\delta} \\ \sin \gamma = \sin \varphi \cos \hat{\delta} - \cos \varphi \sin \hat{\delta} \end{array}\right\} \tag{9.161}$$

where

$$\cos \varphi = \frac{v_{xy\alpha}}{V_{xy}}, \ \sin \varphi = \frac{v_{xy\beta}}{V_{xy}}, \ \cos \hat{\delta} = \frac{(\hat{V}_s \cos \hat{\delta})}{\hat{V}_s}, \ \sin \hat{\delta} = \frac{(\hat{V}_s \sin \hat{\delta})}{\hat{V}_s}$$

where

$$V_{xy} = \sqrt{(v_{xy\alpha})^2 + (v_{xy\beta})^2}, \hat{V}_s = \sqrt{(\hat{V}_s \cos \hat{\delta})^2 + (\hat{V}_s \sin \hat{\delta})^2}$$

Unit vectors $\cos \gamma$ and $\sin \gamma$ in (9.161) are the same as the unit vectors generated from a traditional PLL structure. Finally, the estimated grid voltage can be expressed as

$$\hat{v}_s = \hat{V}_s \cos \gamma. \tag{9.162}$$

9.7.2.2 Reference Current Generation

In this study, the reference current generation is based on the energy equilibrium between AC-side and DC-side instead of the PI controller. Due to the reduced blocking voltage, the middle switch in the T-type topology shows very low switching losses, and the input impedance is low. Therefore, the inductor core loss and switching loss can be very small. By neglecting power losses in the rectifier, the power equations of the AC-side and DC-side can be expressed as

$$\underbrace{\frac{V_s I_s^*}{2} - r_s \left(\frac{I_s^*}{\sqrt{2}}\right)^2}_{P_{ac}} = \underbrace{\frac{V_{dc}^2}{R_L}}_{P_{dc}} \tag{9.163}$$

The reference line current (I_s) can be derived from (9.163) as

$$I_s^* = \frac{V_s}{2r_s} - \sqrt{\left(\frac{V_s}{2r_s}\right)^2 - \frac{2V_{dc}^2}{r_s R_L}} \tag{9.164}$$

In (9.164), the grid voltage V_s is estimated from (9.162), the DC-link voltage V_{dc} is measured by the voltage sensor, and the internal resistance r_s value can be identified by LC meter. On the other hand, the output resistance RL is unknown in this equation since this value depends on the output load. To estimate the output resistance RL, Ohm's law in (9.165) is employed due to its simplicity.

$$\hat{R}_L = \frac{V_{dc}}{I_L} \tag{9.165}$$

However, the estimation in (9.165) goes to infinity when I_L is zero at the start-up. For that reason, supposing that at steady-state operating conditions, the rectifier maintains the target DC-link voltage with unity power factor, it is crucial to determine the initial value for R_L.

9.7.2.3 MPC for the T-Type Rectifier

It is well known that the main aim of the PWM rectifier controller is to regulate the amplitude of the output voltage and ensure the unity power factor by forcing semiconductor switches to follow the grid voltage using the PLL loop. Apart from these control requirements, the T-type rectifier requires to maintain the DC-link capacitors voltages balance ($V_{C1} \approx V_{C2} \approx V_{dc}/2$) that play a critical role in the success of the overall control. Therefore, the MPC is used for control of T-type rectifier since it is a multivariable control algorithm and tackles multiple control objectives in a single cost function.

The line current tracking and DC-link voltage regulation are two control objectives for the presented MPC technique. These control objectives can be formulated through the cost function to minimize the error between the control variables and their references. Considering that the two controlled variables (line current and

DC-link voltage) are tightly coupled. The cost function of the T-type rectifier consists of the following components:

$$g_1 = \left| i_s^*(k+1) - i_s(k+1) \right|^2 + \lambda \left| V_{C1}(k+1) - V_{C2}(k+1) \right| \tag{9.166}$$

where $i_s^*(k+1)$, $i_s(k+1)$, and $V_{C1}(k+1)$ and $V_{C2}(k+1)$ indicate the reference line current, the predicted line current, and the predicted DC-link capacitor voltages at an instant $(k+1)$, respectively. In (9.166), the first term is the current tracking error, and the second term is the voltage deviation of the DC-link capacitors. The weighting factor λ is used to handle the relationship between reference current tracking and midpoint voltage balance.

Because of its simplicity, the future behavior of the control variables is obtained by the forward Euler technique as

$$\frac{dx}{dt} \approx \frac{x(k+1) - x(k)}{T_s} \tag{9.167}$$

where T_s is sampling time.

By approximating (9.158) with (9.167), the discrete-time model of the line current can be obtained as

$$i_s(k+1) = i_s(k) + \frac{T_s}{L} \left(\hat{v}_s(k) - v_{xy}(k) - r_s i_s(k) \right) \tag{9.168}$$

Similarly, the discrete-time model of the DC-link capacitors voltages can be expressed as

$$\left. \begin{array}{l} V_{C1}(k+1) = V_{C1}(k) + \dfrac{T_s}{2C_1} \left[S_1(k) i_s(k) - S_2(k) i_s(k) \right] \\[2mm] V_{C2}(k+1) = V_{C2}(k) + \dfrac{T_s}{2C_2} \left[S_2(k) i_s(k) - S_1(k) i_s(k) \right] \end{array} \right\} \tag{9.169}$$

The algorithm of the MPC for the main converter is given in Algorithm 9.3. The first step is to set required initial parameters such as L_s, r_s, C_1, C_2, and then the switching states and pole voltages were defined for using into the predictive block (for-loop). The second step is to predict the future line current is $(k+1)$ and DC-link capacitor voltages $V_{C1}(k+1)$ and $V_{C2}(k+1)$ using the measured values of is, V_{C2}, V_{dc}, and I_L at the instant k and the estimated grid voltage $\hat{v}_s$. After that, the final step is to determine the switching states in Table 9.1 that can minimize the cost function (g_1).

Algorithm 9.3   MPC for the Main Converter

```
Input: i_s, is, v_s, V_C2 , V_dc, I_L
Output: S_1x , S_2x , S_1y, S_2y
set initial values;
states = [1111; 0111; 0011;…….; 1100; 0100; 0000];
v = [0;-0.5V_dc;-V_dc;0.5V_dc;0;-0.5V_dc;V_dc;0.5V_dc;0];
V_C1 = V_dc - V_C2;
```

```
for j = length(v) do
    is(k + 1) = is(k) + ((Ts/Ls) * (v_s - v(j) - (rs * is)));
    Vc1(k + 1) = Vc1(k) + ((states(i, 2) - states(i; 1) -
states(i, 4) + states(i, 3))(Ts(-is - IL))/(2C1));
    Vc2(k + 1) = Vc1(k) + ((states(i, 2) - states(i; 1) -
states(i, 4) + states(i, 3))(Ts(-is - IL))=(2C2));
    g2 = abs(i_s(k + 1) - is(k + 1))² + λ * abs(Vc1(k +1) - Vc2(k + 1));
    if g1 ≤ gopt then
        gopt = g1;
        xopt = j;
    end if
end for
S1x = states(xopt, 1);
S2x = states(xopt, 2);
S1y = states(xopt, 3);
S2y = states(xopt, 4);
```

9.7.2.4 MPC for the Power Decoupling Circuit

The control scheme of the MPC for the power decoupling circuit is shown in Figure 9.59. It is worth mentioning that the control of the power decoupling circuit is independent of the control of T-type rectifier. On the other hand, both control algorithms run as parallel to ensure the control objectives for the main converter and the power decoupling circuit.

The voltage on the switch S_b can take three different values $+0.5V_{dc}$, 0, and $-0.5V_{dc}$ depending on the switching state. Furthermore, by neglecting internal resistance of L_a, the voltage on the switch S_b can be expressed as

$$v_{S_b} = L_a \frac{di_{C_a}}{dt} + v_{C_a} \tag{9.170}$$

By substituting (9.167) into (9.170), the auxiliary capacitor current (i_{Ca}) can be predicted by

$$i_{C_a}(k + 1) = i_{C_a}(k) + \frac{T_s}{L_a}[v_{S_b}(k) - v_{C_a}(k)] \tag{9.171}$$

The auxiliary capacitor power at the $(k + 1)$ instant can be expressed by

$$P_{C_a}(k + 1) = v_{C_a}(k + 1)i_{C_a}(k + 1)$$
$$= v_{C_a}(k + 1)\left[i_{C_a}(k) + \frac{T_s}{L_a}[v_{S_b}(k) - v_{C_a}(k)]\right] \tag{9.172}$$

Since the sampling time T_s is very small and it is smaller than the time constant of LC branch, we can assume that the capacitor voltage at $(k + 1)$ instant is almost equal to the capacitor voltage at k instant $v_{Ca} = (k+1) = v_{Ca}(k)$. Therefore, the discrete-time model of the auxiliary capacitor power can be written as

$$P_{C_a}(k+1) = v_{C_a}(k)\left[i_{C_a}(k) + \frac{T_s}{L_a}\left[v_{S_b}(k) - v_{C_a}(k)\right]\right] \tag{9.173}$$

The instantaneous power from the ac source is

$$P_{in} = v_s(t)i_s(t) = \frac{V_sI_s}{2}\cos\theta - \frac{V_sI_s}{2}\cos(2\omega t - \theta) \tag{9.174}$$

Assuming the active power is P_{dc}, then, the instantaneous ripple power is

$$P_r = v_s(k)i_s(k) - P_{dc} \tag{9.175}$$

Finally, the cost function for the power decoupling circuit is defined as

$$g_2(k) = |P_r(k+1) - P_{C_a}(k+1)| \tag{9.176}$$

The developed MPC algorithm for the power decoupling circuit is given in Algorithm 9.4. The first step is to set required initial parameters such as L_a and C_a, and then the switching states and the voltage vector were defined for using into the predictive block (for loop). The second step is to predict the future ripple power $P_r(k+1)$ and the auxiliary capacitor power $P_{Ca}(k+1)$ using the measured values of i_s, i_{ca}, v_{ca}, V_{dc}, and I_L at the instant k and the estimated grid voltage $\hat{v}_s$. After that, as it was mentioned before, the final step is to determine the switching combination that can minimize the cost function g_2.

Algorithm 9.4   MPC for the Power Decoupling Circuit

```
Input: is, v_s, iCa, vCa, IL, Vdc
Output: Sa, Sb
set initial values;
states = [01; 10; 00; 11];
vSb = [-0.5Vdc; 0.5Vdc; 0; 0];
Pdc = VdcIdc;
for j=length(v) do
        Pr (k + 1) = (is * v_s) - Pdc;
        Pc(k + 1) = vCa*[iCa + (Ts/La)*(vSb (j) - vCa )];
        g₂ = abs(Pr (k + 1) - Pc(k + 1));
        if g₂ ≤ gopt then
                gopt = g₂;
                xopt = j;
        end if
end for
Sa = states(xopt, 1);
Sb = states(xopt, 2);
```

9.7.3 Experimental Studies

9.7.3.1 Steady-State Analysis

Figures 9.61–9.64 illustrate the steady-state results of the system. To highlight the reduction of 2ω ripple power, the presented solution was compared with the single-phase T-type rectifier without decoupling functionality. For all cases, the reference DC-link voltage is set at 300 V and the output load is set at 30 Ω.

Figure 9.61a shows the steady-state performance of the system when the decoupling function is enabled. It is clearly seen that the grid voltage and the grid current are in the same phase. In other words, the single-phase T-type rectifier operates with unity power factor. It is also shown that the DC-link voltage tracks its reference with high accuracy and nearly zero steady-state error. Furthermore, the capacitor voltages are balanced at 150 V that is half of the DC-link voltage. In this operation, the decoupling circuit is active, and almost all the ripple energy is stored in the auxiliary capacitor as shown in Figure 9.62. Figure 9.61b shows the steady-state performance when the decoupling function is disabled. Although, the unity power factor objective is satisfied in this operation mode, a large DC-link voltage ripple (approximately 50 V) was observed since all the 2ω ripple energy goes to the small DC-link capacitor.

Figure 9.63a,b show the harmonic analysis of the output voltage and the input current with and without decoupling function. It is observed that the current harmonic levels in both cases are significantly lower than the limits mentioned by international power quality standards like IEEE Std 519. However, there is 2ω frequency components on the DC-link voltage when the decoupling function is not active as shown in Figure 9.63b.

To prove the superiority of the grid voltage sensorless control algorithm, the tests were conducted under normal and distorted grid voltage conditions. The estimated grid voltage for a purely sinusoidal grid voltage signal is shown in Figure 9.64a. The grid voltage signal is then mixed with harmonic components and the estimation under distorted grid voltage is reported in Figure 9.64b. It can be seen that the voltage sensorless estimation technique is able to extract the fundamental voltage signal in both cases. Furthermore, the estimated voltage angle tracks the actual voltage angle with high accuracy for both cases. It is clear that the error in estimating grid voltage angle results in both cases is almost zero, showing an excellent position tracking performance.

9.7.3.2 Robustness Analysis

To verify the robustness of the control strategy, several tests have been conducted under various working conditions. The results are given in Figures 9.65–9.66.

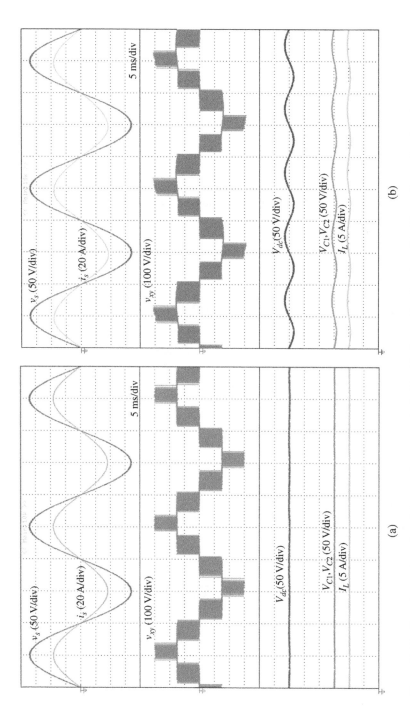

Figure 9.61 Experimental results of the steady-state analysis (a) with decoupling function; (b) without decoupling function.

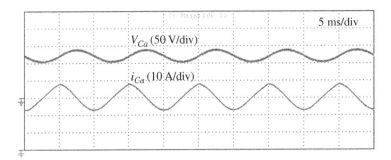

Figure 9.62 Experimental results of the steady-state analysis of auxiliary circuit voltage (V_{Ca}) and current (i_{Ca}) when the decoupling function is active.

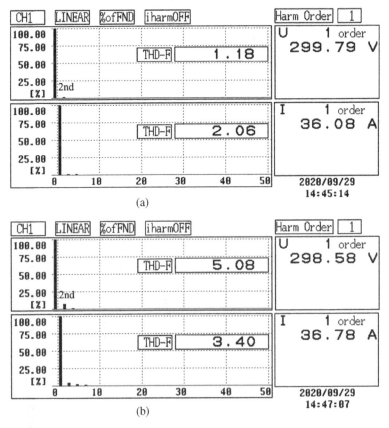

Figure 9.63 Experimental results of FFT analysis (a) with decoupling function; (b) without decoupling function.

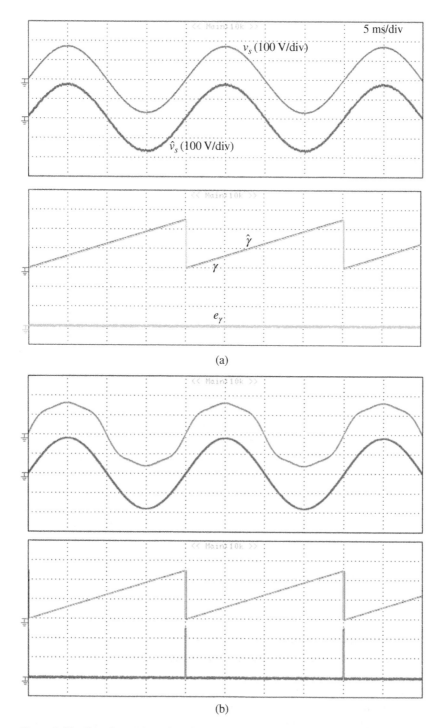

Figure 9.64 Experimental results of the actual grid voltage, the actual grid voltage angle, the estimated grid voltage, the estimated grid voltage angle, and the error between the actual and estimated voltage angles; (a) under normal grid conditions; (b) under distorted grid conditions.

i) *Load Variation:*

In this test, the programmable load is operated in CRL mode and the output resistance (*RL*) is reduced from 30 to 15 Ω. During this test, the reference DC-link voltage is set at 300 V. Figure 9.65a shows the dynamic response of the system when the decoupling function is active. The output voltage is regulated at 300 V and is almost not affected by this sudden 100% load increase. Although the amplitude of the 2ω ripple increased, the DC-link voltage ripple is within the required 2% limit. Thus, *RL* variation seems not to affect the control technique proposed in [22]. The performance of the decoupling circuit is shown in Figure 9.65b under the same operating conditions. The system has excellent ripple power tracking performance in both conditions and the ripple power is almost fully buffered in the decoupling circuit with the control method even during the step change of load power. Furthermore, it can be seen from Figure 9.65c that grid voltage angle was estimated accurately by the technique, even though under load step.

ii) *DC-Link Voltage Variation:*

In this experiment, the programmable load is operated in CRL and CPL modes. For the CRL mode, the output resistance is set at 30 Ω. For the CPL mode, the output power is set at 3 kW. A step change of 50 V in V_{dc} is applied (V_{dc} passes from 300 to 350 V).

Figure 9.66a shows the dynamic performance of the system under CRL mode. It is clear that V_{dc} is regulated at 350 V while VC1 and VC2 are balanced at 175 V. Furthermore, the grid voltage and the grid current are in the same phase for both conditions. Please note that the output power increased from 3 kW to approximately 4 kW since the load is in CRL mode.

Figure 9.66b shows the dynamic performance of the system under CPL mode. It is obvious that similar to previous case, V_{dc} tracks its reference with high accuracy and nearly zero steady-state error and the single-phase rectifier operates with unity power factor. It is clearly seen that the output current reduces from 10 to 8.33 A, while the DC-link voltage V_{dc} increases from 300 to 350 V. In this case, the output power is maintained at 3 kW since the load is in CPL mode.

9.8 Predictive Torque Control of Brushless Doubly Fed Induction Generator Fed by a Matrix Converter

This section presents predictive torque control (PTC) approach for brushless doubly fed induction generators (BDFIGs) fed by matrix converter is developed and investigated. The control technique presented in this section is based on the controller proposed in [24]. Unlike the traditional DFIG, the BDFIG needs lower maintenance cost because it operates without slip rings and brushes. Furthermore,

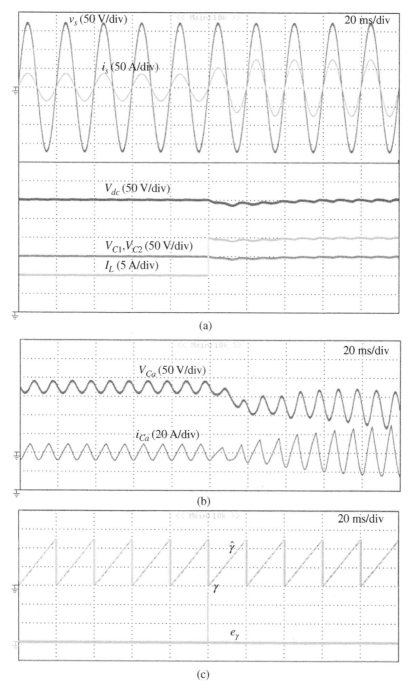

(a)

(b)

(c)

Figure 9.65 Experimental results of the dynamic response (a) main circuit; (b) auxiliary circuit when the decoupling function is enabled; (c) voltage estimation.

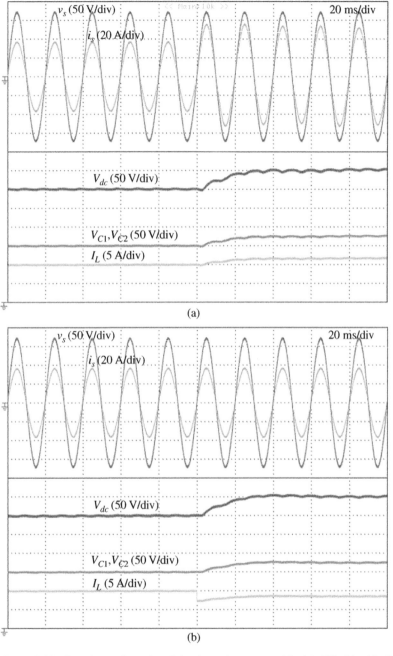

Figure 9.66 Experimental results of the dynamic response (a) with CRL; (b) with CPL.

as a promising alternative to the traditional back-to-back inverter, a matrix converter is employed owing to its high-power density. Thus, the advantages of BDFIG and matrix converter have been brought together to obtain a wind energy conversion system that has high power density as well as high reliability against mechanical faults. To overcome the drawbacks of the cascaded control structure, such as slow dynamic response and reduced control bandwidth, the torque of the BDFIG is controlled by a PTC method. Therefore, the discrete-time model of the BDFIG supplied by the matrix converter is first derived ad inserted into the predictive control routine. For examining the performance of the developed control strategy, simulations based on MATLAB/Simulink are carried out under steady-state and transient conditions. The test results indicate that the presented system exhibits good steady-state and dynamic performance.

9.8.1 Overview of the System Model

9.8.1.1 Topology Overview

In this section, an overview of the investigated system is presented. To simplify the system analysis without affecting the performance of the BDFIG, two DFIGs are electrically and mechanically coupled as shown in Figure 9.67. This configuration is called in the literature as cascaded doubly fed induction generator (CDFIG). In this configuration, the one DFIG is the power machine (PM) and the other DFIG is the control machine (CM). The stator windings of the PM are directly supplied by the utility grid, whereas the windings of the control machine (CM) are connected to a bidirectional power converter. To ensure the coupling of the stator windings for both machines, the windings of the rotor are electrically connected.

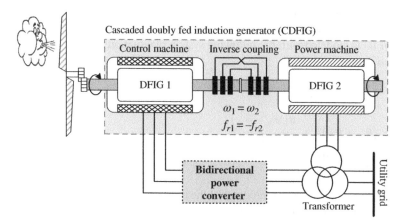

Figure 9.67 System configuration of the CDFIG in wind power generation.

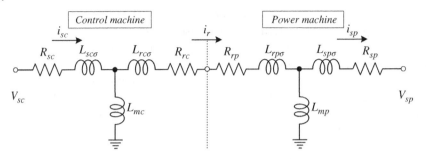

Figure 9.68 Equivalent circuit of the CDFIG based on the single-phase diagram.

9.8.1.2 Mathematical Model of the CDFIG

Figure 9.68 depicts the single-phase equivalent circuit of the CDFIG. Moreover, for the representation of each machine, individual rotor and stator resistances (R_{sc}, R_{rc}, R_{sp}, R_{rp}), leakage inductances ($L_{sc\sigma}$, $L_{rc\sigma}$, $L_{sp\sigma}$, $L_{rp\sigma}$), and magnetizing inductances (L_{mc}, L_{mp}) are considered. It can be seen that the control machine is coupled to the power machine through the rotor windings. The leakage inductances of both machines, control and power, can be written as

$$\left.\begin{aligned}
L_{sc} &= L_{mc} + L_{sc\sigma} \\
L_{rc} &= L_{mc} + L_{rc\sigma} \\
L_{sp} &= L_{mp} + L_{sp\sigma} \\
L_{rp} &= L_{mp} + L_{rp\sigma}
\end{aligned}\right\} \tag{9.177}$$

where L_{mc} and L_{mp} are the magnetization inductances of the machines, and $L_{sc\sigma}$, $L_{rc\sigma}$, $L_{sp\sigma}$, $L_{rp\sigma}$ are the leakage inductances of the stator and rotor windings. Furthermore, the subscripts p and c are used to indicate a variable of the power and control machine, respectively.

The CDFIG is modeled with the following system of equations in the $\alpha - \beta$ representation frame:

$$\bar{v}_{sp} = R_{sp}\bar{i}_{sp} + \frac{d\bar{\psi}_{sp}}{dt} \tag{9.178}$$

$$0 = \left(R_{rc} + R_{rp}\right)\bar{i}_r + \frac{d\left(\bar{\psi}_{rp} - \bar{\psi}_{rc}\right)}{dt} \tag{9.179}$$

$$\bar{v}_{sc} = R_{sc}\bar{i}_{sc} + \frac{d\bar{\psi}_{sc}}{dt} \tag{9.180}$$

where $\bar{v}_{sp}$ is the voltage vector of the stator of the power machine and $\bar{v}_{sc}$ the respective voltage vector of the control machine. Moreover, the stator and rotor flux vectors are $\bar{\psi}_s$ and $\bar{\psi}_r$, respectively.

The angular frequency of the rotor is expressed as

$$\omega_{SL} = \omega_{ec} + p_c \omega_r = \omega_{ep} - p_p \omega_r \tag{9.181}$$

where the angular frequencies of the stator are ω_{ep} and ω_{ec} for both machines, power and control. Furthermore, the power machine pole pair number of the power machine is p_p, whereas p_c denotes the pole pairs of the control machine. The electrical frequency of the control machine stator can be defined as

$$\omega_{ec} = \omega_{ep} - \left(p_p + p_c\right)\omega_r \tag{9.182}$$

The dq model of the CDFIG can be expressed by using the synchronous rotating frame. The power machine stator flux is considered aligned with the frame as next:

$$\bar{v}_{sp} = R_{sp}\bar{i}_{sp} + \frac{d\bar{\psi}_{sp}}{dt} + j\omega_{ep}\bar{\psi}_{sp} \tag{9.183}$$

$$0 = \left(R_{rc} + R_{rp}\right)\bar{i}_r + \frac{d\left(\bar{\psi}_{rp} - \bar{\psi}_{rc}\right)}{dt} + j\omega_{SL}\left(\bar{\psi}_{rp} - \bar{\psi}_{rc}\right) \tag{9.184}$$

$$\bar{v}_{sc} = R_{sc}\bar{i}_{sc} + \frac{d\bar{\psi}_{sc}}{dt} + j\omega_{ec}\bar{\psi}_{sc} \tag{9.185}$$

where the flux vectors of the stator ($\bar{\psi}_s$) and rotor ($\bar{\psi}_r$) are defined as

$$\left.\begin{aligned}
\bar{\psi}_{sp} &= L_{sp}\bar{i}_{sp} + L_{mp}\bar{i}_r \\
\bar{\psi}_{sc} &= L_{sc}\bar{i}_{sc} - L_{mc}\bar{i}_r \\
\bar{\psi}_{rp} &= L_{rp}\bar{i}_r + L_{mp}\bar{i}_{sp} \\
\bar{\psi}_{rc} &= -L_{rc}\bar{i}_r + L_{mp}\bar{i}_{sc}
\end{aligned}\right\} \tag{9.186}$$

Additionally, $\bar{\psi}_r$ r can be calculated as

$$\bar{\psi}_r = \left(\bar{\psi}_{rp} - \bar{\psi}_{rc}\right) = L_{mp}\bar{i}_{sp} - L_{mc}\bar{i}_{sc} + L_r\bar{i}_r \tag{9.187}$$

where $L_r = L_{rp} + L_{rc}$. The electromagnetic torque of the motor is determined as

$$T_e = \frac{3}{2}p_p\text{Im}\left(\bar{\psi}_{sp}^*\bar{i}_{sp}\right) + \frac{3}{2}p_c\text{Im}\left(\bar{\psi}_{sc}\bar{i}_{sc}^*\right) \tag{9.188}$$

$$T_e - T_L = \left(J_p + J_c\right)\frac{d\omega_r}{dt} + \left(F_p + F_c\right)\omega_r \tag{9.189}$$

where T_L is the load torque, J_p and J_c are the frictions of control and power machines, respectively. Moreover, F_p and F_c are the inertias of control and power machines, respectively.

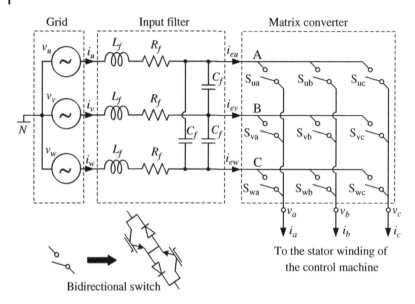

Figure 9.69 Equivalent circuit of the 3 × 3 matrix converter.

9.8.1.3 Mathematical Model of the Matrix Converter

In this study, a 3 × 3 matrix converter is employed as a bidirectional converter to control the CDFIG. The equivalent circuit model of the 3 × 3 matrix converter is illustrated in Figure 9.69.

It can be seen that load side is connected to the input side through the bidirectional switches without using DC-link capacitor. However, the matrix converter requires the use of an input filter for the following purposes: (i) for avoiding the generation of over-voltages because of the fast commutation of the currents i_{eu}, i_{ev}, i_{ew}, (ii) for eliminating the high-frequency harmonics of the input currents i_u, i_v, i_w [17].

It can be seen in Figure 9.69 that each bidirectional switch can be represented by using the variable S_{xy} where $x = u, v, w$ and $y = a, b, c$. When having $S_{xy} = 1$, the switch xy is considered on, whereas when $S_{xy} = 0$ that switch is in the off state. The relationship that connects both the load and input voltages can be defined by using the instantaneous transfer matrix $\mathbf{A}$. In this equation, both voltages are referenced to the neutral point N.

$$\begin{bmatrix} v_a(t) \\ v_b(t) \\ v_c(t) \end{bmatrix} = \underbrace{\begin{bmatrix} S_{ua} & S_{va} & S_{wa} \\ S_{ub} & S_{vb} & S_{wb} \\ S_{uc} & S_{vc} & S_{wc} \end{bmatrix}}_{A} \cdot \begin{bmatrix} v_{eu}(t) \\ v_{ev}(t) \\ v_{ew}(t) \end{bmatrix} \tag{9.190}$$

Additionally, the voltages at both terminals, input and load, are also defined with following vectors

$$
\mathbf{v}_o = \begin{bmatrix} v_a(t) \\ v_b(t) \\ v_c(t) \end{bmatrix} \qquad \mathbf{v}_i = \begin{bmatrix} v_{eu}(t) \\ v_{ev}(t) \\ v_{ew}(t) \end{bmatrix} \tag{9.191}
$$

Finally, (9.190) can be expressed as

$$
\mathbf{v}_o = A.\mathbf{v}_i \tag{9.192}
$$

By applying Kirchhoff's current law to connection points of the inverter switches, the following system of equations is obtained:

$$
\begin{bmatrix} i_{eu}(t) \\ i_{ev}(t) \\ i_{ew}(t) \end{bmatrix} = \underbrace{\begin{bmatrix} S_{ua} & S_{va} & S_{wa} \\ S_{ub} & S_{vb} & S_{wb} \\ S_{uc} & S_{vc} & S_{wc} \end{bmatrix}}_{A^T} \cdot \begin{bmatrix} i_a(t) \\ i_b(t) \\ i_c(t) \end{bmatrix} \tag{9.193}
$$

The currents at the input and load terminals are defined with the following vectors:

$$
\mathbf{i}_i = \begin{bmatrix} i_{eu}(t) \\ i_{ev}(t) \\ i_{ew}(t) \end{bmatrix} \qquad \mathbf{i}_o = \begin{bmatrix} i_a(t) \\ i_b(t) \\ i_c(t) \end{bmatrix} \tag{9.194}
$$

whereas the equation for calculating the current is

$$
\mathbf{i}_i = A^T.\mathbf{i}_o \tag{9.195}
$$

where A^T is the transposed matrix A.

9.8.2 Predictive Torque Control of CDFIG

The algorithm uses model predictive control, which regulates the state variables of the system by using real-time optimization over a finite-step horizon. Figure 9.70 depicts the block diagram of the PTC technique, which consists of the speed controller, the flux estimator, the predictive model and the cost function minimization block. The algorithm first computes the future trajectories of the controlled variables by using the model of the system along with the actual values of the current, voltage, and speed. Then, by minimizing the objective function in real-time over a finite prediction horizon, the optimal switching state for the next time step is chosen. Here, the PTC scheme is described in the following steps.

9.8.2.1 Outer Loop

It can be observed in Figure 9.70 that the computation of the reference torque is based on an outer speed control loop. The speed controller's parameters (K_p and K_i) have been chosen by Ziegler–Nichols tuning method. This reference torque signal is used in the cost function minimization block of the predictive controller.

9.8.2.2 Internal Model of the Controller

In this section, the continuous-time model of the CDFIG is discretized and the discrete-time model of the system is determined. The purpose of this extracted model is the prediction of values of the electromagnetic torque and the stator flux over several sampling intervals. Due to its computational simplicity, the forward Euler approximation method is adopted for obtaining a discrete-time model. The considered sampling period is T_s.

$$\frac{dx}{dt} = \frac{x(k) - x(k-1)}{T_s} \tag{9.196}$$

For calculating discrete-time model of the stator flux of the control and power machines, the stator voltage equations, which are given in (9.178) and (9.180), can

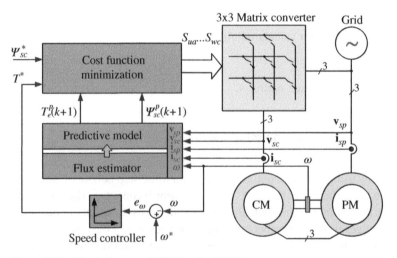

Figure 9.70 Block diagram of PTC for the CDFIG.

be used. By substituting (9.196) into (9.178) and (9.180), the discrete-time model of the stator fluxes is expressed as:

$$\overline{\psi}_{sp}^p(k+1) = \overline{\psi}_{sp}(k) + T_s \overline{v}_{sp}(k+1) - R_{sp}T_s \overline{i}_{sp}(k+1) \tag{9.197}$$

$$\overline{\psi}_{sc}^p(k+1) = \overline{\psi}_{sc}(k) + T_s \overline{v}_{sc}(k+1) - R_{sc}T_s \overline{i}_{sc}(k+1) \tag{9.198}$$

It is worth noticing that since the control machine is fed by the output of the matrix converter according to Figure 9.69, it is $v_{sc} = v_o$ and $i_{sc} = i_o$. Therefore, the voltage vector to be applied to the stator windings of the control machine at $(k+1)$ instant is $\mathbf{v}_{sc}(k+1)$. The voltage vector is defined by the state of the switches of the matrix converter as well as the input voltage vector according to (9.190).

To predict the CDFIG torque, a predictions of the stator currents $\overline{i}_{sp}^p(k+1)$ and $\overline{i}_{sc}^p(k+1)$ are required. By using (9.178), (9.179), (9.180), (9.187), and (9.194), it is possible to predict the stator currents ($\overline{i}_{sp}^p(k+1)$ and $\overline{i}_{sc}^p(k+1)$) at the $k+1$ instant:

$$
\left.
\begin{aligned}
i_{sp}^p(k+1) &= \left(1 + \frac{T_s}{\tau_\sigma}\right) i_{sp}(k) + \\
&\frac{T_s}{\tau_\sigma + T_s} \left[\frac{1}{r_\sigma}\left(\left(\frac{k_r}{\tau_r} - k_r j\omega_{ep}\right)\overline{\psi}_{rp}(k) + \overline{v}_{sp}(k)\right)\right] \\
i_{sc}^p(k+1) &= \left(1 + \frac{T_s}{\tau_\sigma}\right) i_{sc}(k) + \\
&\frac{T_s}{\tau_\sigma + T_s} \left[\frac{1}{r_\sigma}\left(\left(\frac{k_r}{\tau_r} - k_r j\omega_{ec}\right)\overline{\psi}_{rc}(k) + \overline{v}_{sc}(k)\right)\right]
\end{aligned}
\right\}
\tag{9.199}
$$

where r_σ, k_r, τ_σ, and τ_r are the machine coefficients that can be calculated by using the machine parameters. The rotor flux is computed as follows:

$$
\left.
\begin{aligned}
\overline{\psi}_{rp}(k) &= \frac{L_{rp}}{L_{mp}}\overline{\psi}_{sp}(k) + \left(L_{mp} - \frac{L_{rp}L_{sp}}{L_{mp}}\right)\overline{i}_{sp} \\
\overline{\psi}_{rc}(k) &= \frac{L_{rc}}{L_{mc}}\overline{\psi}_{sc}(k) + \left(L_{mc} - \frac{L_{rc}L_{sc}}{L_{mc}}\right)\overline{i}_{sc}
\end{aligned}
\right\}
\tag{9.200}
$$

According to the (9.188), the torque prediction directly depends on the stator fluxes and currents of the CDFIG. Thus, the torque prediction is defined as

$$
\begin{aligned}
T_e^p &= \frac{3}{2}p_p \mathrm{Im}\left(\overline{\psi}_{sp}^p(k+1)\overline{i}_{sp}^p(k+1)\right) \\
&+ \frac{3}{2}p_c \mathrm{Im}\left(\overline{\psi}_{sc}^p(k+1)\overline{i}_{sc}^p(k+1)\right)
\end{aligned}
\tag{9.201}
$$

9.8.2.3 Cost Function Minimization

For tracking the reference values of the torque and the stator flux, the cost function of the predictive control strategy is formulated as

$$g = |T^* - T^p(k+1)| + \lambda_\psi |\overline{\psi}^*_{sc} - \overline{\psi}^p_{sc}(k+1)| \tag{9.202}$$

where the reference torque is T^* and designated as the output of speed controller, whereas the reference flux is ψ^*_{sc} and is the control machine rated flux. The first term in (9.202) achieves the reference torque tracking whereas the second term ensures the stator flux tracking with high accuracy. The weighting factor for normalizing the units of the torque and flux is λ_ψ. The flowchart of the devised control strategy is depicted in Figure 9.71.

9.8.3 Simulation Results

For demonstrating the effectiveness of the developed control strategy, the results of the simulations are presented and discussed in this section. The simulations were performed by using MATLAB/Simulink environment. Table 9.11 summarizes the parameters of the considered CDFIG.

Figure 9.72 shows the results of the simulated electromagnetic torque, the winding flux and currents of the control machine, and the stator currents of the power machine under steady-state conditions. It can be seen that the CDFIG is tracking the reference torque (10 *Nm*) with high accuracy and the flux of control winding is 0.71 Wb. It can be seen that the frequencies of control and power currents are 10 and 50 Hz, respectively. It is clear that the suggested system exhibits satisfactory steady-state response characteristics.

In order to verify the response of the system under speed reference variations, the mechanical speed is changed from 600 to 900 rpm. The simulation results under this speed variation are shown in Figure 9.73. During the acceleration, the machine is tracking the reference torque (20 Nm) and the flux of the control winding is 0.71 Wb. It is obviously seen that the developed control strategy manages to operate properly in a broad speed range with good performance when encountering speed variations. From this perspective, the system operation can be considered robust against speed variations.

The controller has been also tested under load torque changes. More specifically, the reference torque rapidly changes from 5 to 15 Nm. The simulation results of this test case are shown in Figure 9.74. It can be observed that the actual torque accurately follows the reference torque in a very short time without overshooting. Both the torque and flux of the machine have been precisely tracked by the developed algorithm.

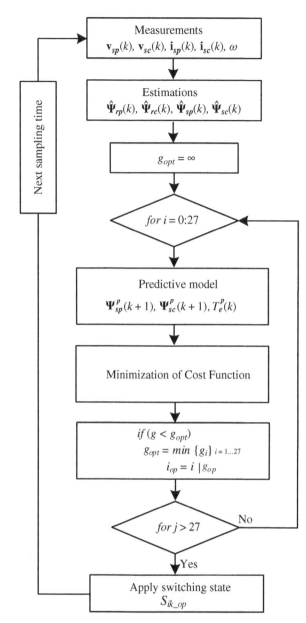

Figure 9.71 Flowchart of the MPC algorithm.

Table 9.11 CDFIG electrical parameters.

Parameter	PM	CM
Stator resistance (Ω)	0.7	0.4
Rotor resistance (Ω)	0.4	0.3
Stator inductance (mH)	12.25	8.25
Rotor inductance (mH)	18.4	12.4
Mutual inductance (mH)	48.4	38.4

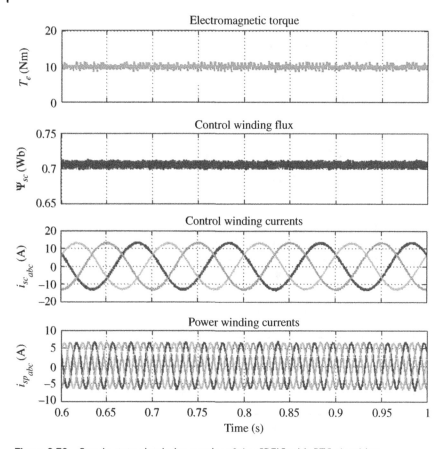

Figure 9.72 Steady-state simulation results of the CDFIG with PTC algorithm.

9.9 An Enhanced Finite Control Set Model Predictive Control Method with Self-Balancing Capacitor Voltages for Three-Level T-Type Rectifiers

The control technique presented in this section is based on the controller proposed in [25]. An effective finite control set model predictive control (FCS-MPC) is introduced for single-phase three-level T-type rectifiers supplying resistive as well as constant power loads (CPL). The main problem of CPL is the negative resistance phenomenon that endangers the rectifier's stability. Hence, the presented FCS-MPC method is based on Lyapunov's stability theory such that the stability of rectifier is guaranteed under all operating points. Unlike the existing FCS-MPC

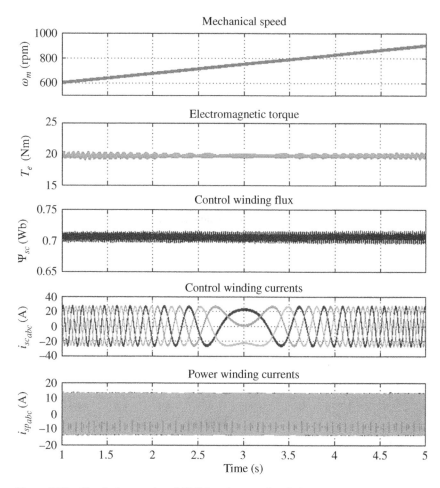

Figure 9.73 Simulation results of CDFIG under speed variation.

methods, the cost function design in the presented control method is formulated on the rectifier's stability. According to Lyapunov's stability theory, the rectifier stays stable provided that the rate of change of Lyapunov function is negative. In this case, the derivative of Lyapunov function can be used as the cost function without utilizing any weighting factor. Therefore, contrary to the existing FCS-MPC methods, the weighting factor requirement is eliminated, which leads to easiness in the design and implementation of the controller. Experimental results reveal that the presented control approach exhibits very good performance with undistorted and distorted grid voltage conditions when the rectifier feeds resistive and CPL loads.

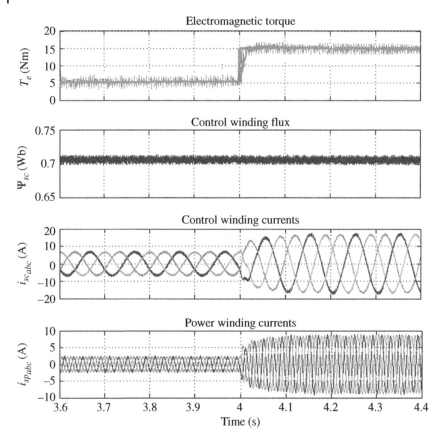

Figure 9.74 Simulation results of CDFIG under torque step.

9.9.1 Overview of the System Model

The schematic diagram of a single-phase three-level T-type rectifier supplying DC load bus is presented in Figure 9.75. It is apparent from Figure 9.75, there are eight switches whose on and off states can be represented mathematically as follows

$$S_{jm} = \begin{cases} 1 & \text{closed} \\ 0 & \text{open} \end{cases}, \quad j = 1, 2, 3, 4, \quad m = x, y \tag{9.203}$$

Due to the multi-level ability of the system, the rectifier can produce three distinct pole voltages with respect to the neutral point O. These pole voltages can be produced when the midpoint of each rectifier leg is connected to positive (P), neutral (O) and negative (N) points by means of appropriate switching. The operation states, switching device states, and produced pole voltage levels are depicted in

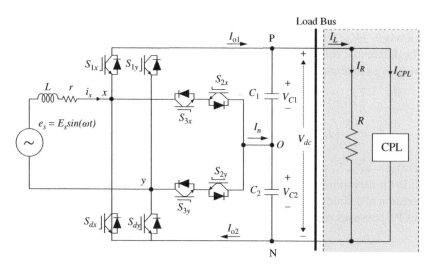

Figure 9.75 Schematic diagram of a single-phase three-level T-type rectifier supplying a DC load bus.

Table 9.12. In the P operation state, the rectifier generates $v_{mO} = +V_{dc}/2$ when S_{1m} and S_{2m} are closed (ON) and S_{3m} and S_{4m} are open (OFF). In the O operation state, the rectifier produces $v_{mO} = 0V$ when S_{2m} and S_{3m} are ON and S_{1m} and S_{4m} are OFF. Finally, in the N state, the rectifier produces $v_{mO} = -V_{dc}/2$ when S_{1m} and S_{2m} are OFF and S_{3m} and S_{4m} are ON. For the sake of expressing the multi-level voltage (v_{xy}) and DC-side currents, the following control input switching functions are defined

$$S_1 = S_{1x} - S_{1y} \tag{9.204}$$

$$S_2 = S_{2x} - S_{2y} \tag{9.205}$$

Analyzing Table 9.12, one can obtain the multi-level voltage and neutral current shown in Figure 9.75 in terms of S_1 and S_2 as

$$v_{xy} = S_1 V_{C1} + S_2 V_{C2} \tag{9.206}$$

$$I_n = I_{C2} - I_{C1} = (S_2 - S_1)i_s \tag{9.207}$$

Table 9.12 Operation states, switching states and pole voltages.

Operation states	S_{1m}	S_{2m}	S_{3m}	S_{4m}	v_{mO}
P	1	1	0	0	$+V_{dc}/2$
O	0	1	1	0	0
N	0	0	1	1	$-V_{dc}/2s$

Substituting switching state values into (9.206), one can obtain five distinct voltage levels 0, $\pm V_{dc}/2$, and $\pm V_{dc}$.

The differential equations of the rectifier feeding a resistive load and a CPL can be written as follows:

$$\frac{di_s}{dt} = \frac{1}{L}\left(e_s - v_{xy} - ri_s\right) \tag{9.208}$$

$$\frac{dV_{dc}}{dt} = \frac{2}{C_1}\left(I_{o1} - \frac{V_{dc}}{R} - \frac{P_{CPL}}{V_{dc}}\right) \tag{9.209}$$

where $v_{xy} = uV_{dc}$, $I_{o1} = ui_g$, u is the switching function and P_{CPL} is the power of the CPL. In the derivation of (9.209), it is assumed that $V_{C1} = V_{dc}/2$ and $I_{CPL} = P_{CPL}/V_{dc}$. The reason of defining CPL as a voltage controlled current source is due to the fact that CPL possesses negative incremental resistance described as

$$R_{CPL} = \frac{dV_{dc}}{dI_{CPL}} = \frac{d\left(\frac{P_{CPL}}{I_{CPL}}\right)}{dI_{CPL}} = -\frac{P_{CPL}}{I_{CPL}^2} = -R_{inc} \tag{9.210}$$

Similarly, I_{C1} and I_{C2} can be written in terms of the grid current (i_s), S_1 and S_2 as shown below:

$$I_{C1} = \frac{1}{2}\left(S_1 i_s - S_2 i_s\right) \tag{9.211}$$

$$I_{C2} = \frac{1}{2}\left(S_2 i_s - S_1 i_s\right) \tag{9.212}$$

9.9.2 Problem Definition

As determined in (9.210), the incremental resistance of CPL is negative, which endangers the stability of the rectifier. Now, consider small perturbations in the following variables

$$
\begin{aligned}
e_s &= e_{so} + \tilde{e}_s, \quad u = u_o + \tilde{u} \\
i_s &= i_{so} + \tilde{i}_s, \quad V_{dc} = V_{dco} + \tilde{V}_{dc}
\end{aligned}
\tag{9.213}
$$

where e_{so}, u_o, i_{so}, and V_{dco} represent the steady-state values of e_s, u, i_s, and V_{dc}, respectively. Substituting (9.213) into (9.208) and (9.209), one can obtain the following small-signal equations:

$$\frac{d\tilde{i}_s}{dt} = \frac{1}{L}\left(\tilde{e}_s - r\tilde{i}_s - u_o\tilde{V}_{dc} - V_{dco}\tilde{u}\right) \tag{9.214}$$

$$\frac{d\tilde{V}_{dc}}{dt} = \frac{2}{C_1}\left(u_o\tilde{i}_s + i_{so}\tilde{u} - \frac{\left(P_{CPL}R + 2V_{dco}\tilde{V}_{dc}\right)}{RV_{dco}}\right) \tag{9.215}$$

The derivative of (9.215) can be written as

$$\frac{d^2 \tilde{V}_{dc}}{dt^2} = \frac{2}{C_1}\left(u_o \frac{d\tilde{i}_s}{dt} + i_{so}\frac{d\tilde{u}}{dt} + \tilde{u}\frac{di_{so}}{dt} - \frac{2}{R}\frac{d\tilde{V}_{dc}}{dt}\right) \tag{9.216}$$

Now, substituting (9.214) into (9.216) yields

$$\frac{d^2 \tilde{V}_{dc}}{dt^2} = \frac{2}{C_1}\left(\frac{u_o}{L}\left(\tilde{e}_s - r\tilde{i}_s - u_o\tilde{V}_{dc} - V_{dco}\tilde{u}\right) + i_{so}\frac{d\tilde{u}}{dt} + \tilde{u}\frac{di_{so}}{dt} - \frac{2}{R}\frac{d\tilde{V}_{dc}}{dt}\right) \tag{9.217}$$

The transfer functions from ac grid voltage to DC load voltage can be obtained as

$$H_1(s) = \frac{\tilde{V}_{dc}(s)}{\tilde{e}_s(s)} = \frac{2u_o}{LC_1\left(s^2 + \dfrac{4}{RC_1}s + \dfrac{2u_o^2}{LC_1}\right)} \tag{9.218}$$

$$H_2(s) = \frac{\tilde{V}_{dc}(s)}{\tilde{e}_s(s)} = \frac{2u_o}{LC_1 s^2 + 2u_o^2} \tag{9.219}$$

While $H_1(s)$ is derived when resistive load (R) and a CPL are connected to the load bus, $H_2(s)$ is derived when the load bus contains a CPL only. It is obvious that the system is stable, while the rectifier supplies resistive load together with a CPL. However, the system becomes unstable when the rectifier feeds a CPL only. According to the Routh–Hurwitz stability criterion, this instability happens due to the missing s term in the denominator of $H_2(s)$. Therefore, a control method, which would guarantee the stability under all operating conditions, is needed.

9.9.3 Derivation of Lyapunov-Energy Function

The essential control objectives set for the rectifiers include regulation of DC voltage (load voltage), unity power factor operation, reasonably small total harmonic distortion (THD) in the grid currents, fast transient response to sudden load variations and stabilized operation with different loads. It should be mentioned that the control of capacitor voltages (V_{C1} and V_{C2}) can be accomplished when the control of load voltage (V_{dc}) is secured. In addition, for the sake of achieving unity power factor operation, the grid voltage and grid current reference must have zero phase difference as shown below:

$$e_s = E_s \sin(\omega t), i_g^* = I_s^* \sin(\omega t) \tag{9.220}$$

where I_s^* is the grid current reference amplitude produced by a proportional-integral (PI) controller by using the load voltage error as follows:

$$I_s^* = K_p\left(V_{dc}^* - V_{dc}\right) + K_i \int \left(V_{dc}^* - V_{dc}\right)dt \tag{9.221}$$

where V_{dc}^* is reference of V_{dc}, K_p and K_i denote the proportional and integral gains, respectively. By making use of (9.208), one can write the derivative of i_s^* as

$$\frac{di_s^*}{dt} = \frac{1}{L}\left(e_s - v_{xy}^* - ri_s^*\right) \tag{9.222}$$

Unlike the conventional MPC methods that employ a cost function, the developed MPC method uses a Lyapunov-energy function. As it is clear from Figure 9.75, the passive components which store energy are C_1, C_2 and L. Since capacitors can store the energy rather than dissipating it, the continuous-time Lyapunov-energy function should be formulated as by using x_1 and x_2 as follows:

$$V(x) = \frac{1}{2}\beta_1 x_1^2 + \frac{1}{2}\beta_2 x_2^2 \tag{9.223}$$

where $\beta_1 > 0$ and $\beta_2 > 0$ are the positive constants and x_1 and x_2 are the error variables, which are described as follows:

$$x_1 = V_{C1} - V_{C2}, \quad x_2 = i_s - i_s^* \tag{9.224}$$

Clearly, x_2 in (9.224) includes grid current reference while x_1 is defined as the error between V_{C1} and V_{C2}. The control objectives in DC-side are to control load voltage (V_{dc}) and capacitor voltages (V_{C1} and V_{C2}). While the first objective is accomplished by the PI controller in (9.221), the latter can be achieved if x_1 is forced to zero. The only possibility to have $x_1 = 0$ is when $V_{C1} = V_{C2}$. Therefore, the capacitor voltages are self-balanced without using their references. Taking derivative of x_1 and x_2 and using (9.208), (9.211), (9.212), and (9.222), one can obtain

$$\dot{x}_1 = \frac{1}{2C_1}(S_1 i_s - S_2 i_s) - \frac{1}{2C_2}(S_2 i_s - S_1 i_s) \tag{9.225}$$

$$\dot{x}_2 = \frac{1}{L}\left(v_{xy}^* - v_{xy} - rx_2\right) \tag{9.226}$$

Lyapunov-energy function has to hold the following features [24], [25]:

i) $V(x) > 0$ for $x_1 \neq 0$ and $x_2 \neq 0$
ii) $V(x) \rightarrow \infty$ for $\|x_1\| \rightarrow \infty$ and $\|x_2\| \rightarrow \infty$
iii) $\dot{V}(x) < 0$

Obviously, while the first two features hold, the third feature ($\dot{V}(x) < 0$) that ensures the stability of rectifier should also be met [31]. Since Lyapunov function in (9.223) is an energy function, the main goal is to force x_1 and x_2 to zero as shown in Figure 9.76. Thus, when $x_1 = 0$ and $x_2 = 0$ converge to zero, $V(x)$ becomes zero as

Figure 9.76 The behavior of energy function versus error variables.

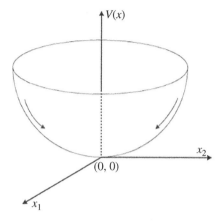

well. For this reason, the main aim is to minimize $V(x)$ as well as guarantee the negative definiteness of its derivative $(\dot{V}(x) < 0)$.

Taking derivative of $V(x)$ yields

$$\dot{V}(x) = \beta_1 \dot{x}_1 x_1 + \beta_2 \dot{x}_2 x_2 \tag{9.227}$$

Substitution of (9.206), (9.225) and (9.226) into (9.227) results in

$$\dot{V}(x) = \frac{\beta_1}{2C_1}(S_1 i_s - S_2 i_s)x_1 - \frac{\beta_1}{2C_2}(S_2 i_s - S_1 i_s)x_1 + \frac{\beta_2}{L}\left(v_{xy}^* - S_1 V_{C1} - S_2 V_{C2} - rx_2\right)x_2 \tag{9.228}$$

Substituting $i_s = x_2 + i_s^*$ into (9.228) and assuming that $C_1 = C_2 = C$, one can obtain

$$\dot{V}(x) = S_1 x_1 x_2\left(\frac{\beta_1}{C} - \frac{\beta_2}{L}\right) + S_2 x_1 x_2\left(\frac{\beta_2}{L} - \frac{\beta_1}{C}\right) + \frac{\beta_1}{C}(S_1 - S_2)i_s^* x_1$$
$$+ \frac{\beta_2}{L}\left(v_{xy}^* x_2 - S_1 V_{C1} x_2 - S_2 V_{C2} x_2 - rx_2^2\right) \tag{9.229}$$

Equation (9.229) can be simplified further if β_1 is selected as $\beta_1 = C\beta_2/L$. In this case, the terms $S_1 x_1 x_2$ and $S_2 x_1 x_2$ are eliminated resulting in

$$\dot{V}(x) = \frac{\beta_2}{L}\left((S_1 - S_2)i_s^* x_1 + v_{xy}^* x_2 - S_1 V_{C1} x_2 - S_2 V_{C2} x_2 - rx_2^2\right) \tag{9.230}$$

Hence, the stability of rectifier can be assured if $\dot{V}(x) < 0$.

9.9.4 Discrete-Time Model

In order to be able to design the FCS-MPC in discrete-time, the derivative of Lyapunov-energy function in (9.230) should be written in discrete-time. Hence, one can define $\dot{V}_x^{(n)}(k+1)$ at sampling instant $(k+1)^{\text{th}}$ as follows:

$$
\dot{V}_x^{(n)}(k+1) = \frac{\beta_2}{L} \left(\left(S_1^{(n)}(k) - S_2^{(n)}(k) \right) i_s^*(k+1) x_1(k+1) + v_{xy}^*(k+1) x_2(k+1) \right.
$$
$$
- S_1^{(n)}(k) V_{C1}(k+1) x_2(k+1)
$$
$$
\left. - S_2^{(n)}(k) V_{C2}(k+1) x_2(k+1) - r x_2(k+1)^2 \right)
$$

$$(9.231)$$

where $n = 1, \ldots, 9$ denotes the index which takes values from 1 to 9 since T-type inverter has nine switching states. The control law can be based on by selecting the values of appropriate switching functions ($S_1(k)$ and $S_2(k)$) which would make $\dot{V}_x(k+1)$ negative. In this case, the rectifier's stability is not endangered as long as $\beta_2 > 0$. This means that different positive β_2 values have no influence on the performance of the controller. Hence, $\beta_2 = 1$ is used in this study. This implies that, contrary to the conventional MPC techniques in which the weighting factor is essential and the performance is subject to the weighting factor value, β_2 used in the developed FCS-MPC technique does not affect the performance at all. Using the definitions in (9.224), the error variables in discrete-time are described as

$$
x_1(k+1) = V_{C1}(k+1) - V_{C2}(k+1) \tag{9.232}
$$

$$
x_2(k+1) = i_s(k+1) - i_s^*(k+1) \tag{9.233}
$$

The values of $i_s(k)$, $V_{C1}(k)$, and $V_{C2}(k)$ at $(k+1)^{\text{th}}$ sampling interval are easily obtained by making use of Euler's forward approximation to (9.208), (9.211), and (9.212) as follows:

$$
i_s(k+1) = \left(1 - \frac{r}{L} T_s \right) i_s(k) + \frac{T_s}{L} \left(e_s(k) - v_{xy}(k) \right) \tag{9.234}
$$

$$
V_{C1}(k+1) = V_{C1}(k) + \frac{T_s}{2C_1} \left(S_1(k) i_s(k) - S_2(k) i_s(k) \right) \tag{9.235}
$$

$$
V_{C2}(k+1) = V_{C2}(k) + \frac{T_s}{2C_2} \left(S_2(k) i_s(k) + S_1(k) i_s(k) \right) \tag{9.236}
$$

where T_s represents the sampling period and $v_{xy}(k)$ is defined as

$$v_{xy}(k) = S_1(k)V_{C1}(k) + S_2(k)V_{C2}(k) \tag{9.237}$$

The prediction of $e_s(k+1)$, $i_s^*(k+1)$ and $v_{xy}^*(k+1)$ are essential in (9.231). These predictions can be achieved as follows:

$$e_s(k+1) = \frac{3}{2}e_s(k) - \frac{1}{2}e_s(k-1) \tag{9.238}$$

$$i_s^*(k+1) = \frac{3}{2}i_s^*(k) - \frac{1}{2}i_s^*(k-1) \tag{9.239}$$

$$v_{xy}^*(k+1) = e_s(k+1) - \frac{L}{T_s}\left(i_s^*(k+1) - i_s^*(k)\right) - ri_s^*(k+1) \tag{9.240}$$

The steps of algorithm regarding the developed FCS-MPC approach are presented below.

Algorithm 9.5   FCS-MPC Based on Lyapunov-Energy Function

```
1: Measure e_s(k), i_s(k), V_C1(k), and V_C2(k).
2: Calculate i_s*(k).
3: Calculate (36)-(38).
4:        for n=1,....., 9 do
5: Calculate (30) and (31).
6: Evaluate (29).
7: end for
8: return minimum V̇_x(k + 1).
9: Select switching states that cause minimum V̇_x(k + 1).
```

The complete block diagram of the enhanced FCS-MPC method is given in Figure 9.77.

9.9.5   Experimental Studies

Figure 9.78 reveals the grid voltage (e_s), grid current (i_s) and grid current reference (i_s^*), five-level voltage (v_{xy}), load voltage (V_{dc}), load voltage reference (V_{dc}^*) and DC capacitor voltages (V_{C1}, V_{C2}) under resistive load $(R_L = 25\,\Omega)$. Obviously, there is no phase difference between e_s and i_s that clearly shows that the rectifier operates at unity power factor. In addition, since i_s and i_s^* are superimposed, the tracking

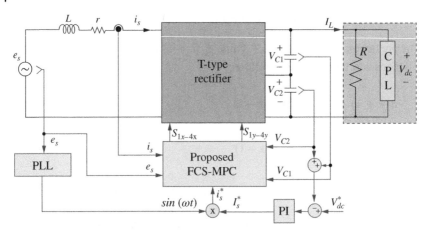

Figure 9.77 Block diagram of the enhanced FCS-MPC with T-type rectifier.

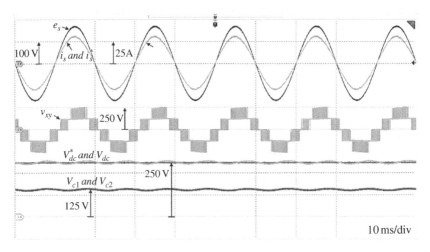

Figure 9.78 Steady-state results of grid-voltage (e_s), grid current (i_s) and its reference (i_s^*), five-level voltage (v_{xy}), load voltage (V_{dc}) and its reference (V_{dc}^*), and DC capacitor voltages (V_{C1} and V_{C2}) under undistorted grid voltage and resistive load.

error between these signals is zero. The total harmonic distortion of grid current is 1.3% which is reasonable. The control of DC-side voltages (V_{dc}, V_{C1}, V_{C2}) is accomplished at $V_{dc}^* = 250$ V and $V_{dc}^*/2 = 125$ V, respectively. Also, as mentioned before, the voltage v_{xy} contains five distinct levels (+250 V, +125 V, 0 V, −125 V and −250 V).

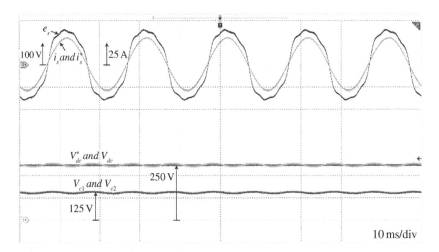

Figure 9.79 Steady-state results of grid-voltage (e_s), grid current (i_s) and its reference (i_s^*), load voltage (V_{dc}) and its reference (V_{dc}^*), and DC capacitor voltages (V_{C1} and V_{C2}) under distorted grid voltage and resistive load.

Figure 9.79 reveals the results with distorted grid utility voltage for $R_L = 25\,\Omega$. The components that cause distortion of grid voltage are 3rd, 5th, and 7th, which have $15\sqrt{2}\,\text{V}$, $7\sqrt{2}\,\text{V}$, and $5\sqrt{2}\,\text{V}$ amplitudes, respectively. In spite of the distorted grid voltage, the distortion in the grid current is reasonably low with THD equal to 2.4%. It can be noticed that the rectifier operates at unity power under distorted grid voltage. Again, the regulation of V_{dc}, V_{C1}, and V_{C2} is accomplished at 250 and 125 V, respectively.

References

1 R. Guzman, L. G. de Vicuña, A. Camacho, J. Miret, and J. M. Rey, "Receding-horizon model-predictive control for a three-phase VSI with an LCL filter," *IEEE Trans. Ind. Electron.*, vol. 66, no. 9, pp. 6671–6680, Sept. 2019.

2 A. Yazdani and R. Iravani, *Voltage-Sourced Converters in Power Systems: Modeling, Control, and Applications.* UK: Wiley-IEEE Press, 2010.

3 L. Wang, *Model Predictive Control System Design and Implementation Using MATLAB.* New York, NY, USA: Springer, 2009.

4 H. Komurcugil, S. Bayhan, N. Guler, and F. Blaabjerg, "An effective model predictive control method with self-balanced capacitor voltages for single-phase three-level shunt active filters," *IEEE Access*, vol. 9, pp. 103811–103821, Jul. 2021, doi: 10.1109/ ACCESS.2021.3097812.

5 Y. Liu, H. Abu-Rub, B. Ge, F. Blaabjerg, O. Ellabban, and P. C. Loh, *Impedance Source Power Electronic Converters*. UK: Wiley-IEEE Press, Oct. 2016. ISBN: 978-1-119-03707-1.

6 A. H. Budhrani, K. J. Bhayani, and A. R. Pathak, "Design parameters of shunt active filter for harmonics current mitigation," *PDPU J. Energy Manage.*, vol. 2, no. 2, pp. 59–65, Mar. 2018.

7 M. A. A. M. Zainuri, M. A. M. Radzi, A. C. Soh, N. Mariun, and N. A. Rahim, "DC-link capacitor voltage control for single-phase shunt active power filter with step size error cancellation in self-charging algorithm," *IET Power Electron.*, vol. 9, no. 2, pp. 323–335, Feb. 2016.

8 S. Bayhan and H. Abu-Rub, "Model predictive control of quasi-Z source three-phase four-leg inverter," in *IECON 2015 - 41st Annu. Conf. IEEE Industrial Electronics Society*, Yokohama, Japan, Jan. 2015, pp. 000362–000367, doi: 10.1109/IECON.2015.7392126.

9 S. Bayhan, H. Abu-Rub, and R. S. Balog, "Model predictive control of quasi-Z-source four-leg inverter," *IEEE Trans. Ind. Electron.*, vol. 63, no. 7, pp. 4506–4516, Jul. 2016, doi: 10.1109/TIE.2016.2535981.

10 Y. Li, J. Anderson, F. Peng, and D. Liu, "Quasi-Z-source inverter for photovoltaic power generation systems," in *Proc. 24th Annu. IEEE Appl. Power Electron. Conf. Expo. (APEC'09)*, Washington DC, USA, Feb. 2009, pp. 918–924.

11 N. Guler, S. Biricik, S. Bayhan, and H. Komurcugil, "Model predictive control of DC–DC SEPIC converters with autotuning weighting factor," *IEEE Trans. Ind. Electron.*, vol. 68, no. 10, pp. 9433–9443, Oct. 2021, doi: 10.1109/TIE.2020.3026301.

12 E. Irmak and N. Güler, "A model predictive control-based hybrid MPPT method for boost converters," *Int. J. Electron.*, vol. 107, no. 1, pp. 1–16, Jan. 2020.

13 L. Cheng et al., "Model predictive control for DC–DC boost converters with reduced-prediction horizon and constant switching frequency," *IEEE Trans. Power Electron.*, vol. 33, no. 10, pp. 9064–9075, Oct. 2018.

14 H. Komurcugil, S. Biricik, and N. Guler, "Indirect sliding mode control for DC-DC SEPIC converters," *IEEE Trans. Ind. Informat.*, vol. 16, no. 6, pp. 4099–4108, Jun. 2019.

15 S. Bayhan and H. Abu-Rub, "Model predictive droop control of distributed generation inverters in islanded AC microgrid," *2017 11th IEEE Int. Conf. Compatibility, Power Electronics and Power Engineering (CPE-POWERENG)*, Cadiz, Spain, 2017, pp. 247–252, doi: 10.1109/CPE.2017.7915177.

16 A. Milczarek, M. Malinowski, and J. M. Guerrero, "Reactive power management in islanded microgrid: proportional power sharing in hierarchical droop control," *IEEE Trans. Smart Grid*, vol. 6, no. 4, pp. 1631–1638, 2015.

17 R. Guzmán, L. García de Vicuña, M. Castilla, J. Miret, and A. Camacho, "Finite control set model predictive control for a three-phase shunt active power filter with a Kalman filter-based estimation," *Energies*, vol. 10, p. 1553, Oct. 2017.

18 Morales, J.; de Vicuña, L.G.; Guzman, R.; Castilla, M.; Miret, J.; Torres-Martínez, J. "Sliding mode control for three-phase unity power factor rectifier with vector operation," in *Proc. IECON 2015 – 41st Annu. Conf. IEEE Ind. Electron. Soc.*, Yokohama, Japan, Nov. 9–12, 2015, pp. 9–12.

19 K. Ahmed, A. Massoud, S. Finney, and B. Williams, "Sensorless current control of three-phase inverter-based distributed generation," *IEEE Trans. Power Delivery*, vol. 24, no. 2, pp. 919–929, Mar. 2009.

20 R. Guzman, L. G. de Vicuna, J. Morales, M. Castilla, and J. Miret, "Model-based active damping control for three-phase voltage source inverters with LCL filter," *IEEE Trans. Power Electron.*, vol. 32, no. 7, pp. 5637–5650, Jul. 2017.

21 R. Guzman, L. G. de Vicuna, J. Morales, M. Castilla, and J. Miret, "Model-based control for a three-phase shunt active power filter," *IEEE Trans. Ind. Electron.*, vol. 63, pp. 3998–4007, Jul. 2016.

22 S. Bayhan, "Grid voltage sensorless model predictive control for a single-phase T-type rectifier with an active power decoupling circuit," *IEEE Access*, vol. 9, pp. 19161–19174, Jan. 2021.

23 S. Bayhan and H. Abu-Rub, "Performance comparison of two sensorless control methods for standalone doubly-fed induction generator," in *Proc. 16th Int. Power Electron. Motion Control Conf. Expo.*, Antalya,Turkey, Sep. 2014, pp. 996–1000.

24 S. Bayhan, P. Kakosimos, M. Rivera, "Predictive torque control of brushless doubly fed induction generator fed by a matrix converter," in *2018 IEEE 12th Int. Conf. Compatibility, Power Electronics and Power Engineering (CPE-POWERENG 2018)*, Doha/Qatar, Apr. 10–12, 2018.

25 S. Bayhan, H. Komurcugil, and N. Guler, "An enhanced finite control set model predictive control method with self-balancing capacitor voltages for three-level T-type rectifiers," *IET Power Electron.*, vol. 15, pp. 504–514, Jan. 2022.

Index

Advanced Control of Power Converters: Techniques and MATLAB/Simulink Implementation,
First Edition. Hasan Komurcugil, Sertac Bayhan, Ramon Guzman, Mariusz Malinowski, and Haitham Abu-Rub.
© 2023 The Institute of Electrical and Electronics Engineers, Inc.
Published 2023 by John Wiley & Sons, Inc.
Companion website: www.wiley.com/go/komurcugil/advancedcontrolofpowerconverters

Books in the IEEE Press Series on Control Systems Theory and Applications

Series Editor: Maria Domenica Di Benedetto, University of l'Aquila, Italy

The series publishes monographs, edited volumes, and textbooks, which are geared for control scientists and engineers, as well as those working in various areas of applied mathematics such as optimization, game theory, and operations.

Printed and bound by CPI Group (UK) Ltd, Croydon, CR0 4YY

27/10/2024

14580673-0004